The Rational Animal

Ludwig Huber

The Rational Animal

In Search of Complex Cognition in Non-Human Animals

Springer

Ludwig Huber
University of Veterinary Medicine
Vienna, Austria

ISBN 978-3-031-60802-5 ISBN 978-3-031-60803-2 (eBook)
https://doi.org/10.1007/978-3-031-60803-2

English translation of the 1st original German edition published by Suhrkamp Verlag, Berlin, 2021

The original submitted manuscript has been translated into English. The translation was done using artificial intelligence. A subsequent revision was performed by the author(s) to further refine the work and to ensure that the translation is appropriate concerning content and scientific correctness. It may, however, read stylistically different from a conventional translation.

This Springer imprint is published by the registered company Springer Nature Switzerland AG
The registered company address is: Gewerbestrasse 11, 6330 Cham, Switzerland

Dedicated to my teacher and mentor
Rupert Riedl

Preface

Whether animals can be rational and whether they are conscious are among the most exciting and controversial questions in biology and comparative psychology [1]. And although Charles Darwin [2, p. 488] prophesied that psychology would be based on a new foundation, namely, the necessary acquisition of every mental power and ability by gradation (and in doing so, also shedding light on the origin of man and his history), biologists and psychologists have long pinned the problem of animal consciousness as outside the realm of serious research, dismissing it as something that one can never truly know and therefore should not think further about, let alone research. Even those scientists who investigate higher forms of cognition in great apes, dolphins, or corvids, and who grant them the ability to think about their physical and social environment, sense of time, tool making, action planning, and perspective taking, resist the question of whether we can ascribe rationality, intentionality, and consciousness to animals. In addition, if we can, to which animal species and to what extent [3, 4]. This has mainly to do with three problems. First, it is still not clear what exactly consciousness, which is related to the other mentioned abilities, is. Even regarding human consciousness, its neural mechanism, and its functions, there are many diverging opinions. This brings us to the second, so-called "hard problem": subjective experience. Consciousness is a multifaceted phenomenon, and subjective experience is its most puzzling aspect. It is not only that our brains collect stimuli and assign them meaning but also that they generate a vivid array of experiences and feelings: seeing green, feeling hungry, or being captivated by philosophical questions. Being oneself is a feeling, and no one else will ever know that feeling as directly as one's self. Third, exploring animal consciousness raises the question of the special role of humans.

As early as 1976, the American biologist Donald Griffin, with his book *The question of animal awareness* [5], broke a previously prevailing taboo regarding research into animal consciousness [6]. In light of the rapidly growing knowledge of today, 48 years after Griffin's book, the time is ripe to illuminate even the most contentious questions of animal cognition, namely, rationality, and consciousness. There are many books about animal thinking, but often, they are anecdotal and superficial. Some of them offer nice stories, beautifully illustrated for the layman, but—garnished with a large portion of anthropocentrism—they lead us to unfounded or unverifiable conclusions. The hot questions and the current approaches to answering them based on solid scientific knowledge are not addressed. These books hardly take us any further in answering the question of animal thinking.

In this book, various aspects of consciousness and rationality will be addressed; however, I will start from a primarily *functional approach*. Griffin was already convinced that the reasons for the traces of human-like consciousness in animals were to be found in the process of evolution. His thesis is that consciousness would not have emerged if it did not have survival-promoting functions. The latest research seems to suggest that the original function of consciousness is to enable voluntary movements. With this function, an organism can better direct its attention and focus on exactly what is important for its fitness and survival. By having cognitive representations of objects and events in the world, the ability to establish relationships between these different representations *(awareness)*, the ability to focus on one *(intention)*, and the ability to plan how to achieve the goal captured by this intention, selected animals possess some of the central functions of human consciousness. That humans have additional abilities, such as being able to become aware of their own conscious processes, having self-reflection, and communicating this in detail to others, is beyond question.

However, even self-monitoring, or "metacognition" in technical terms, occurs not only in children but also in some animal species. Some animals also have the ability to think about the knowledge of others, to anticipate their actions, and thus the ability to partly adopt the other's perspective. Some animals can be innovative and creative—they can spontaneously find solutions to completely new problems by recombining their previous experiences or by selectively copying conspecifics. Some can not only use tools efficiently but also enhance the effect through targeted modification or even manufacture new tools. Animals can act in preparation with the use of tools. For example, some can generally anticipate future events and plan their actions accordingly, suppressing their current need in favor of a future one. Some animals make decisions by weighing goals, choosing the most efficient way to reach them, and considering current and future motivations. With these abilities, they meet (some, though not all) criteria of practical-rational action.

In this book, the major and, at the same time, most difficult questions of comparative cognition research—henceforth, the "sextet of animal intelligence"—will be presented using good scientific examples, discussed in all their facets, and finally answered with necessary caution. Starting from the question of human rationality, which will be historically illuminated at the beginning and presented as a framework concept, the latest research results will be discussed. This approach will either question preferred interpretations or demand new ones. It is particularly important to critically question hastily made categorisations, often derived from the human species and everyday psychological ideas, and to view animal (mis)achievements from the perspective of species-specific, natural requirements. I will also deal extensively with language, both its evolutionary origins and various forms of communication as well as the most complex manifestations found until now in the animal kingdom. Finally, I will highlight forms, origins, degrees, and criteria of consciousness; sketch the underlying neural processes; and end with the ability of humans to be aware of their own consciousness.

Why is it important at all to know whether animals can act rationally, intentionally, or consciously? The answer is that it is of both theoretical and practical relevance. It is theoretically relevant in two respects: because it is good in itself for understanding animals better, but this knowledge is also of great importance for our self-assessment. This results in practical relevance: We are not only surrounded by animals; we live *with* animals; we live *from* animals. Yet we view ourselves as separate from them in important matters. Consciousness, language, and rationality are essential characteristics with which we justify our human uniqueness and thus our superiority and ultimately our right to dominate and use animals in various ways. If an animal can feel pain, we feel obliged to avoid causing unnecessary pain. However, we still do not attribute this to the special intrinsic value and dignity associated with consciousness and rationality. This has far-reaching consequences for human action and the coexistence of humans and animals.

Even the contrast to *the* animal (in the singular) is often a sign of human arrogance. Because of course there is not just one but approximately nine million animal species. Nevertheless, the title of this book says "animal" because, on the one hand, I defend the fact that it is not a mere object of experience but also the name for a construct that becomes comprehensible only within the framework of cultural order patterns. After all, in human language, terms are not only exact descriptions of an object or state of affairs but also abstractions, symbols, and categories.[1] On the other hand, I would like to emphasise that animals are individuals and not simply (nameless) representatives of their species. Modern cognitive biology recognises and appreciates the—often—large individual differences, and one can confidently speak of personalities in non-human[2] animals, many of which we will get to know in the following (including Alex, Betty, Figaro, Guillem, Kanzi, Kermit, and Sarah).

From an ethical perspective, current cognitive research questions traditional attitudes towards animals, especially their instrumentalisation, whether as food or laboratory animals. Here, massive changes occur on the horizon. If great apes have already been withdrawn from the consuming grasp of humans, perhaps all apes will be withdrawn very soon, and then, perhaps all mammals and birds, even fish and cephalopods, could eventually disappear from our diet. The pain and suffering of animals are the main motivations for rethinking traditional attitudes and opinions. They are strongly associated with the attribution of consciousness. At the end of this book, I will outline the ethical consequences and implications that arise from the latest results of comparative cognitive research.

---

[1] I owe to my colleague Judith Benz-Schwarzburg the insight that the construct "animal" is not normatively neutral. We have summarized "the animal" in the singular for centuries and contrasted it with humans, without looking closely at the animals in the plural, at what they are and what they need. It is therefore not simply an order pattern, but a subordination pattern [7].

[2] Throughout the book, I will repeatedly use the attribute "non-human" to occasionally remind us that we humans are also animals in a biological sense.

New perspectives also emerge for human self-understanding. By understanding the ways in which animals can be rational, we are called upon to rethink our own rationality. Perhaps our thinking is much more fragmented and particular, much less theoretical, and more embedded in our environment and conditioned by it than we have previously assumed. The possibility of reducing relatively complex animal behaviour to a number of relatively simple, domain-specific processes might cause us to re-evaluate our presuppositions and prejudices about human rationality. We would have to ask ourselves whether it is right to apply double standards to human and animal rationality. A well-founded and unbiased look at rationality may show that there is no sharp boundary between human reasoning and way of thinking in animals. Comparisons between different animal species and between animals and humans, especially children, in terms of certain talents and thinking styles can reveal to what extent certain abilities are based on common (homologous) or at least convergent (analogous) processes and how these processes reveal rationality. These insights will also increase the understanding of human rationality by shedding light on its evolution.

Against this background, there are three broad themes, each with two domains, that deserve special attention: (1) creative tool making and (2) other forms of technical intelligence (causal understanding); (3) extending thinking into the future (preparation and planning) and (4) into the past (episodic memory); (5) reflecting on one's own knowledge (metacognition) and (6) on the knowledge of others (perspective-taking and theory of mind). These six intellectual achievements are primary candidates for studying rational thinking in an extended sense. Whether these aspects of rationality should be given a different term to save the uniqueness of human rationality, or at least its potential for normative justification and decentration (displacement of thoughts), remains a question for philosophers.[3]

The individual chapters of this book address the cognitive processes of animals that can provide scientific evidence for their rationality. These are thought processes that, in humans, are not possible without consciousness; examples include creative tool use, perspective-taking, insightful problem-solving, and time travel (episodic memory, planning). In the first chapter, I will also cover concept formation, logical reasoning, volitional actions, intentionality, and the so-called executive functions. This list is not exhaustive, but it will leave out those areas where experimental, systematic, and standardised research is still in its infancy, such as rational imitation, creativity, and teaching.

This book is intended not only to present what we know (since the beginning of the third decade of the twenty-first century) but also how we discovered it. Therefore, a multitude of studies of a predominantly experimental nature[4] will be presented. In

---

[3] If I sometimes only mention one gender in the book, (almost) both are always meant.

[4] The attribute "experimental" often implies "invasive" in regard to animal research. Unfortunately, experiments on and with animals have also slipped into this connotation. This is regrettable and misleading, because experiment and trial are good and important tools of empirical research. Not only has my own research remained exclusively non-invasive for more than 30 years, I also only describe results in this book that were obtained non-invasively.

particularly critical cases, where the interpretations and conclusions depend on a precise understanding of the methods used, I will go into detail and describe the experimental setups and the considerations behind them. Sometimes the results must also be presented in more detail and partially illustrated because experimental setups can often be difficult to understand if one does not visually have the involved apparatus and objects, the arrangements, the stimuli, and the animals themselves. Many examples come from my own research; these I have often treated in more detail and provided with personal comments.

I would like to thank the Institute for Advanced Sciences (Wissenschaftskolleg; WIKO) in Berlin, specifically Daniel Schönpflug and Thorsten Wilhelmy, and the Institute for Advanced Sciences in Lincoln, specifically Anna Wilkinson and Stuart Humphries, for their hospitality and support. In the interdisciplinary atmosphere and at a distance from the—sometimes—hectic everyday university life in Vienna, large parts of the book were created. The same applies to my professional home, the Messerli Research Institute, which, with its interdisciplinary focus on questions of human-animal relationships, has provided a great breeding ground for this book. At this point, I would like to especially thank the Messerli Foundation and (posthumously) its founder Herta Messerli for allowing me to hold a chair for the scientific foundations of animal welfare and human-animal relationships. This has enabled me to work with a great team of eager and talented colleagues as well as with many motivated and interested students to trace the intellectual abilities of pigeons, marmosets, dogs, keas, Goffin's cockatoos, poison dart frogs, red-footed tortoises, and Kune Kune pigs (to name only the most important ones). Special thanks go to my friend and colleague Ulrich Kattmann, who read large parts of the book and made many good suggestions for improvement, as well as Alice Auersperg for the careful correction of the Goffin's cockatoo chapter. Much of what I have put to paper here was presented in the philosophical-theological-biological seminar (the "Zwettl seminar") and tested in subsequent discussion rounds with critical students under the wonderful direction of Marianne Popp. I would also like to thank my colleague Judith Benz-Schwarzburg from the ethics unit of the Messerli Research Institute for critical comments and helpful suggestions.

I would like to thank my brother Martin for the introduction to Suhrkamp (for the German edition) and for his good advice on all matters of the book project. Special thanks go to my parents, Aloisia and Ludwig (posthumously), for their everlasting generous support and my wife, Petra, for her inexhaustible patience and great understanding during the drafting of the book.

I would like to thank Louise Mackie for the careful correction and improvement of the English translation, also Karoline Eadie for help with the creation of the bibliography and index and Nadja Kavcik-Graumann for the Illustration of the sometimes abstract and complex experiments. Finally, I would like to thank Suhrkamp for the efficient and constructive cooperation, especially Jan-Erik Strasser, whose careful and critical editing has contributed much to the precision of terms and clarity of argumentation, and Springer for the publication of the English edition.

Vienna, Austria Ludwig Huber

Contents

Chapter 1
Historical Outline of the Basic Questions of Animal Cognition

A cat runs across the meadow because a dog is chasing it. Arriving at an oak tree, it hooks at the last moment and saves itself on a neighbouring maple tree. The panting dog is inattentive for a moment, overlooks the cat's quick manoeuvre, and barks at the oak. The cat is safe. This thoroughly realistic example has been used by philosophers such as Norman Malcolm [8] and Donald Davidson [4] to discuss the question of whether animals can be mistaken since the dog, at least at first glance, is mistaken when it barks at the wrong tree. But does this really occur? Or is an error possible only when someone is able to distinguish between true and false? Only someone who is convinced of something, has an opinion, or can form a judgement can then err in it.

Already in antiquity and during the Middle Ages, philosophers (and later also theologians, jurists, and natural scientists) found varying answers to the questions of whether animals can think, make decisions, have goals, plan actions, and finally, whether animals do something in full consciousness. Often, the question of thinking in animals was posed in such a categorical and general way that no consideration was given to the possible differences between animal species or even individuals or to the possibility of different categories of thinking. The question "Do animals think?" was posed in such a general way because one wanted to point out a decisive, categorical difference from thinking humans. In the so-called "*anthropological difference*", it was and is a question of all or nothing. In the question of human nature, in the sense of human self-knowledge, animals are still important yardsticks.

L. Huber, *The Rational Animal*, https://doi.org/10.1007/978-3-031-60803-2_1

To Err Is Human?

According to the German idealist Hegel, one cannot be reminded often enough that what distinguishes humans from animals is thinking.[1] Accordingly, animals cannot err. The dog has overlooked the cat's quick turn; he is missing a crucial piece of information, but he is not mistaken. He reacts to those stimuli he has perceived, but he does not ponder whether his beliefs are true or false. While he barks at the oak tree, he does not think about possible alternatives. Nor does he form any concept or judgement about what he has seen and knows. Thus, he could not report anything to another dog, not so much because he lacked any particular means of expression but because he lacked the concepts to describe the fact "The cat is sitting on the oak tree".

Donald Davidson [4] calls this particular form of mental content—a belief, a wish, or an intention—a *propositional attitude*. On the one hand, this content has a certain form, the so-called subject–predicate structure; on the other hand, it also has truth value. Similarly, just as a malfunctioning blood pressure monitor does not indicate false values in the actual sense, it is mistaken only in a figurative sense; but in reality, there is only a technical failure. The dog is also mistaken or deceived only in a figurative sense because, in reality, it is only subject to a perceptive (based on mere perception) or cognitive (based on mental processes) failure.[2]

The question now arises whether one must have propositional attitudes to be able to err, i.e., whether to err is solely human, as the philosopher Anselm Oelze (University of Helsinki) explains [10]. In its extreme form, this may be true because—as my professor at the University of Vienna, Rupert Riedl, always emphasised—believing pure nonsense is a privilege of humans [11]. However, could we not simply act unreasonably instead of going completely absurd? Aristotle, in *De Anima*, already made a preliminary decision in this regard because he only meant man in possession of reason (*ratio*) and understanding (*intellectus*). Not only did ancient schools of philosophy, such as the Stoics, follow him in this regard but also many philosophers of the Middle Ages [10]. The inability to intellectually correct sensual misjudgements goes hand in hand with the absence of an intellect, according to Avicenna Latinus. In addition, in *De animalibus,* Albertus Magnus, while conceding to apes the formation of "incomplete conclusions"—simple practical

[1] G. F. W. Hegel repeatedly emphasised in the *Science of Logic* (1812–16) as well as in the *Encyclopaedia* (1817) that thinking distinguishes man from the animal. See [9].

[2] Anselm Oelze explains it this way: "But can we truly only speak of an error when it is propositionally realised? This, too, may at first seem like a trivial or, to be more precise, an overly pedantic question. However, it does have its philosophical justification. After all, if we claimed that it is not necessary to have propositional attitudes in order to be able to err, then we could just as well speak of our thermometer being wrong when it tells us the wrong temperature. Certainly, we often make statements of this kind in everyday life and claim that our thermometer is wrong. But if asked, we would probably always explain that we are only using the term 'wrong' in a figurative sense. After all, the supposed 'error' of our thermometer is ultimately nothing more than a simple technical and not a cognitive failure." [10].

syllogisms and inductions, did not believe ape imagination to be connected with the intellect. Simple distinctions and even generalisations do not require judgements, at least not those that go beyond those of the inner and outer senses. New judgements in the sense of an assessment of a state of affairs cannot be strictly corrected but can be replaced only by new judgements. In the case of humans, corrections are made by the intellect. On the other hand, as Thomas Aquinas aptly put it in *Quaestiones disputatae De veritatae,* animals are not able to judge ("*nec iudicant de suo iudicio*").

Similarly, Adam Wodeham argued, although he pointed out the epistemological problem that we can observe only the behaviour of an animal or certain effects and movements ("*effectus et motus*") but not what causes this behaviour. We cannot see the thoughts of animals, if they have any at all, from the outside. It would therefore be possible in principle for animals to make a "complex objective judgement" (*iudicium complexe obiective*), but we do not know for sure. If we did know, we would have to attribute practical reason (*ratio practica*) to the animal and consequently classify it as a rational creature (*animal rationale*). This would lead to a shift in the boundary between humans and non-human animals. To avoid this, possibly out of caution towards the Church and other anthropocentrists, Wodeham tended to explain the behaviour of animals on the basis of simple acts of apprehension without intellectual judgement. Certain stimuli trigger instinctive reactions as a result of a natural impulse, a natural instinct. However, the animal does not think or judge what it sees.

Another philosopher and theologian of the Middle Ages, the Augustinian hermit Gregory of Rimini, saw the matter quite differently in his *Lectura* and rejected a simple stimulus–response model to explain animal behaviour, pointing out that it cannot satisfactorily explain *all* animal behaviour. How else could it be explained that animals may react differently to one and the same stimulus? Different motivations would only inadequately explain this. Consequently, he attributed a certain capacity for judgement to animals. Even if judgements cannot be expressed in language, they nevertheless reflect certain convictions. Errors are even evidence of judgement, not in the sense of propositional attitudes but in the sense of beliefs that can be corrected by new facts. A lack of concepts or linguistic expressions, such as "the cat is sitting on the oak tree", must not lead to the degradation of animals to stimulus-response machines or to separating them from us more than necessary in terms of their cognitive abilities. Unlike Adam Wodeham, Gregory had no problem viewing animals as rational beings.

Based on these examples from antiquity and the Middle Ages, we can already see that we are dealing with at least two problems in regard to the question of whether animals can think, have minds, or err. On the one hand, there is a problem with the meaning of the concepts and terms with which we describe complex facts, and on the other hand, there is a problem with the methodological difficulty of uncovering the mental processes that underlie the actions of animals. Whether animals can err thus depends on whether we apply a strict concept of judgement and error. If we do, we then are concluding with Aristotle, Adam Wodeham, or Donald Davidson that animals, because they cannot form propositional attitudes, can neither judge nor err. Or we can take the two concepts somewhat more broadly, following Gregory of

Rimini, Thomas Aquinas, Albertus Magnus and Avicenna, so we can speak of animals judging and erring. The other, methodological, problem stems from a general lack of knowledge of how animals act, solve problems and, more specifically, at what level animals make decisions or correct them. Thinkers of antiquity and the Middle Ages already had good knowledge (through self-reflection and exchange of opinions) about the mental abilities of their own species but only extremely rudimentary knowledge about the "the life of the soul" of other species.

Can Animals Think?

Whether animals can think was also the subject of many controversies among philosophers of the modern era. Of all of them, René Descartes stands out for his provocative theses in the *Treatise on the Method* (1637)—he speaks of animals as "wound-up clocks"—drawing an absolute line between humans and animals.[3] Although he did not deny animals' life-sustaining functions, he considered both a vegetative and a sensitive soul to be superfluous, contradicting even Aristotle. The automaton concept of animal behaviour stems from Descartes' mechanistic view of all living things (even humans would be mere "machines with limbs" without a spirit—he speaks of a *soul* in the *Treatise On Man* (1662)). Although he granted animals perceptions and sensations, he insisted on a radical anthropological difference between animals and humans. Well-designed machines would be indistinguishable from animals; they could also perform movements *as if* they had a soul, and both behave predictably and inflexibly. Following an ideal mechanistic and scientistic explanation, he rejected analogies from humans to the intelligent behaviour of animals as unreflective prejudices, for example, the reasonable-seeming behaviour of foxes moving cautiously and testingly over frozen ice. This example comes from another early modern French philosopher, Michel de Montaigne, who, following ancient sceptics such as Sextus Empiricus, sought in the *Essais* (1580) to narrow the difference from both sides by curing human arrogance and "humanising" the mental capacities of animals [12].

English empiricists, such as John Locke, also considered the differences in mental abilities between humans and animals to be much smaller than Descartes did. In Locke's *An Essay Concerning Human Understanding* (1690), he stated that the difference between clever animals and stupid humans could be far less than that between stupid and clever humans. In addition, for David Hume, no truth seems more evident than that animals, just like humans, are endowed with thought and reason. He even goes a step further in *A Treatise of Human Nature* (1739) and says that the arguments for this case are so obvious that they do not escape even the most stupid and ignorant [12].

[3] See the excellent explanations by Markus Wild in Chap. II.2 of his *Tierphilosophie* [12].

Therefore, how is it that some of the greatest thinkers in history are so divided on the question of animal thought? That for some, the question itself is unreasonable because it is based on a categorical error; for others, not only is the question justified but also the (affirmative) answer cannot escape even the most stupid and ignorant? Quite obviously, this discrepancy is not based on the use of different facts and observations but on the different use of terms and varying methodological views. Leaving aside the Cartesian extreme position of animal automata, which denies animals any mental activity, the question remains as to which terms and concepts best describe animal behaviour and its underlying mental processes. Most certainly, the task of empirical research is not to explain animal actions or utterances with careless crossing over to "thinking" but to recognise the actual cognitive performances of animals in circumstances peculiar to them [13]. In doing so, it is essential not to become involved in the mere aggravation of "Do animals think, yes or no?". This question would not only reduce the wealth of the cognitive abilities of animals to a single characteristic but also fuel the unfortunate desire of many philosophers to argue about mere definitions. Just as it is unreasonable to specify a number at which a quantity of grains becomes a "heap" or to determine the height at which a "hill" becomes a "mountain", one will not be able to satisfactorily grasp the gradual transitions between simpler and more complex forms of thinking, deciding, and acting by arbitrarily drawing boundaries. The absolutist's determination must give way here to description by the accumulation of characteristic features for the respective levels. For, if the performances of animals are "often strikingly close to or similar to thought" [13, p. 31], if the amount of similarity is greater than the amount of difference, and if—above all—any determination can be merely deficient and provisional due to the lack of important data, we should beware of exaggerated and hasty judgements.

Of course, starting from human thinking and approaching animal performance from this relatively safe position is justified and even purposeful. After all, we know both from the "I" perspective—through introspection—and from the "you" perspective of mutual linguistic mediation—how we make certain decisions, what the motives of our actions are, what our goals are, and how we feel about them. This experience aspect is of central importance. Beyond this, however, we as scientists also adopt the third-person perspective and make human thought and action an objective matter of investigation. In compliance with established methods, in mutual discourse and diverse reflections, we create an objective and, namely, an intersubjective world in which animals and their behaviour are also the subject of discussion.

Conceptualisation

A central aspect of human thought is concept formation. Some philosophers, such as Davidson [4], think, as I have argued, that without concepts, there is no thinking. Indeed, whether we speak or not, it seems that we think in terms or, more generally, in concepts. First of all, concepts are condensed (mental) representations[4] of reality. They are concepts in the sense that they are abstractions. That is, details, contingencies, or situational arbitrariness are excluded. Concepts are also summaries of typical characteristics and thus enable generalisations. An essential function of these concepts is categorisation—the quick classification of objects or events into one's own mental order structures. Fundamental to this is the ability to treat similar but not identical things as somehow equivalent on the basis of their shared characteristics by placing them in the same categories and reacting to them in the same way.

People assign linguistic markers or labels to these concepts or terms by assigning them words or symbols, a process called naming. A concept can be represented by several of these labels, both by words in different languages and in one language (synonyms). Crucially, however, the ability to form concepts and to categorise them does not *depend* on language, nor did it arise with or even through language.[5] Philosophers developed different theories to explain the emergence of concepts before human language evolved. We can first ask ourselves whether the lion has a concept of wildebeest. Undoubtedly, it reacts very accurately and very purposefully to the sight of a wildebeest. Apparently, the lion—actually we should say: the lioness—has some kind of conception, a mental image so to speak, of this target object. In this case, cognitive scientists like to speak of *mental representations*. Detailed investigations have shown that the lioness makes very precise distinctions and that she distinguishes wildbeest well from other animals, for example, cheetahs, because she hunts the former and not the latter. This "wildebeest concept" is also immensely flexible because the perceptions of wildbeest are highly variable—colour and brightness change with light, size with distance, shape with perspective, shape in tall grass or behind bushes, and so on. Depending on hunger and other motives, the use of the concept also changes in terms of the reaction to its contents. The lioness can charge off at first sight, but she can also wait or refrain from reacting at all if she is not in the mood to hunt. Therefore, we see that not only the content of the lioness' wildebeest concept is variable but also its use. This fulfils essential criteria of a concept beyond language, as postulated by contemporary philosophers. The lioness makes systematic distinctions [19]—these distinctions are rule-governed because they follow certain criteria [20] and are purposeful because they can be used to control important prey-catching behaviour. Which leaves the question of whether

[4] The term "representation" is a problematic one, but nevertheless an unavoidable one in the cognitive sciences. Mostly, the term is used as a synonym for mental image or as a synonym for neuronal pattern. Neurobiologist Antonio Damasio explains it this way: representations are patterns that are consistently related to something else [14]. It is unclear how accurately neuronal patterns or representations reflect the objects they denote; in any case, they are not mere images or facsimiles of reality, but remain constructs of the brain.

[5] I took this view in my dissertation [15]. In doing so, I referred to prominent psychologists. David Premack considered sorting or preverbal classification as an immediate precursor of verbal classification [16, p. 215], and Richard Herrnstein added in 1984: "Human language may depend on categorization, but the evidence shows clearly that does not depend on language" [17, p. 257]. I describe some examples of categorisation and concept formation in animals a little later, and in Chap. 9 I go into this again in detail. See also [18].

this concept arose evolutionarily in the sense of an innate prey schema or whether it was acquired individually by the lioness through learning or other cognitive processes.

In so-called teleosemantics [21–23], the example of the prey-catching behaviour of the frog is often discussed [24]. The frog snaps its tongue at a moving fly; thus, the sight of the fly activates a neuronal pattern that corresponds to a mental representation because it triggers snapping. This particular state in the frog's nervous system, it is argued, accurately represents the presence of a fly because its biological function is to indicate its presence. This display function corresponds to an evolutionarily generated meaning. It has a selection advantage and is thus adaptive, functional, or (biologically) purposeful. Although the fly schema is modifiable—as shown above all by the toad studies of Jörg-Peter Ewert at the University of Kassel[6]—it nevertheless remains far removed from the concepts of language that constitute thinking for Davidson and many other analytical philosophers. Even the notion of wildbeest in lions, for all its variability and flexibility, has a strong genetic component; after all, we find hardly any serious differences between lionesses across Africa in terms of wildbeest hunting. Let us therefore turn to the question of whether animals can learn complex object classes and whether they are able to "grasp" initially completely unknown patterns—in other words, whether they can form adequate concepts.

Categorisation and Concept Formation

In my doctoral thesis, I described experiments with pigeons that made categorisations using perceptual concepts [15, 26–28]. I was able to show that these birds were able to distinguish very quickly between complex stimulus classes according to the criteria I had specified and to classify them according to the categorisation rules I had established. In doing so, they could not fall back on innate categorisations but had to determine the features of the images that were important for the given rule and then link them in the right way to the appropriate categorisation rule (Fig. 1.1).

As I have been able to show in many other experiments with pigeons, they have several possibilities for class formation. Feature learning is only one of these

[6] Contrary to what was assumed at the beginning of the last century, visuo-motor behaviours related to amphibian prey capture and feeding are not reflexive, but relatively variable and flexibly adaptable to the corresponding stimulus situations. The neuronal circuitry that mediates between prey stimulus and prey capture behaviour consists of retina-fed pathways of the midbrain and corresponding nuclei of the rhomboid brain. The distinction between prey and non-prey (mostly predators) is based on the interactions of these pathways with structures (tectum, tegmentum) in the diencephalon. Directed attention, i.e., the translation of perception into action, is based on modulating forebrain loops of the telencephalon (esp. striatum). Dopaminergic modulations in the macronetwork can alter prey capture strategies in terms of hunting and waiting for prey. In addition, various modulatory forebrain loops involving parts of the telencephalon (esp. the hippocampus) allow modifications of prey recognition through visual and olfactory learning [see 25].

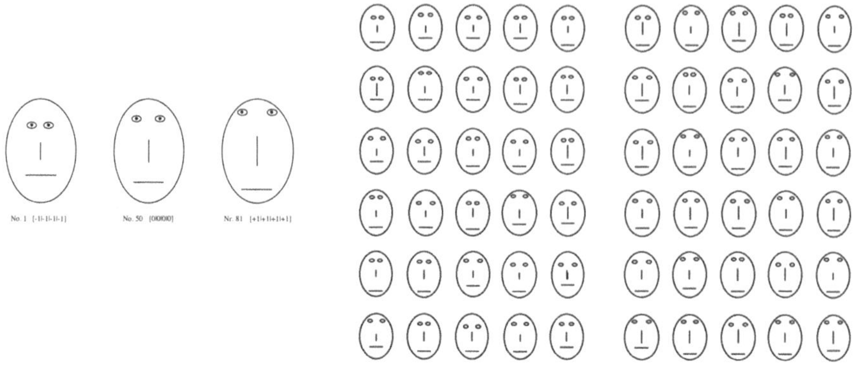

Fig. 1.1 After a few training sessions, pigeons were able to distinguish between two face classes, each with 30 instances, according to a polymorphic feature rule. To do this, they had to correctly combine the four feature dimensions: eye, nose, mouth, and chin. The short eye distance, the deep eye position, the short nose, and the high mouth position received (arbitrarily) the attributes −1. The contrary characteristics received the attributes +1 and the middle ones 0. By summing these values, the class criterion (the sum) was obtained, which varied from −4 to +4. All faces with a negative sum (<0) formed the left class, and all faces with a positive sum (>0) formed the right class [15]. The faces with a sum of 0 were not used. (From [15])

methods. In addition, they can also proceed by extracting prototypes but also in an emergency by just using exemplars or by remembering many copies of the stimulus classes [26, 28–30]. This classification competence of pigeons based on the so-called polymorphic (multiform) feature rules is a characteristic of perception-based categorisation but is only one of the several strategies for prelinguistic concept formation. The important issue here is that the characteristic features can be distinguished from the non-essential features. The basic principle is thus the *abstraction* of unimportant stimulus features.

Categorisation is not unreasonably seen as the basis of cognition, it is a process by which the individual assigns meaning to objects or events by treating them as equivalents [29, 31]. Considering, on the one hand, the enormous amount of information arriving in the perceptual systems of mobile organisms and the few possible behaviours in non-human animals on the other hand, categorisation can therefore be considered an adequate solution to this "information bottleneck". A drastic reduction of information is a basic principle of cognitive economy and is therefore widespread in the animal kingdom. The evolutionary pressure to minimise processing requirements in small information-processing systems, such as bees and pigeons, is obvious. Nevertheless, the specific solutions used by a variety of animal species to compress the amount of information to be processed and stored vary at different levels, from purely perceptual (i.e., selective attention) to the level of learning or representation to the eventual level of thinking.

To behave appropriately, an animal needs a description of the stimuli in each environment that results in a small variation of signals for meaningful behavioural consequences. Somehow, the animal must select certain aspects of her experience for use as a basis for judgement, grouping, and splitting. Then, a balance must be struck between grouping and splitting. Different occasions require different descriptions and different dimensions of invariance. Categories are neither purely perceptual nor purely functional, but both. Categorisation tasks are jointly determined by the physical variation of stimuli and the consequences of behaviour. The available evidence suggests that natural selection has endowed animals with considerable adaptations to deal with the categorisation problem in this sense [32, 33].

Animals can represent or encode stimulus classes in (at least) six different ways [18]: (1) *pictorially* (imaginably) as arrays of features or elements defined in their own absolute values, thus being a matter of assessing similarity; (2) *associatively* as collections of objects or events signalling the same consequence or follow-up event; (3) *functionally* as collections of items with the same inherent function; (4) *abstractly* as relations between two or more objects or events; (5) *analogically* as a relationship between two or more other relationships; and (6) *symbolically* as relations to other classes. The categorisation literature contains a wealth of examples of animals' capacity for one or more of these coding mechanisms [34].

An essential criterion for categorisation is the establishment of connections, not only between stimulus elements but also between members of a class. The most advanced form is based on the establishment of relationships between relationships: Analogies. Many animals tested so far have failed on such tasks, but great apes and grey parrots have succeeded if they have received appropriate (explicit) training beforehand [35]. They use particularly efficient forms of associative learning. Associative concepts provide a basis for human language so that spoken and written words and the objects they represent become a class of interchangeable stimuli.

Unlike perceptual classes, which share common physical properties, the basis for the common response to members of an associative class is completely arbitrary and must be acquired through training. It is precisely the property of arbitrariness that gives associative classes a flexibility that is evident in the extent to which they are used in human language. An associative class can consist of an object and its various symbolic representations. For example, the object itself, the chair, the spoken word "chair", and the written word CHAIR are all members of a common associative class. However, the mechanisms of associative concept learning and associated behavioural flexibility are also evident in the adaptive behaviours of animals without language [36].

The ability to form symbolic representations enables humans to group or combine objects and events in countless ways. This "combinatorial explosion" is certainly one of the most outstanding features of human language. Furthermore, human concept formation differs from the forms of categorisation mentioned so far, especially in terms of the degree of abstraction. All the concepts are formed by abstracting and generalising. In this respect, concrete concepts are also a result of abstraction, even if they designate a specific, given object or a specific class of objects. However, the degree of abstraction can still be significantly increased by humans, either to represent a relation between individual objects or to refer to non-objective entities such as freedom or justice. We (now living) humans are thought to have a "metaphorical capacity for abstraction" that allows us to co-opt skills originally developed for physical problem solving and social coordination to abstract objects and combine them productively [37, 38]. This psycholinguistic phenomenon refers to constructions associated with concrete scenarios but extends by analogy to more abstract concepts. Such metaphors are ubiquitous in human languages—for example, the "sea of flowers"—but do not exist outside languages.

As we shall see, animal cognition research is devoted, among other things, to the question of how far animals can refrain from concrete and accidental conditions and how they can grasp the general, the permanent, or the regular in their conceptions of things and events. The content of animal concepts is questionable, as is their use. How well can they be combined and how are they organised—are they merely strung like pearls or perhaps hierarchically ordered? By using words or symbols, humans create concepts and, with them, their own extremely diverse combinatorics. Through this, and because the concepts are represented by names, it is possible to talk about classes of things or properties of properties. Finally, new meanings can emerge from the structure of concepts, or sentences, which can go far beyond those of the underlying concepts. However, let us stay with the nonstructured concepts for the time being.

John Locke distinguished between simple and complex concepts. Simple concepts are obtained through immediate experience, whereas complex concepts are a product of abstraction and formed by the mind. Both had been equated by English empiricists, including George Berkeley, who had ideas or images in the mind.

Immanuel Kant also still had concepts of experience and understanding, which essentially still corresponded to the categories or concepts described above. Beyond that, however, there are concepts of reason (ideas) that only the cognising subject can have, because laws would never be able to exist in appearances but always only in the cognising subject. Concepts, according to Kant, serve the understanding in their activity of judging and going beyond the merely sensual.

As early as the nineteenth century, the linguistic component of concept formation increasingly came to the forefront. Johann Gottfried Herder and Wilhelm von Humboldt deserve special mention. Concepts in the sense of word, name, or expression are elementary components of language. They replace meaning as the smallest unit of thought, as an idea, or as a concept. Conversely, many think that the fundamental function of language is to enable conceptual thinking. In fact, the way a person thinks is strongly influenced by the semantic structure and vocabulary of their native language—the so-called Sapir–Whorf hypothesis [39–41]. Nevertheless, it is questionable whether this gives rise to certain thoughts that cannot be understood by someone speaking another language. Concepts divide and order extralinguistic reality, and this happens in different ways in many individual languages. So here we are dealing with conventions that have a certain degree of arbitrariness but that do not change one objective reality or recognisability.

The principle of this "linguistic relativity", which was already anticipated in the nineteenth century by Wilhelm von Humboldt with the concept of the inner form of language [42], is far removed from a "linguistic or language-related relativism" that goes beyond and postulates the complete dependence of thought on language. The semantic structure of a language may influence and limit the possibility of concept formation. Our view of the world may be influenced by our language, and we may also arrive at different views of the world, but our fundamental perception and cognition are not completely determined by this.

Thoughts, Judgements, and the Search for Truth

At some points in the course of hominid evolution, our ancestors developed the ability to isolate things, their relationships, and processes and to understand them in their abstract meaning. In doing so, humans can not only "think about something" but also "think something" or "think it" [43]. The human thinking process is characterised by the fact that thinking and what is thought intrinsically belong together. As a result, thoughts can be fixed and passed on. They become an intersubjective common property.

In the philosophy of mind, the concept of thought has changed accordingly. Put simply, one could regard the thought as the result of a thought process. It contains what has been thought of, but subsequently also an opinion, a view, or an idea. Increasingly, however, this subjective, cognitive, and individual understanding of thought was replaced by an understanding in the abstract, objective, permanent, and true sense. With the emphasis on the general, the essential, and the lawful, the validity aspect of thought came to the fore over the origin aspect of it. Logicians such as Gottlob Frege and Edmund Husserl understood thought as a product of the thinking process in the form of a judgement, a concept, or a combination of both. This brings us, after the things of the external world and (subjective) ideas, to the realm of logic. An essential characteristic of this new form of mindless and timeless thought is that something like truth comes into play. "A thought, however, is something of which it is true or false; there is no third thing" [44, p. 36]. This refers to scientific knowledge, such as the Pythagorean theorem, thus moving far away from the original concept of the (psychological) thought in the form of an idea or conception. The question of validity is not about perception but about truth, quite independent of whether or not it is thought of by this or that person. In sum, thought is "the state of affairs asserted by a proposition" [45, p. 223]. A being can therefore only have thoughts if it can know things or be mistaken and if it can have certain intentions.

With the focus on the objective content of a thought, the fact or the so-called proposition, the process of thinking in modern logic also undergoes a fundamental shift in emphasis. According to Frege, thinking is not to be regarded as the production of the thought but as its "grasping". This is followed by judging as the recognition of the truth of a thought. Finally, thought expresses itself in asserting—the manifestation of judgement.

As a result of this rationalist conception, thinking and language become intertwined, mediated via judgement and assertion. However, formal-logical judgemental thinking is not new but has been regarded as the rationality characteristic *par* excellence since antiquity [43]. Logical thinking is not spontaneous, subjective, or able to be psychologised, for it follows the laws of thought (the law of contradiction, the law of identity, the law of the excluded third party, and the law of the sufficient reason) and strives for truth and validity. The focus is on the statement or the assertion as a consequence of formally correct conclusions. However, which *formally correct conclusions* are correct and which are not, in terms of *content* in different areas of human life, is disputed. The sentence "I love and hate you at the same time" may be contradictory, but is it not nevertheless part of our everyday reality of life [43]?

Over time, the meaning of thoughts and concepts has changed in philosophical discourse, essentially by referring to different aspects than in biology and (cognitive) psychology, where the focus is on the subject having thoughts, namely, in the sense of individual representations relevant to the causation of actions. Here, thoughts are primarily thought operations or conscious mental actions. Two important philosophical trends have moved away from this *subjective* dimension. While some focus on the *objective* dimension, in that concepts are related to the world and to its extension, others, following the so-called linguistic turn, focus on the *intersubjective* relation to the language community, which maintains a practice of attributing beliefs. They ask how the word in question is used in everyday speech and what the speaker wants to express with it. In Ludwig Wittgenstein's late work [46] and in the work of the philosophers of language who followed him in the process, it was no longer the reference to extrasubjective reality that was decisive but the internal rules of a language game, i.e., the rules of the way individual words and sentences are used.

Both schools of thought have in common that they attest humans the ability to grasp reality intersubjectively. This approach overcomes the subjective, egocentric perspective and creates an objective reference to the world. In the triangle of speaker, interpreter, and object, something like a common world emerges. Whether (initially) through pointing gestures or (later) through signs, symbols or words, people refer to something outside themselves and thus establish an objectifying reference.

From the first year of life, humans communicate in complex ways through gestures, which can be defined as *intentional signals* in that they are directed at a receiver with the aim of influencing their behaviour in a particular way [47]. Human children use gestures such as pointing (for example, with the outstretched hand and index finger) or offering (of objects) to draw another's attention to a particular object in the environment, especially to share interest in that object or to achieve a more specific goal, such as retrieving the object [48]. These signals are mechanically ineffective—apart from a change in the receiver's behaviour—and are *referential* because they are used to direct the attention of others to an external entity. As *triadic interactions*—between sender, receiver, and object—gestures represent early key elements in human language development.

However, there are now an increasing number of ethological studies on referential communication in animals. Not only in laboratory experiments chimpanzees use pointing gestures to communicate [for example 49], but also in the wild they seem to use targeted scratching to indicate to conspecifics areas of the body where they want to be groomed [50]. Moreover, bonobos appear to use pointing gestures to indicate potential danger to conspecifics [51]. Furthermore, recent studies with birds and fish suggest that referential communication, relating thoughts to objects, is not limited to humans and apes but is widespread in the animal kingdom. Common ravens have been observed to use referential gestures in the context of social bonding between mating partners [52]. Even fishes such as the Roving coral grouper and the Leopard coral grouper are not inferior to large-brained mammals or birds; both use referential signals to indicate hidden prey to cooperating hunting partners who are not even the same species [53]. All of these case studies exhibit five essential features of referential signals: they are directed towards an object, as well as towards a potential recipient; they are mechanically ineffective; they receive a voluntary response; and they demonstrate hallmarks of *intentionality* [52].

Shared Intentionality

Comparative psychologists recognise the deliberate clue signs of the examples given—especially in the cases of apes that are repeatedly used for comparisons with children—but qualify that the animals are concerned with their own individual needs and goals in doing so, taking advantage of or even exploiting others [54]. As animals compete for important resources, they gather information from others to manipulate them as "social tools", and even when they coordinate their actions with them, this usually occurs for their own benefit. In contrast, human children are often busy exchanging information with others by giving them helpful hints, sharing common intentions and attention, and learning from demonstrations produced for their own benefit.

Out of an altruistic and cooperative mindset, infants, the argument goes, would put their already present sociocognitive skills of referential, manipulative communication, social learning, cooperation, group coordination (think of chimpanzees hunting together), and others such as gaze following to the service of the culturally shaped community [54, 55]. Unlike chimpanzees, humans live not only in communities but also in cultures consisting of social practices, norms, and institutions whose very fabric consists of the cooperation and division of labour for a common goal. In the course of human evolution, the sociocognitive capacities and motives for *shared intentionality* have been strengthened and partly transformed, enabling children to participate in the cultural practices around them.

Towards the end of the first year of life, human children develop a strong motivation for shared attention and the transmission of information, emotions, interests, and their own mental states. This motivation extends and transforms the phylogenetically acquired sociocognitive abilities that for 30 years have been described in a wide variety of animal species under the term "Machiavellian intelligence" [56]. Infants already reciprocally exchange not only glances but also emotional states with others. This strong need to communicate (prelinguistically) and exchange information and emotions with others enables growing children to later respond to and even internalise different kinds of social norms, collective beliefs, and cultural institutions [54].

In this early phase, it is predominantly the caregiver, usually the mother or father, who is the target of so-called social referencing, the orientation on the target's emotional body language, especially facial expression, which is important for the evaluation of an action or an event [57, 58]. In addition to cognitive maturation, this kind of reassurance forms an important part of the child's emotional development. Cognitively, social referencing is the beginning of the ability to assume feelings, needs, intentions, expectations, and opinions of other people. When the insecure child reassures itself of the mother's affects, the child already knows that the mother has intentions directed towards the outside world. This attitude can therefore be seen as a preliminary stage in the development of perspective taking or mind reading (see Chap. 8).

Four types of transformations from individualistic to sociocultural tendencies have been studied in chimpanzees and infants [54]: from gaze following to joint attention, from social manipulation to cooperative communication, from group activity to cooperation, and from social learning to instructed learning. Human infants not only follow each other's gaze to external targets and not only want to know what the other sees but also seek to share attention with others. They want to engage the other in their own psychological world and share in their interests, needs, and desires. With pointing gestures, toddlers want to satisfy their own needs (for example, hunger) and to inform others about something they do not know. This kind of referential communication is thus in the service of cooperation. When chimpanzees hunt together, on the other hand, it is not certain whether in doing so they are committed to a common goal or are merely coordinating their own interests [59]. Children, however, seem to do this very early on, and even more, they often cooperate for the sake of cooperation or out of a tendency to conform. For example, they continue a collaborative activity after its (instrumental) goal has already been achieved. Finally, chimpanzees imitate others to solve a task but not with the motivation to show the other that they are "in tune" with the current situation [60]. Children dress like adults, imitate their occupational activities and copy ("overimitate") recognisably useless actions, quite obviously to be like their adult role models. As has already been demonstrated in several studies, great apes do not do this [61] (but dogs do[7]).

It is said that being human is not only to be social but also ultrasocial, that is, cultural. Only humans have developed the "We-intentionality", which is important for collaborative interactions in which individuals share psychological states with one another [64–66]. As the Russian developmental psychologist Lev Semonovich Vygotskiĭ noted almost a century ago [67], the underlying idea is that what makes human cognition unique is not individual intelligence but the ability to learn through other people and their artefacts and to collaborate with others in collective activities. Social interaction between learners and teachers plays an important role, and human

[7] It must be said that dogs are also prone to overimitation when observing ostensively displayed behaviour of the person to whom they have developed a close bond (usually the owner), as we have recently been able to demonstrate [62, 63].

knowledge is largely socially constructed knowledge (so-called social constructivism [68]).

In this context, cultural, pedagogical, and especially normative influence must not be overlooked [54]. Humans not only learn from others but also need to teach others. Children therefore not only learn passively through silent observation, as do most animal species studied so far, but they also learn from instruction by being taught by adults. In doing so, they understand certain *ostensive* signs as requests to pay attention and learn—a form of *natural* pedagogy [69]. Finally, there is a normative aspect of human learning, as children imitate adults "because that is the way it should be". Growing up in a world of symbols, artefacts, and conventions, not everything is understood as functional, but also as culturally acquired. However, norms also mean that there are sanctions, up to and including exclusion from the group, if a child does something harmful to the community. Exclusion is imposed by adults and even by the children playing along, who refer to supraindividual social or moral norms.

Early human cognitive development in the cultural environment seems to have a snowball effect because there are only small, quantitative differences in many basic aspects (including common, shared intentionality) between humans and apes, but the sum of these small differences yields a remarkable qualitative difference in collective and cooperative capacities for culture and cognition.

Images in Your Head

An essential question in attributing mental phenomena is how relevant behavioural changes are in the process. With suitable behavioural tasks, one should be able to detect the effect of certain thought processes. The so-called Turing test, named after its inventor Alan Turing, became famous for this [70]. This test is supposed to be able to determine whether a computer can think or at least has a thinking capacity equivalent to that of a human being. In this test, one would ask any questions but not be able to see who answers them; only the answer should allow one to conclude the source behind them. If one cannot decide after numerous questions whether one is dealing with a machine or a human being, then the computer has passed the Turing test. In the same way, one would be able to draw conclusions about the functional similarity to human thinking in animals on the basis of the answers in specially devised tasks.

Once we have established criteria for the consequences of certain mental activities, specially devised behavioural tests or neurophysiological analyses, to a limited extent, should help to identify the underlying thought processes. In contrast, reference to "images in the head" or mental representations is only an aid to the hypothesis that animal cognition goes beyond direct sensory experience and mere stimulus-response schemata. A separate "*mental language*", distinct from a public language, has even been proposed [71]. However, the danger is that this obscures more than it reveals. As long as we have a mid-level functionalist language between neuron discharges (eliminative materialists) and stimulus–response associations (behaviourists) on the one hand and mental representations (cognitive representationalists) and speech acts (lingualists) on the other hand, we should strive for everyday language and common sense [46].

In most cases, talk of mental representations can be replaced by talk of higher mental abilities, such as special memories or the ability to plan with foresight [20]. In this context, concepts are, as explained earlier, classification rules composed of previously abstracted features. With experimentally testable classification tasks, the feature rules can be identified.

Which concepts we can ascribe to animals also depends on empirical research into animal classification and discrimination abilities. We attribute concepts to animals by uncovering the parameters that determine their discriminative behaviour. Whereby conceptual comprehension requires more than mere discrimination. In fact, the formation of complex perceptual classes is based on the use of several properties of the members, on a distinction between typical and accidental properties, on the appropriate and regular combination of these, and on a situation-dependent, flexible modification and adaptation. Being able to classify in this extended sense also means capturing the normative dimension of terms because following (self-recognised) rules implies the recognition of errors. Through the feedback of application errors, incorrect classifications can be corrected. For example, young meercats learn to narrow the class of ground predators to snakes and to neither warn nor flee from lizards [72]. Learning complex class rules, i.e., rules that encompass several features, and applying them flexibly to the situation satisfy the claim of being intentionally valid and thus subject to normative distinctions. Taken together, the learned classification rules fulfil the criteria of non-linguistic terms: Classification is rule-driven, rule-driven behaviour is intentional, and intentional behaviour is flexible [20].

At this point, at the latest, the philosopher will object that it is not enough to observe animal behaviour. This has to be interpreted, and this interpretation in turn depends on the terms we use for it. The attribution of certain mental states and cognitive abilities depends very much on *which* terms we use or *how* we use them. Speaking, thinking, feeling, even more so having beliefs, being conscious or acting—these things (not to mention terms such as reason and understanding) need to be defined or at least explained before they are used to describe empirical facts. It is therefore not enough to simply collect empirical data. How we interpret these data depends largely on which term we bring to bear and consider appropriate [73].

Clarification of what it means to have mental qualities must be accompanied by very basic considerations about consciousness, language, and thinking. In addition to the purely cognitive ability to gain knowledge and solve problems, there are the conative abilities to demand or intend something. Only then do desires, intentions, and convictions come into view; everything that is intentional, that is, everything that has an object or content [74]. If we now allow animals to act intentionally to a certain extent, do they already act rationally?

Chapter 2
Can Animals Be Rational?

The question of why an animal acts this or that way invokes the further question of whether it acts for specific reasons. In a more sophisticated sense, the capacity for intentional action—not merely giving in to a desire and aiming for a goal but also knowing why, which means acting for specific reasons, which in turn means that the decision taken is based on a reasonable conclusion. Can animals reason, can they draw practical conclusions? More generally, can they be rational? In his famous essay on rational creatures, Donald Davidson [4], as mentioned above, made a clear differentiation about the possession of propositional attitudes. Only creatures that can have beliefs, desires, and intentions, which in turn presupposes the possession of language, are rational. Snails are therefore not, and according to Davidson, human infants are not yet rational either. The point here is that communication partners can exchange ideas about external things, think about the same things, thus sharing their world and objectifying it in the process. They produce an *intersubjective truth* in the sense of objectivity via a process of *triangulation* (2 subjects and one object). Rationality is thus a social construct.

However, this language-based rationality is not the classical perspective. The traditional answer since antiquity is that we humans are rational when our emotions and actions are controlled by our thinking and when they are reasoned, that is, justified by reasons. Going back to Aristotle is syllogism, an argument formed by three propositions, where the subject of the first proposition is the predicate of the second (called premises) and the third proposition is composed of the remaining terms (called conclusion); thus, the truth of the conclusion is guaranteed by the truth of the premises. For the modern philosopher Gottfried Leibniz, logic was the foundation of rationality, which was no longer just a tool for convincing arguments but rather a universal system of rules of thought, according to which even God's thoughts are necessarily logical.

With Gottlob Frege and others, the modern science of logic became propositional logic, which was developed in connection with the problem of founding mathematics and solving language problems, as mentioned above. However, this

L. Huber, *The Rational Animal*, https://doi.org/10.1007/978-3-031-60803-2_2

conception of so-called formal logic is useless for describing cognitive processes in nonlinguistic beings. However, could not more modest logics be devised that are applicable to animals—and perhaps also to humans in their less intellectual moments? Even for those who argue that logic requires language, in the sense of a formal operation on whole propositions or thoughts, analogies to familiar reasoning processes in the absence of a linguistic structure are possible [22, 75]. Otto Koehler [76] described such behaviour as "*unnamed thinking*". A well-known joke illustrates a minimum rationality such as this:

> Clyde is looking for his keys under a street lamp on a dark night. Jane: What are you doing? Clyde: Looking for my keys. Jane: Did you lose them here? Clyde: No, I lost them over there (pointing to a dark alley). Jane: Why, then, are you looking here? Clyde: The light is so much better here. [77, p. 108]

Obviously, reasoning as such is not yet a guarantor of what we intuitively understand by the term *rationality* [43]. In fact, logic underwent a major paradigm shift in the twentieth century when advocates of the classical view had difficulty describing the way we think. After abandoning classical classification logic, many followed Rudolf Carnap's principle of tolerance, according to which there is not one but many logics. One consequence of this was the creation of nonclassical logics, such as fuzzy logic, which provided better tools for describing processes such as pattern or object recognition [78]. Increasingly, studies in economics and human decision-making have documented cases in which not only lay people but also experts strayed far from the rational ideal and followed intuition more than reason [79]. Systematic analysis of these "cognitive biases" has revealed that the patterns of our deviations from being the ideal logician are due to certain competencies and heuristics that are alternative sources of intuitive judgments and decisions [80]. From Sigmund Freud to modern cognitive science, there is agreement that the rationality of men and women is fragile and that many different unconscious motives and simple reasoning underlie our "reason". Logical omniscience is arguably appropriate only for a deity, but it contradicts what human psychology teaches us.

Rationality, it is said, is "*bound*" because it is exercised by actors with limited capabilities and because it is far from being a tool for optimisation or maximisation [81]. Instead, it serves primarily to satisfy the actor's expected utility by selecting those choices that are good enough, but not necessarily perfect, according to his or her desired outcomes. Such moderate conceptions of rationality leave room for widely observed phenomena of suboptimal human thinking. Rather than being discarded as "unintelligent behaviour", these psychologically more realistic models of human decision-making can explain the countless deviations from perfect rationality as a symptom of the fact that we use formally imperfect, but often more efficient, "*fast and frugal*" heuristic procedures in everyday life [82].

Full-fledged rational thinking may indeed be a purely human capacity, but it might be limited to rare situations, such as philosophising and writing books. Everyday thinking might be better described as "*ratiomorphic*", a term coined by Egon Brunswik [83], proposed to characterise processes that are very similar to rational behaviour in both formal and functional terms but have little to do with conscious reason. Both Konrad Lorenz [84] and my teacher Rupert Riedl [11] distinguished a system of "*innate forms of ideation*" that allows us to form space, time, comparability, causality, finality, and a form of subjective probability together into a system of unconscious, subpersonal human thought. When we confronted people with a series of events and, as they became more apparent, asked whether the sequence appeared regular or stochastic, we were able to identify two problem-solving strategies for people of different ages and education levels [85, 86]. While on one side of the spectrum, mathematicians with enough time to make a decision displayed a classically rational strategy based on calculating probabilities, on the other side of the spectrum, children and nonmathematicians who were under pressure to make a decision revealed a ratiomorphic strategy. The latter is strikingly similar to the behaviour of pigeons or rats in the "Skinner box" [87], where the number of confirmations or punishments, or rewards versus disappointments, determines decisions and choice behaviour. However, as we shall see, modern conditioning theory can explain incomparably more complex behaviour. Its basic elements go back to the principles of the learning theory of Ivan Petrovich Pavlov and Edward Lee Thorndike [88].

This brings us to a central point in the debate about animal thinking. Are all, even the seemingly "intelligent" and human-like thinking abilities of non-human beings simply due to associative learning? Are we dealing with animal cognition only with (innate or learned) abilities instead of reason? Are even the decisions and actions of animals, which in humans are unhesitatingly explained by logical, deductive reasoning, also only a consequence of associative learning? A famous and often reviewed task is "transitive reasoning", which ultimately goes back to Aristotelian logic. If Rudolf is greater than Klaus and the latter in turn is greater than Robert, who is greater, Rudolf or Robert? The Swiss developmental psychologist Jean Piaget had already confronted children of different ages with such questions to test their ability to deduce a relationship between things that had never before been presented together [89]. Interestingly, Piaget did not find this ability of basal logical thinking in the so-called *preoperational phase* between two and seven years of age, but only in children in the stage of "*concrete-operational intelligence*", i.e. between seven and 11 years of age. Before that, children can infer inductively, i.e. generalise from individual cases, but they are not able to deductively infer the logically compelling consequences from given premises. In cognitive psychology, this is referred to as emergent relationships and the use of prior knowledge to determine the missing piece of the puzzle with the help of basic logic, i.e. with the inferences of certain premises.

Much of the experimental work on human reasoning has focused on deduction. This is perhaps partly because many researchers believe that the ability to perform valid deductions is the sine qua non of rationality [90]. Later, researchers in comparative cognition developed a nonlinguistic method, replacing pairs of premises (or propositions) with successive discrimination tasks [91]. They trained squirrel monkeys sequentially with four stimulus pairs: A+ B−, B+ C−, C+ D−, and D+ E−. The letters represent different arbitrary stimuli, and the signs indicate whether the choice of stimulus in each pairing was rewarded (+) or not rewarded (−). In the crucial test (on the understanding of transitivity), the animals were presented with the novel pair B D. Although the two stimuli were never presented together before but were only together with other stimuli once rewarded and not once rewarded, the animals almost always chose stimulus B. The authors interpreted the preference for B over D in their monkeys as forming a linear representation of the pattern series and as consistent with the principle of transitive logic [92]. Not only chimpanzees but also rats, pigeons [reviewed in 93], and 6-year-old children have shown very similar response profiles to squirrel monkeys [94]. Meanwhile, many more animal species have been tested with similar serial-pair tasks, and indeed, with the exception of honeybees, all species tested thus far exhibit transitive-like behaviour [95]. However, whether this overwhelming frequency means that animals truly draw logical conclusions is the subject of debate that continues to this day. With the current state of knowledge, the arguments for cognitive-logical models and those for reinforcement models based on associative conditioning balance each other.

However, as a caveat, it must be said that laboratory experiments with arbitrary, artificial pairs of patterns do not adequately reflect the ecological conditions for intelligent behaviour. Cognitive ethologists also reject the claim that associative learning is the most parsimonious and therefore the best assumption against which to test cognitive models, arguing instead that transitive thinking is an evolutionary adaptation to life in large, stable social groups.

How can the evolutionary emergence of transitive thinking be explained functionally? Social dominance relationships typically do not manifest themselves in externally perceptible features on the individual. If animal A dominates animal B and animal B dominates animal C and no direct comparison of animals A and C is possible to reveal the relationship between them, transitive reasoning would be very helpful, especially if the loss of a fight could lead to injury and reduced access to food and other resources, with potentially severe consequences for fitness [96].

Since the number of possible pairwise interactions increases rapidly with group size, members of large social groups benefit if they can infer relationships between them based on indirect evidence by observing the interactions of others. For example, social dominance hierarchies in troops of bear baboons (or chakma) consist of 80 or more individuals, each of which theoretically faces 3160 different dyadic combinations and 82,160 different triadic combinations [97]. In birds, too, there are large, stable groups with linear dominance relationships, for example, in the Pinyon jays, which are very gregarious members of the corvid family. In a very sophisticated experiment, individuals were allowed to observe interactions between conspecifics from other groups and later interacted with some of these birds themselves [98].

The interactions consisted of controlled encounters in an experimental aviary after raising partitions and the resulting competition for a central food source with unshelled peanuts (Fig. 2.1a). In such encounters, the dominant bird keeps the upper hand, and the lower-ranking bird gives way to it. The greater the difference in rank, the more submissive the latter behaves, while quarrels do occur with opponents who have small differences in rank. The experiment was intended to test whether the jays could draw conclusions about their own dominance status in relation to that of strangers they had observed interacting with known individuals.

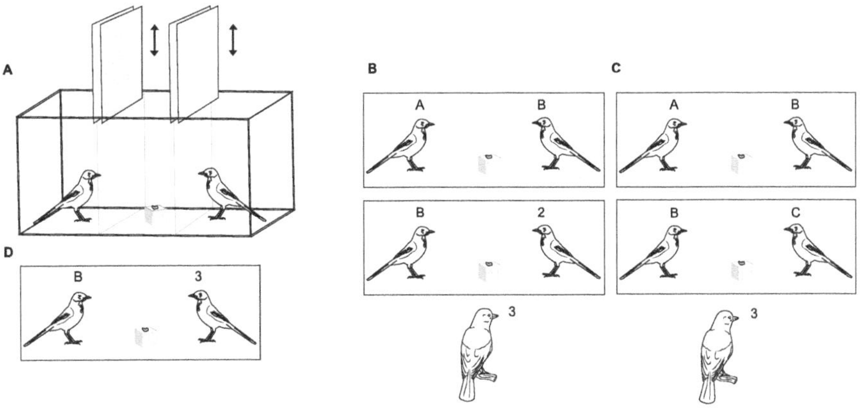

Fig. 2.1 Experiment to test the ability of Pinyon jays to draw conclusions about their dominance status compared to that of strangers whom they observed interacting with known individuals. In a chamber divided into three compartments by opaque and transparent dividers, a small box was filled with a single unshelled peanut in the central area. (**a**) By opening the dividers at the same time and thus creating competition for the food, the dyads were tested for relative dominance. (**b**) A third bird whose prior knowledge of the opponents was systematically controlled could observe their interactions. Explanation in the text. (From [98])

For this purpose, the prior knowledge of the observer (3) about one of the opponents of a demonstrator (A) was systematically varied. In the experimental situation, the demonstrator (A) and an opponent (B) were foreign to the observer (3), but the opponent (2), who repeatedly lost to the demonstrator (A), was familiar to the observer (3); the opponent (2) had dominated the observer (3) in previous staged encounters. In the control situation, on the other hand, observer (3) had no prior knowledge of the observed opponents. This should rule out alternative explanations, such as direct recognition of the rank of a conspecific through status traits or specific dominance behaviour.

In the experimental situation, observer (3) witnesses how, on the one hand, demonstrator (A) is dominant over opponent (B)—A gets the food, B holds back—and, on the other hand, opponent (B) is dominant over opponent (2) (in simplified terms: A > B and B > 2 (Fig. 2.1b)). Demonstrator (A) and opponent (B) are unfamiliar to observer (3), but in previous intergroup tests, opponent (B) was found to be dominant over opponent 2; therefore, observer (3), if endowed with the capacity for transitive reasoning, should be able to anticipate being subordinate to B. Specifically, observer (3) should make inferences about his rank and show a graded, quantitative response based on his observations. He should behave submissively towards B and not try to obtain food (Fig. 2.1d).

In contrast, in the control trials, where both the demonstrator (A) and its two counterparts are unknown to it, observer (3) cannot use transitive logic to predict its status relative to B, having observed the interactions A > B and B > C (Fig. 2.1c).

Indeed, the jays drew inferences about their own dominance status relative to that of strangers they had observed interacting with known individuals. For the authors, the results are fully consistent with the hypothesis that the jays used transitive reasoning to make inferences about their own relative dominance, supporting the hypothesis that social complexity is a crucial cause for the development of cognitive abilities.

The example of the jays reveals a case of practical rationality. It does not tell us what thoughts and considerations the jays have during dominance encounters, but their behaviour certainly cannot be explained by rigid, instinctive behaviour, rigid stimulus–response mechanisms or step-by-step trial-and-error learning. Quite obviously, animals process observations very quickly in a way that would hardly be possible with the previously mentioned mechanisms. Apart from ready-made solutions based on innate stimulus–response mechanisms, the two major opponents on the stage of animal cognition remain for solving novel problems: learning and thinking. This distinction has its roots in the debates between the German Gestalt psychologist Wolfgang Köhler and the American psychologist of learning Edward Lee Thorndike that took place in the first quarter of the twentieth century. While Thorndike [99], with his *law of effect*, advocated learning on the basis of the increasing, positive effect of *reinforcement,* Köhler [100] contrasted this with the possibility of sudden solutions through mental reorganisation of problem elements. However, while gradual learning based on association formation became accepted as the standard for animal problem solving in learning theory dominated by behaviourism, the concept of "*insight*"—with only a few exceptions—fell into disuse [101]—it was either forgotten or misunderstood [102].

Three quarters of a century later, the distinction between learning and thinking was addressed again, this time in connection with the question of symbol processing, connectionist models, and theoretical considerations [103]. Although "the psychology of thinking seems to create dichotomies" [104], one is only as valuable as the explanatory power of the hypothetical systems it distinguishes. Unlike associative learning, where perceptual similarities and temporal relations are important, "insightful" thinking is "rule-based" because it operates on symbolic structures with logical content and variables and because its solutions have properties that are normally assigned to rules. It seems that these two information processing systems serve complementary functions, but can they simultaneously generate different solutions to the same "thinking problem" [103]?

Logic and Reasoning

It would be wrong to dismiss thinking as nonassociative information processing. Instead, one can distinguish learning and thinking by the type of information that the individual associates with. Whereas learning involves the association of two perceptual events, thinking is based on the association of a perceptual and an imaginary event—that is, perceptually unavailable event [105]. Accordingly, the task of learning is to identify regularities of either purely external events (classical or Pavlovian conditioning) or associations between one's own actions and external events (instrumental or Thorndikean conditioning). In contrast, reasoning can be understood as the search for the causes of events that cannot be explained and as a certain understanding of the observed phenomena. In addition, in learning, linkage usually occurs slowly after several to many repetitions of the events, namely, in the form of Hebbian reinforcement of the synaptic connections between the neurons representing these events in the brain, whereas inference can occur upon a single presentation of the premises.

One way of testing whether animals can spontaneously make associations between perceptual and imaginary events is by confronting them with tasks in which information is missing. Suppose you fill two boxes in front of the animal's eyes, one with a banana and the other with an apple. After a short pause, you eat the apple and then let the animal choose a box. Will it often or always choose the box that still has the banana in it? If this task had never been given before, so that the subject could not learn, would she still act correctly immediately? Would she understand that the box with the fruit that the experimenter did not eat (the one with the banana) is still filled?

Thinking by exclusion or thinking about alternatives is a relatively seldom-noticed phenomenon in comparative cognition research, as it was originally considered a hallmark of human cognition and related to language acquisition. Chimpanzees solved the task without hesitation, not only after a phase of trial and error [106]. The consequence of eating an apple in terms of the subsequent states of the two boxes was not revealed before the correct action, so this information was not available but had to be inferred. According to the authors, the monkeys showed "reasoning by exclusion". Not only could these results be replicated [107], but also they were further extended in subsequent studies. First, the task could also be solved spontaneously by other great apes—bonobos, orangutans, and gorillas—and they were able to fill in the missing information in both the visual and auditory domains [108]. The animals chose the correct cup when only the empty cup was either revealed or shaken silently, meaning that the apes chose correctly without having seen or heard the food. The statistically significant correct choice behaviour[1] in the very first trials proved that they did not simply learn quickly to use sound information as cues. In addition, in special control tests, it was ultimately shown that the animals were not merely attracted to noisy cups and were repelled by silent cups. Thus, explanations based on fast learning and innate behavioural tendencies were ruled out.

[1] Statistical significance is an essential characteristic of empirical studies. It is determined by statistical tests. The result of a statistical test is called statistically *significant* if the values found deviate so strongly from a previously established assumption that it is rejected. This basic assumption, the *null hypothesis*, is that a statistically found result has also come about by chance without the presumed effect. According to common practice, a significance level is set beforehand for the said deviation, which is called the *probability of error*. The question is how likely it is that the null hypothesis was rejected by mistake, i.e., rejected although it is correct. The so-called p value is used to estimate whether the probability of error was exceeded. The significance level does not indicate the probability that a hypothesis is correct, but only the probability that the null hypothesis is rejected although it is correct. Such an error is called a type-1 error or α-*error*.

In the past decade, several other nonhuman animals have been found to have the ability to eliminate alternatives. Ricoh, a Border Collie, became famous when he was trained by his owner to associate more than 200 everyday objects with words and to bring them up when the object was named. In front of an audience on a German TV show, he was not only asked to fetch an object with an acoustic name he knew from another room he then had to find an object with a name he did not yet know. In fact, after an initial surprise at the new word, he brought the only object he did not know from the next room [109]. Since the dog had never perceived either the new word or the object before, his choice must have been based on the spontaneous formation of a new link.

Children also learn new vocabulary partly in this inferential way, which is called *fast mapping* in linguistics [110]. Chimpanzees, sea lions, seals (and dolphins) have also spontaneously drawn the right conclusion in a very similar way when confronted with some kind of sign language or computer screen tasks [reviewed in 111]. In doing so, they not only demonstrated reasoning skills but also showed important elements of human language.

The ability of animals to speak has been tested not only in mammals but also in birds. Research by Irene Pepperberg from Harvard University (USA) with Alex, the grey parrot, attracted particular attention. The researcher used human language to set tasks and formulate questions. When Alex was given a tray with wooden dice he knew and metal thimbles he did not, he was asked for the number of "woods;" he then gave the correct answer [112]. However, when Pepperberg wanted to know the number of "thimbles"—a word Alex had never heard before—he gave the total number of items—suggesting that he may have interpreted "thimble" as the word for "total". After being corrected and asked again, Alex gave the correct answer. On the next three trials with new objects, Alex first gave the number of all objects on the tray and then the number of objects with names he did not know. On the fifth attempt, he immediately gave the correct answer.

Now, it can be argued that Alex and his successors in Pepperberg's laboratory were intensively trained animals who had to master linguistic tasks almost daily for several years. Therefore, it made sense to test birds that had no prior experience in this regard. Would they also be able to solve tasks by inference? My colleagues developed an elegant and, above all, ecologically relevant and everyday variant of a test paradigm previously used with children and apes [107] and confronted corvids and parrots with it [113]. Common ravens and New Zealand mountain parrots (Kea) were similarly tested while searching for a piece of food hidden in either a straight or a curved tube. The main focus of these simple tests was not on the number of trials the birds needed to find the hidden food but on the search strategies used. Would the animals proceed deductively and efficiently in the sense discussed earlier, or would they act more exploratively according to simple rules of thumb? This question was investigated with ravens and kea because these birds are known not only for their high intelligence but also for different foraging behaviours and object exploration. Thus, the experiment had an ecological note in addition to a cognitive note.

We can roughly distinguish three approaches to foraging. *Excessive searching* means that a subject continues searching even after she has already seen the food. *Insufficient foraging* occurs when a search is terminated before sufficient information about the location of the hiding place is available. Finally, an *efficient search* for food is terminated when sufficient information to find the food has been acquired. These differences are evident when comparing straight and curved tubes. A single glance into the straight tube, no matter which side, reveals whether there is food inside. In the positive case, there was no need to look further, either on the other side or in the other tube. However, if the first look does not reveal a piece of food, the search must continue. A further examination of the bent tube would make sense, but not a look from the other side of the straight tube. In the experiment, the glances into the tubes were counted alongside the actual choices, i.e. which tube the birds touched to obtain the piece of food. If the correct tube was chosen, the animal received the piece of food hidden in it; otherwise, the animal went away empty-handed.

The experiment yielded interesting and, at first glance, contradictory results: overall, the kea were more successful, but the ravens proceeded more efficiently or more logically to be precise. The latter followed exclusion logic, while the kea inspected the straight pipe from both sides in 35% of those cases in which the first look was negative [113]. This result was explained by the divergent foraging ecology of these two bird species. Kea are hardly exposed to predation pressure, so they exploratively search an area, mainly the ground, for plants, fruits, and roots [114]. Whereas, the carrion-eating ravens must be more cautious in light of the risk of getting in the way of a fox, wolf, or bear. They inspect the situation in detail and, if necessary, warn each other before flying to a carcass on the ground from their safe places in the trees [115]. Therefore, while the raven thinks twice before deciding on a risky action and inspects its surroundings very carefully when hiding food, as we shall see, for the parrot threatened with starvation in winter, it is not only safe but also advisable to look a second time to avoid overlooking a nutritious piece of food. The interconnection of cognition and ecology is—as will be discussed—a basic principle in the evolution and species-specific development of cognitive abilities.

One consequence of this entanglement for the question of what kind or degree of logic animals possess is the change in the method of verification. While everyday situations provoke the use of everyday heuristics, animals might be forced to use the full potential of their cognitive abilities when they have to push their limits in abstract situations. My colleague Ulrike Aust came up with an abstract test for inference by exclusion. This discrimination task performed on the touch screen has two important advantages. First, the animals' choices can be objectively and easily evaluated quantitatively. It is the computer that sets up the preprogrammed tasks and records the animals' choices. Second, the task is so abstract that it allows for a fair comparison between many animal species and also humans.

In the first application of this abstract test design, six pigeons, six dogs, eight children, and six human adults were exposed to the same visual stimuli and the same training and testing procedure [116]. First, Ulrike presented eight pictures, four of which were assigned to the positive (S+) class and four to the negative (S−) class

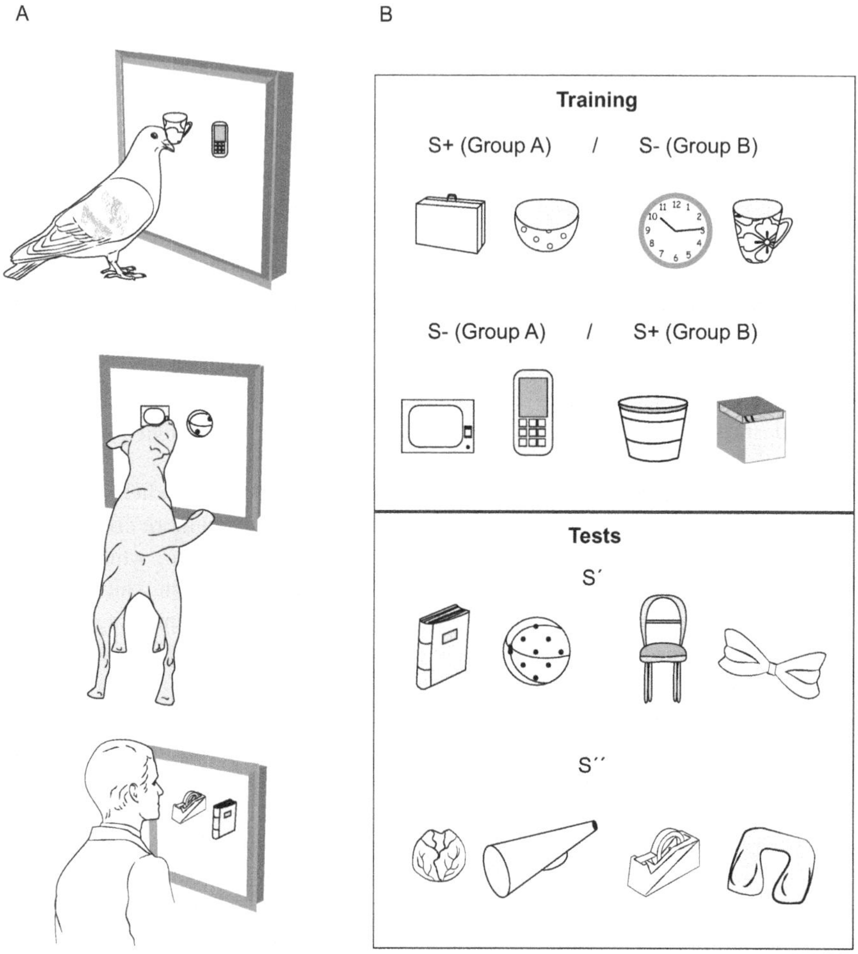

Fig. 2.2 Test for inference by exclusion. A pigeon, a dog, and a student learn to select one of two pictures by touch. The pictures are simple clipart images from the Internet. In training, four pictures were rewarded, and four were not. Four new pictures were used for each of the first (S′) and second tests (S″). (From [116])

(Fig. 2.2, top right). One positive picture and one negative picture were presented for each of the possible combinations during training. The subjects had to learn the assignment of the pictures by trial and error—choosing the positive picture was rewarded with food (pigeons and dogs), sweets (children), or symbols (adults), while choosing the negative picture gave nothing, only the screen turned red. As soon as they reliably solved this task, new pictures (S′) were introduced in the first test (Fig. 2.2, bottom right). These were paired with a negative image known from training, and their choice had no consequences; there were neither rewards nor the red screen as a signal of an incorrect choice. This was to check whether the subjects chose the new picture by exclusion. In fact, all the children, five adults, three dogs, and a single pigeon did so.

So far, so good. However, this procedure allowed for two other explanations for the choice of undefined test images. The new picture could have been chosen because of mere curiosity or avoidance of the known negative picture. To rule out these alternative explanations, we showed the first test pictures (S′) together with completely new test pictures (S″) in a second test (Fig. 2.2, right, bottom). Now, curiosity would lead to the choice of the new test pictures (S″), and the avoidance strategy would not allow any preference. On the other hand, the consistent application of the exclusion principle in the first test would have required the assignment of the test images (S′) shown in the process to the positive category and would have had to keep this assignment in the second test as well, since the completely new test images did not provide any information. Now the wheat was separated from the chaff, as the only pigeon remaining in the test preferred the new test pictures (S″), while all three dogs, three out of five adults, and six out of eight children stayed with the pictures of the first test (S′) and thus showed the ability to reason.

As so often in science, our joy at the result did not last long. Our method has innovative aspects but also has several shortcomings. Since the right choice was not rewarded in the first test, this might have led to a departure from the chosen strategy based on the exclusion principle. Moreover, our test procedure was geared toward a confrontation between inference and curiosity, which must lead to an unnecessarily hard decision in very curious minds. In two other experiments with birds known for their unbridled curiosity, kea and Goffin's cockatoos, my PhD student, Mark O'Hara, eliminated these methodological weaknesses [117, 118]. The essential trick was to satisfy curiosity already in training by replacing the negative class picture twice during each training session with a new picture each time.

In a total of 30 sessions with 20 individual tests each, the familiar training pair (S+, S−) was shown 16 times, with four test pairs (TEST 1–4) interspersed between them. Thus, a total of 120 new test images were shown. In contrast to our earlier study [116], the correct choices according to the exclusion principle were rewarded. Although the four test pairings were always inserted at different positions (at least two training trials between each test), their internal order remained constant (TEST 1, TEST 2, TEST 3, and TEST 4) because it stemmed from a particular logic (Fig. 2.3): TEST 1 contained the negative training image (S−) and a new image (S1); TEST 2, the positive training image (S+) and a second new image (S2); TEST

Fig. 2.3 Modified test for inference by exclusion in Goffin's cockatoos and kea. Shown is the sequence of test images in one test run for the two groups A and B, which differ only in test pairings 2 and 3. Note that the position within the 16 training pairings changes, but the internal order remains the same. (From [118, 119])

3, the image S1 (now already shown once) and another new image (S3); and finally, TEST 4, the new image from TEST 2 (S2) and a fourth new image (S4). Following the exclusion logic, image S1 should be chosen for TEST 1 because it is paired with the known negative training image. In TEST 2, the well-known positive training image should be chosen, and at the same time, the new image S2 should be assigned to the negative category. In TEST 3, for the first time, there is no training image to base one's decision on. Here, picture S1 shown in TEST 1, which was assigned to the positive category, should be chosen, and the completely new picture S3 should be assigned to the negative category. Finally, in TEST 4, the choice should fall on the brand new picture S4 because it appears with picture S2, which should be assigned to the negative category in TEST 2.

To check what effect memory plays on the solution of this difficult task, a second group of animals (B) was confronted with a sequence in which TEST pairings two and three were swapped (Fig. 2.3). This meant that the assignment of a new picture (in the example of Fig. 2.3, the kiwi and the toaster) was immediately checked in the following test (1–2, 3–4), which requires a lower memory performance. As complex as the test design was, as varied are the possible outcomes of the experiment. For group A, the choice of the first test image (the kiwi in Fig. 2.3) follows the logic of our earlier experiment [116]. However, since this time the choice of the test pattern was rewarded, one could explain the correct choice in the third (for group B in the second) test with learning from a single experience (*one-trial learning*). However, this is no longer possible in the case of the toaster (Fig. 2.3) in TEST 4, as it was not rewarded before. Only the correct assignment to the negative category would explain this preference over the heart.

To evaluate the animals' performance with regard to the presence of the exclusion logic, we took a very conservative count. We considered only those test runs successful because of the exclusion logic in which all four individual tests were completed correctly. Since this is only 1 of 16 possible combinations for the four decisions, it results in a random probability of 1/16 (6.25%). Only a statistically significantly greater frequency would justify the assumption of inferential reasoning. In fact, this was the case in the majority of the animals; in 8 out of 14 kea and in 8 out of 12 Goffin's cockatoos [117, 118]. As a group, the kea were successful on average in 16% of the cases, and Goffin's cockatoos were successful in 20% of the cases, which exceeded the chance probability by a factor of 3. In runs in which they did not pass all four tests, they preferred other strategies, such as learning from a single experience or avoiding the negative training image; choosing based on curiosity, on the other hand, hardly ever occurred. Thus, we were inclined to assume the presence of inferential thinking—after chimpanzees, sea lions, seals, dolphins, grey parrots, and dogs—is also common in kea and Goffin's cockatoos. The abstractness of the task, the considerable number of alternative reaction patterns, the impossibility of associative strategies, and the large number of new test images substantiate the conclusion that thinking without language is common, at least in birds and mammals.

Act Willingly

Experiments on inferential thinking are important for determining whether animals are capable of spontaneously associating a perceptible and an imaginary event. Can animals make mental connections and thus control their actions even without language? As we have seen, comparative cognition researchers tend toward the now well-founded assumption that some animal species mentally fill in missing or incomplete information and subsequently act on the basis of what they merely thought. This represents two major achievements in the evolution of the mind in terms of increased behavioural flexibility. On the one hand, behaviour can be decoupled from the here and now, i.e. at least partially freed from incoming stimuli; on the other hand, a completely new problem can be solved without first having to go through a lengthy learning process.

This brings us to a central problem of animal thinking. The problem consists of the key questions of which functions animal thinking must fulfil in principle and for what purpose certain cognitive performances have developed in the course of evolution. There can be no doubt that the most fundamental problem of any living being is deciding what to do next. According to Thomas Goschke, a motivation psychologist from the TU Dresden, the ultimate function of cognitive processes such as perception, attention, and memory is enabling living beings to select, under changing and only partially predictable environmental conditions, from the multitude of possible behaviours those that promote survival, reproduction, and well-being [119].

It is therefore advantageous not only for humans but also for animals not only to be bound to the current stimulus situation, the current needs, and to what has once been learned but also to gain control over one's own actions to be able to choose from the available alternative actions the best ones not at the moment but in the long term.

One can understand this type of action control as an extended form of "cognitive control", which describes the ability to restrict attention to goal-relevant information and to inhibit or suppress irrelevant distractions [120]. We humans can do more than react reflexively to sensory information that is immediate and salient. We can sometimes direct our behaviour towards distant goals. To this end, neural mechanisms have evolved in human evolution that can either attenuate or amplify reflexive and habitual responses to orchestrate behaviour in accordance with our intentions. Their function is primarily to control lower-level sensory, memory, and/or motor processes for a common purpose. Cognitive control is thus essential to what we mean intelligent behaviour.

Cognitive control is the collective name for a set of regulatory and executive processes that attract and focus attention, manipulate and evaluate information, initiate the search for new information when needed, remember past events to anticipate and plan future responses, and address distraction and impulsivity when they threaten goal attainment. These are not only characteristic features of our cognitive system but also milestones in our cognitive development. It is precisely in relation to the focusing of attention that complex states of consciousness up to and including self-knowledge go hand-in-hand in humans. In addition, there are the abilities related to time travel, including episodic memories and prospective planning, as well as metacognition, mind reading, and creativity. For many authors, the focus of attention on processes of monitoring, controlling, directing, and regulating is precisely the special function of consciousness [120, 121].

An important assumption for the evolution of cognition is that even before the emergence of modern humans (*Homo sapiens*), different solutions to the problem of extended behavioural control and flexibility were found, but they all share one central feature, namely, the decoupling of behaviour from the immediate stimulus situation and need. This decoupling is not an all-or-nothing feature but happened in the course of evolution in varying strengths and degrees. Even if they reach their highest level of performance (thus far) with humans, brains with neuronal control mechanisms that enable rapid and effective adaptation to changing environmental conditions have proven themselves many millions of years earlier. The decisive factor here was that the old, relatively rigid control mechanisms did not disappear but continued to exist in parallel to cope with unchanging problems in a proven way. While cognitive control, the ability to restrict attention to goal-relevant information, supports the solution of a variety of difficult cognitive problems, there are also everyday functions that require little to no control and others that even benefit from reduced control.

A central component of the instinct theory of classical comparative behavioural research, as advocated by Konrad Lorenz, Niko Tinbergen, and Wolfgang Schleidt [122], is the innate releasing mechanism (IRM). It is a kind of stimulus filter that responds to the so-called key stimuli in biologically relevant environmental situations and triggers species-specific, stereotypical behaviours (motor coordination). These fixed action patterns (FAPs) with very little variance in the manner in which the actions are executed have evolved over the course of the evolution of a species. Such instinctive movements enable individuals to solve many everyday tasks quickly and adequately, as they permanently occur in the environment to which the species is adapted [123]. As a result of neuroethological research to elucidate the neuronal mechanisms of instinctive behaviour, we later speak more generally of trigger systems in which trait filtering, locating, motivational and learning systems interlock, as has been described, for example, for the prey-catching behaviour of the common toad [124].

It is now evident that completely rigid systems do not provide adequate solutions for the behaviour of complex organisms. These solutions include sensory-motor modifications of invertebrates, such as honeybees, which exhibit adaptive capabilities that go beyond simple stimulus–response links [125]. Most animals can modify their behaviour in a variety of ways, beginning with the modification of the IRM by learning additional features characteristic of the releasing stimulus configuration.

According to Lorenz, if one prefers a broad definition, learning can be seen as an adaptive modification of the complex, physiological mechanism responsible for the functions of animal and human behaviour. The central paradigm is associative learning, although there are other nonassociative forms, such as habituation and imprinting [84]. In the cognitivist literature, learning is generally understood as a form of internal representation of the relationship that exists between events in the animal's environment. In other words, learning is the acquisition of knowledge [88]. This associative knowledge is acquired by progressively strengthening the connection with each effective experience of a relationship between events [126].

Behaviour is not only adapted to changing environmental conditions through learning but can also be modified at relatively short notice depending on the need stimulated at the time. Common to all the processes described in this chapter is that the organism reacts to the "here and now", using mechanisms that harness past experience whether acquired in evolutionary periods, as in IRMs, or in shorter periods, as in learning. An essential step in the evolution of cognition now consists of decoupling from the here and now in "*decentration*", that is, the transposition of thoughts in space and time that enables action directed toward the future. In human psychology, or more precisely, in the psychology of motivation and action, the term *voluntary* (or *volitional*) *action* is commonly used [127]. The capacity for voluntary action is considered essential to human nature. Unfortunately, this phenomenon is disputed between philosophers and neuroscientists due to the sometimes heated discussion about "*free will*", which is treated only stepmotherly in animal research [128]. The topic has also traditionally been dismissed as unscientific in neuroscience and behavioural biology, perhaps because the mechanisms that lead to action have long been unclear. However, recent research has identified networks of brain areas that underlie voluntary action [129]. In humans, these areas generate information for upcoming actions and cause a distinct conscious experience of wanting to act and then controlling one's own actions [130]. The group of decisions about whether to act, what action to take, and when to take it is called will or, more technically, volition.

Volition is defined as a category of mental functions that mediate the coordination of a large number of individual subfunctions—perception, attention, cognition, emotion, motivation, activation, and movement control—on the basis of a unified control principle. The control principle is called "intention" or "goal" [125]. The function of voluntary processes is to strengthen the reaction tendencies compatible with an intention so clearly that they can be carried out in place of the initially stronger habitual or impulsive reactions [131, p. 678].

Voluntary actions are characterised by two features: (1) the ability to mentally represent target states, i.e. desired effects and results of one's own actions, and (2) the ability to anticipate future needs [119, 132]. The first aspect, the ability to anticipate effects of action, is initially based—both in the phylogeny and in the individual history (ontogeny) of humans—on the sensory effects of movements. Later, increasingly richer associations between actions and their effects are added; these associations are internalised and increasingly contain sequences of actions further in the future in the imagination. Over time, the human child—and probably many animals as well—forms a whole network of associations between actions and the effects these actions produce under certain conditions [133]. If the effects of actions are associated with positive emotions and are thus increasingly desired or wanted, they take on the attribute of action goals or goal states. One speaks of goal-directed behaviour when the subject is able to select actions that bring about the goal states on the basis of the idea of the goal states and in the absence of direct stimulus influence. Instead of merely achieving the desired effects of actions through associations, voluntary actions are directed toward the attainment of mentally represented, possibly distant goal states [133, 134].

In everyday life, we speak of intention and mean the mental state with which we strive for the realisation of a goal with a certain commitment and consider it achievable; we are also determined to put the necessary action steps into practice. Although stimuli can initiate an action, they no longer determine the action itself because this action is controlled by the idea of the goal to be pursued, which is the priority at the moment. Action control, action monitoring, and action regulation are synonyms for this capacity. The control of actions by mentally represented goal states not only enables a far-reaching decoupling of behaviour from the immediate stimulus situation (*stimulus dependence of* voluntary actions) but also the ability to plan, whereby elementary actions are recombined into new behavioural sequences and "mentally simulated" before they are actually executed [119]. The highest form of voluntary action lies at the end of a continuum that has simple reflexes at the other end. Thus, while reflexes are immediate motor responses whose form is determined by the stimulus situation, the occurrence, timing, and form of a voluntary action are not determined directly, or at best, very indirectly, by a recognisable external stimulus. Such actions are not governed by the immediacy of sensory input, a hallmark of cognition [135].

Action goals not only have to include action effects but also needs. The anticipation of future needs, i.e. the ability to anticipate needs in one's mind that one does not yet have, is a further step toward greater behavioural flexibility [136]. Instead of foraging only when one is hungry, one could also anticipate a future state of hunger and stockpile food accordingly. Of course, many animals do this out of inner drives and genetically preprogrammed behavioural dispositions without having to imagine winter hunger while in autumn. The particular challenge for the subject is not only to anticipate future needs while ignoring the current state of need but also to even tolerate a momentary deterioration of the current state of need in the process (*detachment from needs*).

The ability to anticipate future action effects and needs can greatly increase behavioural flexibility and problem-solving capacity. However, the orientation of behaviour towards a distant goal means that decisions must be made between alternative courses of action, which becomes more difficult the more of them there are. Moreover, conflicts can arise when one simultaneously pursues two or more goals that are incompatible or because they conflict with currently stimulated needs or strong habits. The challenge for cognitive systems is to gain control over various response tendencies, thereby coordinating a large number of sensory, cognitive, and motor subsystems in the sense of higher-level goals and ultimately suppressing inadequate responses. Only in this way is the coherence of behaviour and fast, goal-oriented action possible.

Neuroscientists have made great efforts to decipher the neuronal basis of cognitive control. Great progress has already been made in this regard, as networks in the brain underlying voluntary, goal-directed actions have been identified both in

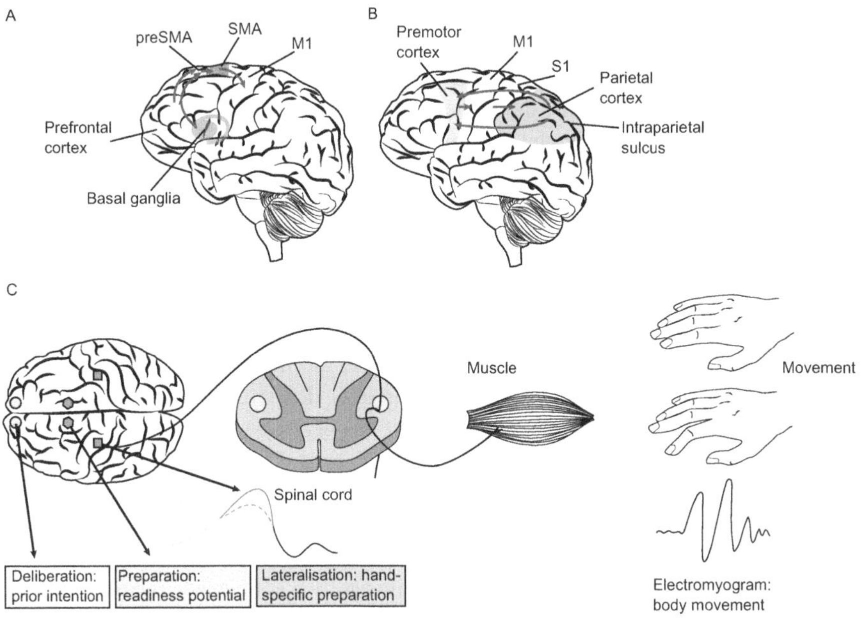

Fig. 2.4 The parts of the human brain that appear to be responsible for the exercise of voluntary action. (**a**) The primary motor cortex (M1) receives two broad classes of input, one from the supplementary motor area (SMA) and the other from the presupplementary motor area (preSMA). The latter receives inputs from the basal ganglia and prefrontal cortex. (**b**) A second, parietal-premotor circuit (early sensory cortices S1 > parietal cortex > lateral part of the premotor cortex > M1) controls object-orientated actions, such as grasping, using sensory information. This also contributes to "voluntary" behaviour. (**c**) Brain activity preceding a voluntary action of the right hand. Long-term plans and intentions are generated by the fronto-polar cortex. The action is prepared in the presupplementary motor area, where the readiness potentials are generated together with other premotor areas. Before the action is performed, the primary motor cortex becomes active, and neuronal signals are sent to the spinal cord and contralateral hand muscles. The difference between the readiness potentials in the two hemispheres (contralateral > ipsilateral; solid and dashed lines) results from their different activities in the later phases of preparation. In the electromyogram, the contraction of the muscles is represented in the form of an electrical signal. (From [129])

humans and in other primates. The primary centres here are the anterior prefrontal cortex, the presupplementary motor cortex, and the parietal cortex [129] (Fig. 2.4).

The prefrontal cortex is an association of neocortical areas that exhibit a unique pattern of connectivity with virtually all sensory neocortical and motor systems and a large number of subcortical structures. Thus, this cortex provides an ideal infrastructure for synthesising the diverse information required for complex behaviour. It also has extensive projections back to these systems, enabling it to exert a "top-down" influence on a wide range of brain processes and thus both control and decision-making [137]. Among the subcortical structures that are important for memory and the processing of emotional information, the limbic systems, especially the hippocampus and the amygdala, are particularly important.

Voluntary action is a special form of decision-making. Decisions to act always allow for several possible solutions. In contrast, in the so-called perceptual decisions, complex stimulus information is reduced to simpler representations [138]. While in perceptual decisions, the (sensory) brain only has to process information that is already available through the stimulus, the (motor) brain has to generate new information to make action decisions [139]. The development of the association cortex in primate evolution—and almost certainly also in other cerebral animal groups—makes it possible for motor responses to be not only direct, as in object-oriented actions such as grasping, but also arbitrary, flexible, and only conditionally responsive to stimuli [140].

Once a decision has been made to strive for a certain goal, this must be kept in focus and safeguarded against interference. Here, the frontal lobes in general and the motor precursors (preSMA) in particular play decisive roles. Using imaging techniques (functional magnetic resonance), it has been shown that different subregions of motor precursors (preSMAs) are active in regard to choosing between alternative tasks and switching between them [141, 142]. Finally, a decision must be made between means to achieve the goal since most goals can be achieved in different ways. Such movement decisions are secondary to decisions about goals. The corresponding motor commands are derived inversely, i.e. quasibackwards from the general goal description, and are made in the parietal and premotor areas.

Overall, the human cortex performs a variety of functions that are important for cognitive or executive control. It enables flexible switching between tasks and goals, active maintenance of goal representations, planning and internal rehearsal, shielding against disruptive stimuli, suppression of automated responses, self-regulation of emotional responses, and reward deferral. Some behavioural performances are also mediated by routine measures and associative mechanisms, but many others require cognitive processes that cannot be captured by these mechanisms.

Self-Control

A fundamental problem of intelligent organisms capable of directing their behaviour towards distant goals is the dynamic regulation of the balance between preservation and flexibility. To ensure goal-directed behaviour in a constantly changing environment, intelligent organisms are generally confronted with two basic requirements. First, they must maintain their current goals over time and protect their intentions against possible distractions. Second, they must be able to switch flexibly between goals and update working memory when significant changes occur. Maintaining a balance here, as well as switching between stable focus and flexible switching depending on the situation, requires antagonistic control processes. For an organism may suffer from constant distraction and impulsivity if it is unable to protect intentions and goals from disruptive stimuli and predominant but inappropriate reactions. Conversely, an organism would suffer from inappropriate persistence, sticking to ideas and behavioural rigidity [132] if it is unable to keep a made decision flexible under changing conditions. Intelligence can be seen as the cognitive solution to such control dilemmas.

One of the best studied, and in the largest number of animal species, subskills of cognitive control is self-control. It is defined as foregoing an immediate reward in the interest of a higher-value but delayed reward. We are therefore dealing with a regulatory process in which individuals have to compare present and future rewards. Because it is often not clear whether an individual is simply hesitating or deliberately waiting to receive the distant reward, choice tasks have been used for the so-called intertemporal decisions, i.e. those that involve a trade-off between the amount of reward and the waiting time. Such time-dependent trade-offs are ubiquitous in human decision-making, from dieting to retirement planning.

The Austrian-born developmental psychologist Walter Mischel studied impulse or inhibitory control in children with a classic test that was entered into the literature as the "Marshmallow test" [143, 144]. With this simple, easily quantifiable test procedure, he was able to check children's ability to turn down the immediate reward of pieces of foam sugar to secure a larger quantity later. This ability not to give in to every temptation can be seen as fundamental to many other cognitive control processes, underlying the slowed, voluntarily controlled allocation of attention, controlled access to memories, and monitoring and control that occur in information-seeking behaviours when uncertainty is experienced. In fact, Mischel was able to demonstrate in follow-up studies that children who were particularly good at controlling themselves in the Marshmallow test at approximately four years of age were more competent as adolescents in both academic and social domains than children with lower self-control. They were also better able to handle frustration and stress and resist other temptations. In general, self-control seems to be based on the ability to willfully distract ourselves from a temptation, which ultimately contributes to a person's academic, emotional, and social success [143]. Later studies also demonstrated that the ability to delay gratification is related to other self-control skills and, to some extent, to general intelligence [145]. These abilities seem to be inherited. However, it seems that the last word has not yet been said about this issue, as a recent study with a very large sample found no correlation between inhibition control in childhood and later behaviour as a teenager when family and parental upbringing were taken into account [146].

In animals, self-control ability has been measured with different experimental setups. Traditionally, researchers have focused on variants of the so-called intertemporal choice tasks, sometimes called delay of gratification tasks, self-control tasks, or intertemporal trade-off tasks, in which small rewards available immediately must be weighed against larger but postponed rewards. The preference for the immediate reward in this dichotomous choice was then taken as an indication of impulsivity and/or a failure of self-control.

The researchers first found low levels of reward deferral in various species of animals—rats, various species of monkeys, lemurs, apes, jays, chickens, and bumblebees—indicating high impulsivity. A typical finding is that animals make no distinction between a small reward offered immediately and a reward offered twice as large a few seconds later [147]. Even the most patient nonhuman species, the chimpanzee, waits only a few minutes [148–150]. This seems to reveal an enormous difference from humans because they can wait up to several months for the greater reward—a difference of several orders of magnitude [151]. These results support the common view that animals are much more impatient than humans [147, 152].

However, this conclusion might be premature. There are now similar findings in averagely patient people with low inhibitory control, suggesting that the known patience in humans may not be quite so robust [151, 153]. Moreover, the results vary widely and show great context dependence, such as the influence of attention, sleep deprivation, social context, and time frame.

Overall, there is very little evidence to support the intuitive assumption that behaviour in intertemporal choice tasks reveals self-control and impulsivity. In contrast, it seems that the apparent lack of self-control reflects a limitedly rational, but not necessarily impulsive, reward maximisation strategy [154]. This is because there is growing evidence that animals even adopt a long-term rate maximisation strategy without fully incorporating reward delay into their decisions. Tasks that measure animal impulsivity could greatly overestimate impulsivity and "contaminate" it with processes that have nothing to do with time preferences. In ecologically valid tasks, animals may be much more patient than abstract laboratory tasks suggest. Finally, the intertemporal choices of humans and animals could be based on different mental operations.

Several other tasks have been devised to describe the time-dependent decision-making behaviour of animals more reliably and more realistically. One particularly intuitive task, which notably requires no explicit training, is to load a rotating food tray with two food rewards of different values or amounts [155]. As long as the animals could restrain themselves, they let less attractive food pass by until they finally grabbed a later but better food. Since another piece of food was placed on the tray after each turn, the reward could be continuously increased by waiting. This so-called "accumulation test" has also proven successful with children [156].

Another more subtle test showed that capuchin monkeys would forgo eating a readily available but less valuable food (celery sticks and pretzel sticks) in exchange for an out-of-reach, higher-value food reward (peanut butter) using the less preferred food as a tool [157]. Finally, exchange tasks were devised in which animals could exchange an object of lower value for a higher value and thus a more desirable option [148, 158, 159]. In all these studies, in addition to primates, corvids [160], Goffin's cockatoos [161, 162], and dogs [163] have demonstrated patience [reviewed in 164, 165]. These findings contradict the previous view that only humans possess this ability and that they correspond to the ecological needs of animals; for example, they have the chance to eat or mate in the face of a higher-ranking group member but also to share food with relatives or to search for food in a new area. In all these cases, a high degree of self-control is advantageous.

Comparisons between animal species using simple tests of inhibitory control also pointed in the direction of a phylogenetic link. If self-control is indeed a result of higher cognitive abilities and is representative of the executive functions involved in various decision-making processes, one should find systematic tendencies toward larger brains or greater general intelligence. Indeed, the latter has been demonstrated in chimpanzees [166]. In a large-scale study, 567 individuals from a total of

36 animal species, albeit 23 of which were primates, were compared for self-control using the so-called cylinder test [167]. In this test, the animals must avoid reaching directly for a food reward located in the middle of a Plexiglas tube and instead choose a detour via one of the two lateral entrances to the tube [168]. Even though, in primates, performance on a series of nonsocial self-control tasks has been associated with variability in social systems, performance on the cylinder task across all species yielded a surprising result that absolute brain size best predicted inhibitory control ability. Only within primates was variation in diet a good indicator of species differences, not group size or brain size relative to body size.

However, these correlations cannot be generalised to other animal groups. Among birds, corvids have some of the largest relative brain sizes, although they are small in absolute terms compared to large mammals. Indeed, the performance (in the cylinder task) of common ravens, straight-billed crows and jackdaws differed little from that of large-brained monkeys [169]. Common ravens, whose brains are approximately 25 times smaller, performed similarly to chimpanzees, and jackdaws outperformed capuchin monkeys. This is astonishing not only with regard to the differences in brain size but also with regard to the enormous temporal divergence since the phylogenetic lineages of birds and mammals separated almost 300 million years ago; thus, evolutionary proximity to humans cannot be an explanation for the good performance of chimpanzees (Fig. 2.5). Although, within corvids, brain

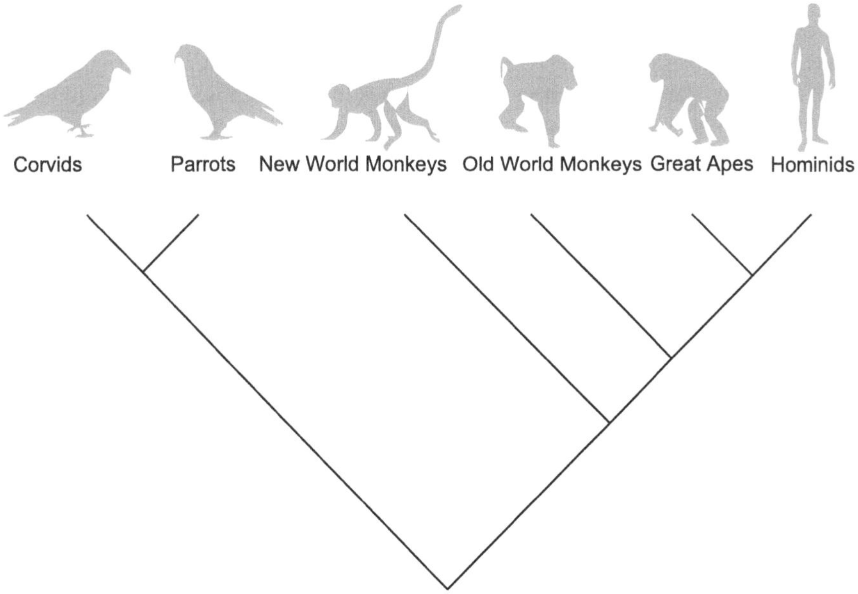

Fig. 2.5 Schematic representation of a hypothetical phylogenetic tree showing the convergence of cognitive abilities between large-brained representatives of birds and mammals. Birds and mammals split 296 million years ago, and parrots and corvids diverged 96.4 million years ago. From the lineage to humans, the New World monkeys split 35 million years ago, the Old World monkeys split 25 million years ago, and the great apes split between 12 and six million years ago

volume, both absolute and relative, seems to be a reliable indicator of inhibitory control, it remains difficult to establish general relationships between brain growth and inhibitory control.

In summary, delaying gratification may be cognitively challenging but still need not be automatically beneficial [170]. In contrast, it may be adaptive not to delay and to consider the costs of waiting. Thus, there are opportunity costs when individuals lose time that could be used for other activities; the risk that the reward will no longer be available before it can be collected; an interruption risk when, for example, another animal gets to the food first; and, finally, an abandonment risk when the chance to get the reward is prevented by, say, a predator and foraging has to be aborted [171]. Thus, the central challenge for the cognitive apparatus remains to decide what to do next, whether to act quickly and effectively or to wait and aim for a long-term goal. After all, under changing and only partially predictable environmental conditions, living beings should be able to choose from the multitude of possible behaviours the ones most conducive to their survival, reproduction, and well-being.

Executive Functions

The human cortex facilitates a variety of functions that are important for the cognitive or executive control of actions. It designs goals, weighs them, considers them, plans and rehearses their achievement, selects the necessary action steps, anticipates their outcomes, keeps the goal in focus, pursues it with purpose, switches to other tasks and goals only when necessary, resists external interference and internal drives, suppresses habits and momentary needs, and shields its actions from competing response tendencies. Although a large part of our behavioural performance is mediated by routine actions and stored associations, the aforementioned elements of goal-oriented actions require the existence of additional cognitive processes.

All three components of executive functions—working memory, cognitive flexibility, and inhibitory control—have been extensively studied in humans [172, 173]. Recently, it has been suggested that these findings should be taken as a starting point for comparative studies of the structure of individual differences in cognitive ability [174]. Executive functions have a strong genetic component and correlate not only with the health of individuals and the fitness of species [172] but also with the general intelligence of individuals across different domains. It is a very recent trend to look more closely at the differences between individuals, not to chase a supposed general intelligence factor but rather by applying a larger number of tests, often so-called test batteries, to uncover the inner structure of intelligence in animals. The ultimate questions are what are the decisive factors for intelligence and problem-solving competence, and how are they interrelated? This does not mean that animals perform equally well or poorly in many different tasks but that there are recognisable relationships with socio-environmental factors [175].

For almost a century, the basis for the study of individual differences was the concept of "general intelligence" (*g*) [176], which was based on the observation that people who perform better on one task tend to do better on another. The striking result of intelligence measurements—the so-called psychometric studies—is a matrix of generally positive correlations. According to Spearman's theory, the definition of *g* is abstract, a hypothetical factor responsible for the correlations. However, the neural or cognitive basis of *g* has not been determined. What functions or processes are reflected in *g*?

It is reasonable to assume that *g* reflects the function of a particular information processing system that contributes to the organisation of many different activities and thus establishes a positive correlation between them. For example, it has been suggested that *g* reflects, at least in large part, the action control functions of the frontal lobes of the (human) brain [177]. This is because the active search for a path to the goals of a task is severely impaired by damage to the frontal lobe, and human behaviour becomes passive, inert, stereotyped, or fragmented. There is a loss of coherent, goal-directed behaviour. This is visible in very many areas of behaviour, from categorising stimuli to orienting in a maze, from describing pictures to solving puzzles, and from remembering a story to observing social conventions. The patient is persistent or distractible, behaves rigidly or inappropriately, passively or impulsively and disinhibited.

Additionally, in some attempts to find a general intelligence factor in animals, positive correlations in performance among different tasks have been found [178]. However, the findings are far less clear than those in humans. First, there are no standardised intelligence tests, neither for individual animal species nor for entire animal taxa. Even the test batteries commonly used to measure human IQ are biased in that they contain tasks that are more or less strongly tied to linguistic ability. Furthermore, the tasks are largely performed in the same way, with no or only modest variation in test situation, motivation, or sensory domain. Human IQ tests are often constructed in the manner of a best-case scenario in which tasks are included in the final battery only if they correlate positively with other tasks. That is, the presence of *g* is assumed, and tasks are chosen to verify its presence [179].

The use of test batteries in animals has thus far yielded only modest success in the search for the *g* factor. The only studies to prove general intelligence outside of rodents [180, 181] have been unable to provide any or only very weak evidence for *g*. A large-scale study investigated cross-task correlations in a total of 106 chimpanzees, 32 orangutans, and 105 two-year-old children using a test battery consisting of 16 different tasks in the areas of space, quantity, causality, social learning, communication, and perspective taking. No evidence of *g was* found in any of the three groups [182]. Later studies were similarly disappointing, once using 23 individuals from all four ape species [183], and on another occasion with 99 individuals from these and three monkey species. It is therefore questionable whether intensive thinking or quick grasping of correlations is actually required when solving the tasks posed.

A Sextet of Animal Intelligence

How and where do we find signs of particularly highly developed intelligence in the animal kingdom? What tasks require cognitive abilities that can be solved neither with instinctive behaviour and routine actions nor with stimulus–response mechanisms and even more complex forms of associative learning? According to a definition by Michael Posner and C. R. R. Snyder [120], such processes are controlled by intentions, run consciously, and require a central processing system. However, automatic processes are triggered by stimuli, run unconsciously, and require no processing capacity. Where do we find evidence for such intentional processes, which Posner and Snyder describe for humans, in animals? In what cases can animal behaviour no longer be explained by more parsimonious means than by voluntary, goal-directed behaviour that is detached from current stimuli and needs and based on thoughts that move freely in place and time? Where do animals find completely new, individual solutions, and in doing so, actually show innovative behaviour? Finally, in which cases is it plausible to assume that animals experience mental states that are analogous to human consciousness and the aspect of subjective experience?

Although the task is very difficult, I would like to dedicate this book precisely to this search because I believe that the chances of success are better today than ever before. In doing so, I would like to focus on six areas of cognitive behaviour in animals because I consider them the most promising candidates. Certainly, this selection is subjective because there are several other phenomena that might be worthy of inclusion in this club of the most challenging intelligence phenomena in animals. I am thinking, for example, of the phenomenon of "*selective imitation*", which in humans has even been called "*rational imitation*" [184]. However, compared to the six selected phenomena, there are very few empirical studies on this topic, namely, up to the time of writing this book, only two studies, one in dogs [185] and one in chimpanzees [186]. The sparse data, the lack of confirmations in other laboratories, and the discussion about the necessary controls—especially in the dog study [187, 188]—do not yet allow them to be treated in the same context as the other phenomena, in my opinion.

1. The first phenomenon in the field of animal intelligence that I would like to discuss for the question of voluntary, purposeful, innovative actions in animals is that of *tool production*. While I will also preface this with an overview of the most important and best-documented cases of *tool use*, I will then focus on the so-called *flexible* and *intelligent* tool use, especially on the modification and production of tools. In these cases, behavioural innovations are thought to emerge after prior mental work, manipulation of ideas, and mental play-through of alternatives. Sometimes individuals not only apply old solutions to cope with new problems but can also create new solutions. When it is no longer sufficient to use a known tool, a new tool must be constructed. Similarly, some animals may select tools at a time when they have no feedback at all about their effectiveness. The idea of the distant goal guides the selection of appropriate methods, thereby

suppressing current stimuli and needs and allowing future gain to determine action.

2. Closely related to the phenomenon of tool-making is the *comprehension of causality*. In certain cases, animals can represent certain physical properties in an abstract or principled way, derive rules about more general modes of action and then apply them to new objects or events. Object properties such as hardness, weight, stiffness, and connection (between objects) are not detectable by direct sensory perception and must be inferred from a larger set of experiences. Understanding these object properties, in combination with spatial ideas, plays a major role in obtaining food that is only indirectly obtainable through the manipulation of related objects. The sudden production of a new adaptive action, not achieved by trial-and-error learning or the repetition of learned stimulus–response units, suggests a solution based on insight. Initially, primates dominated the stage of animal tool tinkerers, but now corvids and parrots are also in the spotlight of technical intelligence research.
3. The third component concerns *prospective thinking*, preparatory or future-oriented behaviour. If the chain of actions is not merely routine but new and its end links must be included in the selection of the first actions, this requires an idea of the consequences of these actions for the achievement of the goal. In short, planning actions are needed. Sometimes, it is useful, namely, to save resources, to anticipate future events, situations, and needs and use them to make the current decision. In Chap. 5, "Looking into the Future", I will discuss a number of well-studied examples of preparatory or forward-looking behaviour, planning actions, and prospective thinking.
4. *Episodic memory*: "Time travel" is useful for future-oriented behaviour, but it is also an important skill for making decisions based on past actions and events. People can not only think about past events but also travel through subjective time, reviving past experiences. Sometimes it is important to know what you yourself did and experienced in a specific situation in the past. The question of whether animals also have such episodic memory has been another defining topic of comparative cognition research for approximately 20 years (after spectacular and hotly debated experiments with food-hiding jays). In recent years, rats have taken on an increasingly important role in clarifying the question of subjective memory.
5. *Metacognition*: Knowing what one has done or experienced in the past touches on the ability to know what one knows and what one does not know. Episodic memory has already been associated with a form of self-consciousness, which is even more true now. It is not a matter of thinking about an object or event in the environment but about something in one's own mental world. Knowing what the mind holds could be a prerequisite for knowing whose mind it is, which in turn is important for having a concept of self, self-awareness, and ultimately fully reflective consciousness. It is therefore not surprising that the ability known as metacognition in animals is also highly controversial—not only among philosophers but also among animal researchers—because it is truly no simple undertaking to prove that an animal thinks about its knowledge or memory content.

However, sophisticated experiments with dolphins, rhesus monkeys, and various great apes have produced confirmed findings for three important features of metacognition: showing uncertainty when making difficult decisions, "hesitating" before making a decision, and seeking additional information.

6. *Mind reading*: The study of metacognitive processes in humans has led to a distinction between a simple level, presumably also found in animals, and a higher level found only in humans. The latter is based on the extension of metacognitive control processes to the interpersonal or suprapersonal level. The particular utility and evolutionary advantage of this second system lies in its role in group decision-making. The effectiveness of collective decision-making based on shared perceptions and experiences seems much greater when individuals share (verbally) their metacognitive representations about them. The interpersonal coordination of metacognitive content is further enhanced when participants in such suprapersonal processes think about each other's thoughts. Thus, in addition to developing a theory of others' behaviour, they also develop a "*theory of mind*" of others. Even if animals are incapable of exchanging metacognitive content verbally, they might at least be able to take the perspective of the other and thus gain access to their knowledge, desires, and even thoughts. Since the first thoughts on the possibility of this so-called mind-reading in animals more than 40 years ago, this cognitive ability has been regarded by many as the crowning achievement of the animal mind. Unsurprisingly, this has triggered the most heated debates among philosophers and divides experimental cognitive researchers into two camps to this day. Ever more intricate tests are devised to check whether animals merely use the outwardly perceptible behaviour of their counterparts on the basis of the most subtle details as an indicator of action or whether animals can actually infer what the other person knows and thinks and how she will act in the next moment.

Join me now on a journey through the minds of nonhuman beings, which I will summarise and evaluate again at the end. Before that, I will discuss in detail two other mental phenomena that are perhaps most often mentioned as unique features of the human mind. However, I do not include them here among the sextet of animal intelligence because *language is* is a form of communication and *consciousness* is more of a general mental state. This journey requires rethinking the "*animal mind*" and redefinition of potential capacities; it will inevitably provoke questions about the place of humans in nature, the anthropological difference, and ultimately, our treatment of animals. For as the French philosopher Étienne Bonnot de Contillac pointed out in his *Traité des animaux* [189], it would be of little interest to know what animals are if it were not a means of knowing what we are.

Chapter 3
Tool Use

Specificum Humanum

Not least through the distinction of man as tool maker ("Man the Tool Maker" [190]), which derives from the special ability to use objects to expand the functional possibilities of his body, the use of tools has been regarded as a brilliant achievement of human reason, by some even as its highest form; for some philosophers and anthropologists, this distinction is even the outstanding, defining characteristic of humans. Indeed, there is no doubt that man's technical intelligence crystallises in the production of tools and takes ever higher and more complex forms, just think of rockets flying to the moon or particle accelerators treating cancer [191, 192].

Certainly, humans had already expanded their physical capabilities with very simple tools made of wood and stone,[1] which they found and used unchanged. Since the discovery of the first stone tool in 1964, which the anthropologist Louis Leakey attributed to the first representative of the human genus (*Homo habilis*) [193], it was assumed that the origins of the sharp-edged stone tool production were linked to the emergence of the genus Homo in response to climate change and the spread of savannah grasslands. The premise was that our lineage alone made this cognitive leap, laying the foundation for our evolutionary success. Later discoveries suggested that the age of the first Oldowan stone tools was 2.5 million years [194]. The simplest stone tool is slightly more than a conveniently shaped natural stone and is selected for use as a striking and hammering stone. These naturally shaped stones were probably among the first tools ever used. Every subsequent step of technical innovation in the early days of man required these simple striking and hammering stones.

[1] I thank my friend, the biodidactician Ulrich Kattmann, for pointing out that stones need by no means have been the first tools, they have perhaps only been better preserved fossilised. In gathering and hunting, the most commonly used tools are digging sticks and carrying straps. Wood and fur or cloth, however, are rarely preserved.

L. Huber, *The Rational Animal*, https://doi.org/10.1007/978-3-031-60803-2_3

However, ancestors of men were already using stones as tools more than three million years ago in East Africa. The oldest finds are those from western Turkana (Lomekwi 3), attributed to *Kenyanthropus platytops* [195], and those from Dikika (Ethiopia), which are attributed to *Australopithecus afarensis* [196]. Neither of these belong to the genus Homo, nor is it clear whether they are among our direct ancestors. Even more interesting is that at Dikika, not stone tools but animal bones with unusual cut marks were excavated. They could indicate that someone used sharp-edged stones to cut meat from bones and perhaps get at the marrow. This would not only be the earliest evidence of the consumption of meat and marrow by hominins but would already suggest an intelligent refinement of tools.

It is precisely the functional improvement or adaptation of tools that shows that early humans thought about certain functions and effectiveness, that they specifically selected or exploited physical properties of materials and that they understood certain cause–effect relationships. Otherwise, the cumulative improvement of tools up to today's technical masterpieces would be inconceivable. The findings of Lomekwi and Dikika are also significant for understanding the development of the human brain. Making tools requires a minimum of motor skill use, which indicates that changes in the brain and spinal cord required for such activities took place more than three million years ago. The artefacts found indicate that the hand motor control of their makers must have been considerable; thus, reorganisation and/or expansion of several regions of the cerebral cortex (for example, the somatosensory, visual, premotor and motor cortex), cerebellum, and spinal tract may have occurred [195].

In any case, it is very plausible that the use of stone tools has a significant advantage in obtaining and processing high-quality food. Hard nuts, bone marrow, mussels, and similar enclosed food sources cannot be tapped well with bare hands but very well with hammers made of stones. Raw meat can be cut with sharp-edged stones but not with unworked pebbles. Therefore, it is natural to assume a selective pressure on tool use, improvement, and manufacture that indirectly acted on the necessary cognitive mechanisms. Even if the origin of tool use in humans remains obscure, because the place, time and exact circumstances of the very first occurrence can never be established, it has sufficient plausibility to derive general principles for cognitive evolution.

Against this background, it was obvious that the use of tools in animals would be seen as a special achievement, even as a human-like ability. Not only did the human-unique selling point begin to disappear, but tool-using animals were also contrasted as particularly intelligent with animals that simply use their bodies to change things in their favour. In fact, initial work on the relationship between tool use and brain size showed a positive relationship [197, 198]. Later, however, doubts arose about the validity of brain–behaviour relationships, mainly of a methodological nature but ultimately also because, upon closer examination, only a very small number of animal species used tools and because there were species among them that did not otherwise stand out for their particularly high intelligence. Rarity could of course be understood as an indication of special intelligence, as this was the key feature of human specialisation for a long time. However, comparisons of very closely related species, both with and without tool use, that otherwise had hardly noticeable differences in intelligence, such as chimpanzee and bonobo, Woodpecker finch and Small tree finch, and New Caledonian crow and closely related crows, have made researchers doubtful.

Methodologically, one faced the problem of obtaining sufficiently large and high-quality data sets to be able to calculate correlations on a large scale and for general statements. Often, data on tool use were based on the literature collected with reports from amateur animal observers. These reports were not based on systematic observations with the aim of specifically detecting tool use or innovations. Thus, it sometimes remains unclear whether a species actually uses tools habitually or only facultatively or whether most individuals of a species use tools or only a few. The amount of data depends strongly not only on the observation intention and effort but also on the observation opportunity. It is much more difficult to obtain good data from animals that are rarely seen because they live at sea depths or in the treetops of rainforests than from animals that live in open forests or in the open countryside.

A better approach both to the question of the relationship between brain size and tool use and to the question of which centres are of particular importance would be expected from neuroscience. However, there are no certain brain centres that are particularly involved in the activities surrounding tool use, which would then show a correlation in size. Nor are the brains of mammals and birds, let alone vertebrates and invertebrates, similar enough to compare such brain centres across phylogenetic trees. Moreover, these centres would most likely also be active in other activities, such as object manipulation or exploration.

Thus, the most promising option at the moment is to observe animals using tools, to analyse their dexterity, and, ultimately, to test them specifically with suitable behavioural experiments. Although, as mentioned, there are only a very small number of species with reliable reports of tool use, there are still enough to study this phenomenon with regard to the cognitive processes involved.

First, it is advisable to obtain an overview of the observations and reports of animal tool use. Two recent works have attempted to at least list, if not all, most of the known cases of tool use.

In the great standard work on tool use in animals, *Animal Tool Behavior*, reports from approximately 3,000 articles [updated and extended edition, 199] have been collected and evaluated. The book describes more than 100 cases but does not venture an estimate of the total number of species in which tool use has been observed. Another review was less cautious and included 418 animal species, including 53 insect species, 65 mammal species and, surprisingly, 277 bird species [200]. In addition, there were 23 species of several other taxa. At first glance, these numbers seem surprisingly high, especially when one considers the relatively short duration (60 years) since Jane Goodall caused a sensation with her description of termite-fishing chimpanzees [201]. However, as measured against the number of animal species (estimates speak of approximately nine million) and even against the number of relatively well-known species, one still has to speak of tool use as a very rare phenomenon.

Certainly, the numbers will still change, not only because diligent researchers continue discovering new cases of tool-using animals but also because the classification of what still does and does not count as such behaviour will change. This depends on increasing the understanding of the processes involved and on increasing the sensitivity and accuracy of observation, especially for borderline cases that make a crystal-clear determination difficult. Two recent reports may illustrate this. First, three Visayan warthogs were observed digging in the ground with a piece of bark or sticks [202]. The two adult females exhibited this behaviour during the final stage of nest building, and the adult male apparently wanted to do the same. Although pigs are known to use their proboscis as a digging tool for nest building, especially for extracting food from the ground, the use of wooden objects would correspond to a significant improvement and extension of their digging technique. The second report is less clear. Two Atlantic puffins were observed spontaneously scratching their bodies with a small wooden stick [203]. The puffin is a species of seabird in the auk family that breeds in burrows on cliffs in the northern Atlantic and was not previously known to use tools. This observation is potentially remarkable, according to the authors, because, on the one hand, it would show the use of tools for grooming in wild birds and, on the other hand, for the first time[2] in seabirds, which lived far away (one bird in Wales, the other in Iceland) and could therefore be widespread in this species. The problem, however, is that these random observations can hardly exclude simpler explanations and therefore carry a high risk of a false-positive result [205]. A more likely and parsimonious explanation would be that during courtship, when the birds playfully pick up objects resembling nesting material or demonstrate readiness to breed, they accidentally touch their plumage with the stick while bringing it towards their breast or that they simply try to scratch themselves with their beak while still holding the piece of wood in their beak [206].

We can clearly see from the example of the puffins how important it is to check anecdotes with systematic, repeated observations, if possible with video recordings, but even better with experimental analyses. Notwithstanding this, it is also necessary to have terminological clarity so that rash sensationalist reports, which also, or perhaps especially, find their way into highly respected scientific journals, are stopped. Therefore, let us turn to the question of how best to define tool use.

What Is Tool Use?

When I fetch a hammer from my workshop and use it to hammer in a nail for a picture, the matter seems clear. Because it would be almost impossible and also very painful to drive the nail with the bare hand, the hammer is a very useful instrument. It extends and improves my limited physical abilities with regard to the goal of

[2] However, the authors of the puffin study overlooked the report of a similar behaviour of a double-crested cormorant [204].

fixing the nail deep into the wall. In the context of this simple, generally understandable example from everyday human life, it is surprising that the term "*tool use*" is not only not unambiguous or self-evident but also has generated relatively heated discussions to this day.[3] Therefore, it is useful to define this phenomenon here before we take a closer look at a few interesting cases. One of the earliest definitions was made by Jane Goodall, who, in 1964, reported the sensational discovery that a chimpanzee poked a termite burrow with a thin stick, then carefully removed and pulled the stick through its mouth to strip the termites between its lips and then eat them.

> A tool-using performance in an animal or bird is specified as the use of an external object as a functional extension of mouth or beak, hand or claw, in the attainment of an immediate goal [208, p. 195].

Thus, tool use is a purposeful action with appropriate means. (In contrast, there are also actions that involve the expert use of objects as the means to reach their goal but where one would not call the objects "tools", such as nest building. One can place everything under the umbrella term "technology" [209]).

In most cases, one starts by describing the object called a tool. In very general terms, a tool can be thought of as a part of what Malcolm McCullough [210] calls "applied intelligence", that is, a technology that enables us to extend the limited repertoire of manual and cognitive skills that we have at our disposal. It seems that this extension of the skills we possess distinguishes tools from other artefacts, machines, and technologies [211].

Ethology, on the other hand, tries to look at tool use more from the perspective of behaviour, the actions involved,[4] and the corresponding goals, especially since the objects that animals use as tools are all very simple in nature (twigs, sticks, stones, etc.). For a long time, people stuck to simple, functional definitions until approximately a decade ago. Under the impression of new, sometimes very exotic, cases of tool use in the animal kingdom, a multitude of new definitions were proposed. These became increasingly extensive, complex, and abstract. I will limit myself to two newer attempts at definitions that have been cited most often.

In 2008, a particularly technical definition was proposed by computer scientists:

> Tool use is the exertion of control over a freely manipulable external object (the tool) with the goal of (1) altering the physical properties of another object, substance, surface or medium (the target, which may be the tool user or another organism) via a dynamic mechanical interaction, or (2) mediating the flow of information between the tool user and the environment or other organisms in the environment [213, p. 1203].

In addition to the controllability and manipulability of objects with reference to a clearly recognisable goal, the authors focus on two previously unconsidered features: dynamic mechanical interaction and the mediation of information through the

---

[3] For example, the current proposal from 2021 to consider chewing sticks in dogs as a form of tool use also triggered discussions again [207].

[4] Action is often used in the sense of a series of movements that achieve a goal set by the organism. See [212].

tool. By emphasising dynamic interaction, they want to exclude cases where the actor establishes a merely static relationship between objects, such as in nest-building when stones are placed in specific locations. Conversely, the addition that a tool can also serve to convey information to the actor is inclusive, as it is now intended to cover cases such as those where a gorilla uses a stick to determine the depth of a water obstacle to be crossed [214] or where bottlenose dolphins detach sponges from the seabed and place them over their snouts to protect themselves from injury while foraging [215].

Despite all this subtlety of the new definition, questions have still remained unanswered, so that the authors of the 2011 great standard work felt compelled to make yet another clarification (set in italics below). The modified definition of tool use is as follows:

> The external employment of an unattached or a *manipulable attached* environmental object to alter more efficiently the form, position, or condition of another object, another organism, or the user itself, when the user holds and *directly manipulates* the tool during *or prior* to use and is responsible for the proper and effective orientation of the tool [199, p. 5].

Since both works argued for their respective definitions over many pages, it would go too far here to make a comparative assessment of the definitions and come to a decision.[5] In any case, we are dealing with a polymorphous concept—as Wittgenstein [46] argued—using the example of "play", which does not allow for a clear demarcation. In the context of the question posed in this book, we want to concentrate on those cases in which cognitive involvement is recognisable and for which corresponding experimental work is also available.

Examples of Tool Users in the Wild

Among the most spectacular examples of tool use in the animal kingdom are certainly the following well-known cases among chimpanzees: those in Tanzania and the Republic of Congo that use sticks to fish termites out of holes in the ground and leaves as a substitute for a sponge to collect water from holes in trees; those in the West African states of Côte d'Ivoire and Guinea that use stones or thick pieces of wood to crack open nuts; those in Senegal that hunt galagos with skewers fashioned from branches; and those in Gabon that harvest honey with five-piece tool sets (all cases described in the two recent reference books [199, 217]). In addition to these "world champions of tool use", orangutans have now also been observed using tools in Borneo and Sumatra. Many spectacular cases have been described in zoos or sanctuaries in the past, but orangutans have also been observed in the field using

[5] The primary question is which cases should still be considered as tool use and which should not. However, there are now so many borderline cases, and there will certainly be more, that it is pointless to argue about a precise demarcation. As Beck [216] already noted, tool use is intertwined with so many other behaviours, such as object play and exploration, that it is difficult to assign it a category of its own.

sticks to pick up edible insects, branches as "fly swatters", and leaves as either "umbrellas" or "gloves" to protect their hands from thorny branches [218]. In addition to these two ape species, two other primate species have also come into the limelight as tool users. First, black-striped (or bearded) capuchin monkeys in Brazil are similarly adept as chimpanzees at breaking open nuts with pieces of wood or stones [219]. Second, long-tailed macaques in Myanmar and Thailand, which use stones to break open oysters, crabs, and snails on the coast and further inland, have also started cracking open palm nuts [220].

Tool use in other mammals is much rarer than in primates. The best known examples are sea otters that use their hands to strike shellfish on stones, which they place on their bellies while floating on their backs [221]. There are also elephants that wave twigs with their trunks to drive flies away from their bodies [222], as do the sponge-bearing dolphins mentioned earlier [215].

There are many more cases of tool use in birds than in mammals, but only in a few cases do the reports come close to the previous spectacular examples. Apart from the textbook example of Egyptian vultures hurling stones at ostrich eggs to crack hard shells [223] and the special cases of mangrove herons [224] and burrowing owls [225], which hunt with bait, the Woodpecker finches in Galápagos and the endemic crows of New Caledonia achieved notable publicity. Both poke into knotholes or tree crevices with twigs, thin branches, cactus spines, or thorns to retrieve larvae or other nutritious food. These two birds have been intensively researched and, along with chimpanzees, form the royal class of tool users in the animal kingdom.

As we will discuss in more detail later, there are a considerable number of other well-described or studied cases of tool use in captive birds, but in most cases, it is still unclear whether one would find among them wild counterparts that use tools facultatively or even habitually.

In addition, there are reports from insects, cephalopods, fish, and reptiles, such as digger wasps, ant maids, various ant species, wrasses, rays, and swamp crocodiles. However, in many cases, research into the exact mechanism, function, and, above all, the understanding of the animals involved is still in its infancy. Often, the initial reports are sensational, but either no further research follows or the subsequent analyses lead to somewhat more modest interpretations. I have already mentioned the questionable reporting of tool use in the two puffins [203, 206].

A classic example of original over-interpretation is the American crows that throw nuts on the street in Japan and California [226]. It was assumed that they did this to have the nuts cracked by the cars driving over them. This actually happened, but follow-up studies showed that they do this just as often when no car is coming. This argues against a deliberate act with the idea or intention of opening nuts as a result. Although this could mean that they do this as a precautionary measure and wait, it is equally likely that throwing nuts on roads has simply been rewarded often enough; thus, the action has been reinforced and is performed without conscious thought. Thorndike's law of effect would suffice as an explanation. In any case, the example shows that one must distinguish between an intelligent result and an intelligent action.

We have established that tool use occurs in many varieties and shades, in very different species, with great differences in observation quality and investigation intensity. But one constant runs through the literature: the vast majority of cases concern only the use of objects in the course of foraging. In the vast majority of cases, it is also a single type of use with very constant action (for example, only poking or only hitting), few different utensils (only twigs or only stones), and hardly any modifications. Until proven otherwise, one should therefore assume—following the principle of explanatory parsimony—that these are often not deliberate, flexible, or even creative actions but rather stereotypical, strongly genetically disposed, simple uses of objects. Another peculiarity of tool use in animals is the high number of species where tool use has been clearly demonstrated in captive individuals but never in their wild conspecifics. We will discuss some cases in more detail, especially those where apparently very creative and complex actions with objects to obtain food could be observed. However, we can already say in advance that tool use and even tool making seem to be "dormant" in some animals but emerge only under the most suitable conditions. Apparently, some—perhaps even many—animal species have sufficient potential to use objects for their own purposes, i.e. to improve their limited physical capabilities. However, favourable circumstances must first be present, such as need, because easy-to-tap food sources have dried up or because of reduced enemy pressure, so that there is enough time to deal with the enclosed fruit. Therefore, we see that there must be more factors at play than mere mental capacity. However, at least there is scope for creative or innovative action.

Categorisations of Tool Uses

Now that we have gained a rough overview of the distribution of tool use and its variety of purpose in the animal kingdom, it would be advisable in the next step to distinguish the different types of tool use. Benjamin Beck [216] has distinguished 21 types of tool use. These types are merely operational and descriptive; they do not characterise functions. Essentially, they are distinguished by motor functions, such as throwing, dropping, pulling, rolling, poking, fishing, and wiping. Besides this, there is only one nonmotor type, symbolising (or referencing), by which an object is held aloft or exchanged, which relates to another object, another organism, or even a psychological state. A good example would be a money coin. In the animal kingdom, this type of use is very rare.

Less than the motor details, however, we should be interested in the function of use. For it makes a difference whether an object is used only to generate greater force or greater reach—as in the vast majority of cases—or whether one uses objects to camouflage oneself, to increase comfort (for example, to make good beds for oneself), or even to create abstract means of communication or barter. Beck [216] classified the 21 functions into a total of six general types. However, the vast majority of cases can be described in terms of only four types. These include (1) extending the reach of the actor; (2) amplifying mechanical force; (3) increasing the signal value of a social expression (agonistic display) or producing such a signal; and (4) effectively controlling fluids and small objects (for example, removing or absorbing them). Two other but rarer functions are (5) camouflage (and hiding) and (6) increasing one's bodily comfort. Finally, Robert Shumaker, Kristina Walkup, and Benjamin Beck [199] added a seventh function, (7), to abstract from or represent reality (in the case of symbols, for example, coins).

Tool Use and Tool Manufacture

All the types of use and functions described above concern the use of a single object. However, in addition to these, there are, much more rarely but all the more conspicuously, actions of animals in which two or more objects are involved. These are mostly used in the closest spatial and temporal connections. Besides these, there are cases where one object is assigned several functions. All these cases have been subsumed under the label "associative tool use". However, since "associative" itself is an element of the whole group, I propose the label "complex tool use" instead. It is particularly worthwhile to take a closer look at the terminology, especially because there is considerable terminological ignorance in the literature—terms are used without defining the associated concepts.

In total, there are 11 types of complex tool use, which include tool making. These are named according to the tool itself or the combination of items. Tool making (or manufacturing) is much easier to define:

> Tool manufacture is simply any structural modification of an object or an existing tool so that the object serves, or serves more effectively, as a tool [199, p. 11].

For a better overview, all forms of complex tool use are summarised in a Table 3.1 according to—but slightly modified—Table 1.2 of the book by Shumaker, Walkup, and Beck [199].

I have grouped all the types of tool use listed in the table above under the umbrella term *complex tool use* to suggest that these actions are different from simple, stereotypical actions in the same context, with the same goal, the same function and, above all, the same object and motor action. In the literature, we also like to speak of *flexible tool use*, where features of the tool are adapted or modified according to the specific requirements of the task at hand, depending on necessities or increasing efficiency or functionality. In these tools, the use or production is both flexible and cumulative, that is, based on accumulated experience. As we will see, these actions have one more thing in common: they are acquired, at least in part, through social learning.

In some of the types of use mentioned, the addition of a new action to an existing one is crucial. This is not done randomly but rather to create a new functional combination or a new sequence to achieve a general goal. We then speak of *cumulative tool use* because it involves the functional integration of multiple elements. While

Table 3.1 Different forms of complex tool use

Designation	Description	Example
Associative Tool (also Combinatorial Tool)	Tools used in any combination to achieve an outcome.	Goffin cockatoos use thick pieces of wood to open a wawai fruit, then thin, pointed pieces to pierce the leathery coating, and finally medium-wide pieces to scoop out the inside.
Sequential Tool	A tool used to acquire another tool.	Capuchin monkeys use a small stone to knock out an enclosed large stone, which they then use to crack open nuts.
Serial Tool (also Tool Set)	Two or more tools used sequentially, usually each in a different mode, to achieve a single outcome.	Capuchin monkeys knock a hole in a tree trunk with a stone and then fish out bees with a stick. Chimpanzees use up to five types of sticks and branches to extract honey from beehives.
Tool Composite	Two or more tools used simultaneously, usually each in a different mode, to achieve a single outcome, where the first tool is not used to manufacture the second.	Weaver birds use pieces of bark to apply colour. Sea otters use two stones as a hammer and an anvil to break open shellfish. Chimpanzees use stones as a hammer and an anvil to crack nuts.
Metatool	A tool used simultaneously with a second tool to increase the efficiency or effectiveness of the second tool, where the first tool (the Metatool) acts directly on the second. The second tool could function as a tool on its own; the Metatool makes it a better tool. Every Metatool is part of a Tool Composite. The use of a Metatool is not tool manufacture.	Chimpanzees use up to two small stones to stabilise a large stone, which serves as an anvil for hitting nuts with a fourth stone, the hammer.
Tool Kit (also Toolbox)	All of the various tools known to be used by a single individual, a single population, a species, or a genus.	A number of stones used by capuchin monkeys or chimpanzees to crack nuts.
Tool Assemblage (also Tool Collection)	All of the tools found in the same context at a particular site that are presumed to have been used by the same group or population. Assemblages may accumulate over time.	In Taï, accumulations of 4300-year-old stones have been found that were used by chimpanzees to crack nuts.
Multi-Function Tool	A single, specific tool used in two different modes and/or for two different functions.	A capuchin monkey digs a hole in the ground with a stone and later knocks open a nut with the same stone.

(continued)

Table 3.1 (continued)

Designation	Description	Example
Secondary Tool	A tool used to manufacture (structurally modify) another tool.	A young orangutan knocks off leaves from a stone, which he then uses for cutting.
Tool Crafting	Tool manufacture that involves a number of successive steps and three-dimensional reshaping. Only one object is usually involved, but it could include secondary tool use.	New Caledonian crows build their hooked tools in several finely coordinated steps. As do orangutans and chimpanzees.
Construction	Two or more tools and/or objects physically linked to make a functional, semipermanent thing that, once completed, it is not held or directly manipulated in its entirety. A Construction itself is therefore *not* a tool. Nor is it tool manufacture, because the product is not a tool.	Nests, ant burrows, traps, bee or termite hives, food hiding places.

the goal remains the same, two or more tools are used simultaneously or sequentially, usually in different ways, to achieve it.

In the technical literature, the individual terms are sometimes used very differently, and different terms are formed, which causes confusion. Therefore, we should be precise and, for the sake of damage limitation, first distinguish five relatively similar forms of tool combination: (1) in *sequential tool use*, one tool is used to obtain another tool; (2) in *serial tool use*, for which the term *tool set* has come to be used, two or more tools are used in succession to achieve a goal; (3) in *combined* or *compound tool use*, two or more tools are used simultaneously and in a complementary (supplementary) way, but the first tool is not used to produce a second tool; (4) in *meta-tool use*, one tool is used simultaneously with a second tool to increase its effectiveness but not to make it—the second tool could also be used alone; and (5) only a *secondary* tool is used to make or structurally modify another tool. These distinctions can be illustrated by a well-known example.

Capuchin monkeys knock out an enclosed large stone with a small stone, which they then use to crack open nuts (*sequential tool use*). They also sometimes knock depressions in the ground with a stone and then poke them with a branch (*serial tool use*). In the second case, they use a set of tools. The behaviour of hitting nuts with two different stones (one serving as a hammer and the other serving as an anvil), famous for chimpanzees, is an example of *combined* or *compound tool use* [227–229].

There is a great deal of confusion about the term *metatool*. This term was coined by Tetsuro Matsuzawa [230–232] and later refined by him and his colleagues. It was used to denote the peculiarity of nut-hammering by chimpanzees. Since the anvil is sometimes very wobbly and makes hammering difficult, chimpanzees have pushed a small stone underneath it as a sort of cradle stone, thus stabilising the anvil. Later, one could even observe how two small stones were used for this purpose [233].

Here, one tool (the cradle stone or the *metatool*) is used at the same time as a second tool (the anvil stone) is used to increase its effectiveness. Actually, the anvil is also a metatool because it supports or even improves the striking with the hammer. In this respect, *metatool* use is a form of *combined* tool use, distinguished only by the fact that the metatool acts directly on the other tool, which is not the case with the anvil.

Later, in laboratory experiments with New Caledonian crows [234–238], following a precursor experiment with gorillas and orangutans [239], studies on "metatool use" were published in which one tool was brought in by another. Strictly speaking, this is *sequential* tool use.

The archetypal case of a *secondary* tool is a coarse stone used to strike another stone so that fragments of stone are splintered off, which can then be used for cutting due to a sharp edge. In hominids, the earliest evidence for the creation of specific stone tools, made by chipping and several fine adjustments, was found in the Omo Valley in Ethiopia; their age dated to about two million years ago [240].[6]

Are there similar behaviours in animals? Chimpanzees sometimes also hit bones, causing them to splinter and produce sharp bone flakes [241]. The question is whether this is done intentionally. Some primatologists doubt that cases of *secondary* tools are known in chimpanzees [242]. However, there is an early report of a young orangutan cutting flakes off a rock, which was subsequently used for cutting [243]. Many years later, capuchin monkeys were observed using small stones to dislodge larger quartz pebbles from conglomerate blocks in which they were enclosed, after which they used the pebbles as hammers to pound logs [244]. As if Friedrich Engels had already thought about this in 1876, when he noted that it was not the use but the production of tools that was the decisive difference for humans [245, p. 445]:

> No monkey's hand has ever made the crudest stone knife. With the production of tools, social work begins, through which man creates a second nature, culture.

Even more controversial is the question of whether nonhuman animals are capable of producing *compound* tools. This is where tools are assembled from two or more elements of different types to form a single unit. For example, humans make spears, knives, and scrapers from a handle or shaft, a stone insert (blade), or a binding material. William McGrew from the University of Cambridge (UK) [246] provided three examples of chimpanzees. One is the cradle stone, which chimpanzees slide under the anvil to stabilise it and thus increase its efficiency. However, this is mostly seen as an example of a metatool because it is a very loose connection, not a real composition. The second example is the leaf sponge. Here, several fresh leaves are combined into a single compressed mass that allows water to be extracted from inaccessible tree openings [247]. Here, the composition is already clearer, but the tool is still very simply made (from folding) and consists of one and the same components (leaves). Finally, great apes link branches, twigs, and leaves together daily

---

[6] However, the cut marks on the more than three million year old bones from Dikika could, as already explained above, also have already come from sharp-edged chips [196].

to build platforms for sleeping (beds or nests), creating frames, supports, blankets, and even cushions [248]. These beds, like nests, are described by Shumaker, Backup, and Beck [199] as *constructions* that differ from tools in that they are not held, moved, or manipulated.

Another way in which complexity increases in the use of tools concerns the number of uses of one and the same object. *Multifunctional* tools are again found in chimpanzees and capuchins for the purpose of extractive foraging. Chimpanzees use sticks for both termite fishing and honey collection [249–251]. Capuchin monkeys use sticks to collect water, arthropods (insects, crustaceans), and insect nest contents. Stones are mostly used as hammers, not only to open nuts or reach other trapped foods but also to break up wood and hard soil to expose insects or roots and, as mentioned earlier, to knock out trapped quartz stones from conglomerate boulders [244]. To what extent the same stones are also used for throwing or drumming, or the same sticks for digging, remains open, but on the whole, these two categories of objects find many uses among apes.

Recent reports have confirmed that even in New Caledonian crows, the *multifunctional* tool sticks are used for purposes other than acquiring inaccessible food by poking into knotholes. These birds also generally use them to explore their environment [252]. Apparently, they use it to keep their distance from possible dangers too.

The multifunctional use of objects shows that actors use them as a means to an end, whereby it is not the means but the ends that vary. In any case, this behaviour demonstrates a high degree of flexibility and at least a basic understanding of the physical properties of the objects. The latter also leads directly to an understanding of the functional possibilities—of what one can do with the object; in English, one uses the term "*affordances*" for this purpose. With the repeated use or application of objects, animals gain experience with the relationships among objects, actions and effects. By observing themselves and their actions, such as by observing the movements of objects, their dynamic properties, and spatiotemporal causal relationships, the actor discovers stimulus effects.

This somewhat abstract or unwieldy formulation requires clarification with an example. A young capuchin monkey can learn that soft sandstone breaks when it hits a hard palm nut, while the nut remains intact. In contrast, the opposite occurs for hard quartz stones. Over time, young monkeys will reach for quartz stones more and more often after placing a nut on the anvil. He has thus learned from his activities to pay attention to certain relevant properties of the possible tool. However, we do not yet know whether (a) the monkey avoids sandstone because it is fragile, (b) because it is light, or (c) because it looks like those stones that were broken in earlier actions. In other words, we do not know exactly what the monkey has learned or what he now understands about stones as hammers when he is selective.

I shall now discuss a recently presented theory of tool use, called "*tooling*" [253], which attempts to unify the different aspects of the body, the mind, the task, and the environment. This is because when analysing a possible case of tool use in the animal kingdom, we must address the real-time dynamics of the specific task, which include perceiving and moving the limbs, the head, the eyes, and the grasped object, as well as perceiving the effects of the actions with the grasped object on the target. Dorothy Fragasy and Madhur Mangalam [253] define tool use, which goes beyond mere instrumental problem solving, as intentionally creating a mechanical effect on a target object by first grasping an object, thereby transforming the animal body into a body-plus-object system, which is then used to create (at least) a spatial relationship between a grasped object and a target object, thereby creating a mechanical interface between the two. This theory, which is inspired by the theory of "*embodied cognition*" [254, 255], excludes the throwing of spears because it dissolves the body-plus-object system, which is defined by an increase in degrees of freedom. It is one of the most advanced and neurocognitively demanding forms of object interaction; it requires rapid muscle movements, trajectory calculations, and precise timing within a narrow time window [256]. Targeted throwing therefore involves unique challenges at the cognitive level, particularly in terms of planning, timing, and precision. Although we share some basic throwing techniques with other primates [257], humans are unsurpassed in their ability to throw at a high enough velocity to kill prey or conspecifics [258]. William Calvin (University of Washington, Seattle, USA) has even suggested that these requirements contribute to hominid brain enlargement, as errors in neural timing could be compensated for by an increase in redundant parallel neural circuits.

The tooling theory, which is based on biomechanical and spatial interactions, has the advantage of describing tool use in terms of the action rather than the object (the tool). The locus of control for this shifts from the part of the body, for example, the hand that grasps the object, to the part of the tool that establishes contact with the environment during use. This object at the other end, unlike a body part, does not provide direct feedback to the brain. Thus, the use of tools transforms the body's movement system in a particular way that is not present in other forms of object use. Using a stick to poke a hole changes the representation of the animal's action in a way that, say, dropping a stone does not.

The central hypothesis of tooling theory is that the difficulty of tool use increases with the following five factors: (1) the number of spatial relations, (2) temporal order (with simultaneous relations being more difficult than sequential ones), (3) temporal dynamics (with dynamic relations being more difficult than static ones), (4) frame of reference (with allocentric actions, oriented towards the environment, being more difficult than egocentric actions, oriented towards one's own body), and (5) specificity (depending on placement, orientation and geometric alignment) [253]. Measuring these aspects of tool use not only allows for a more detailed and coherent characterisation of specific tool use phenomena but also provides measures that can be applied for broader comparative purposes across different animal species and tool use types.

A group of young colleagues in my laboratory systematically studied all of Shumaker's, Walkup's, and Beck's [199] work to determine whether or not it qualifies as tool use according to tooling theory [259]. In fact, this was only the case for around 40%. Moreover, tool use in the sense of tooling seems to be far less common in the animal kingdom than tool use according to the definition of Shumaker, Walkup, and Beck [199], where the greatest diversity of types was found in the order of primates. However, it remains unclear whether animals still master other intellectual challenges in addition to the "embodied cognition" of tooling.

Creative Tool Users

Neurophysiological studies of macaques and humans suggest that the use of tools expands the internal representation of the actor's extremities (hand, foot, beak, etc.) [260, 261]. However, the extension of body representation alone cannot explain how the efficacy of an object is represented [262]. A screwdriver, for example, extends not only the reach of the user but also the functional capabilities of the hand. However, the way we skillfully interact with tools and other artefacts is highly contextual, and aspects of this context are not specified by the physical properties of the task or the actor's body alone. Effective control of tools involves both a tool-specific extension of the body schema and specific thinking that captures how movements of the hands (or body or other extremities) are transformed into actions of the ultimately effective tool. The way humans and some animals skillfully use tools is possibly due to adaptations that integrate sensorimotor and cognitive processes [263]. Complex tool use is enabled by conceptual knowledge about objects and their functions, the actor's intended goals, and interpretations of prevailing task demands. Comparative studies suggest that critical differences in tool use are more likely to be found in the mechanisms involved in causal reasoning than in those that perform sensorimotor transformations [264].

Before we turn to the tricky questions about the cognitive aspects of tool use and discuss a selection of recent experiments on this topic, I will justify why I will limit myself in the following to only a handful of animal species that habitually show complex tool use in the wild. These include, in the mammalian class, chimpanzees (with short excursions to other great apes, long-tailed macaques, and capuchin monkeys), and in the avian class, New Caledonian (and Hawaiian) crows and Woodpecker finches.

I limit myself to these species for the following reasons:

1. They use tools habitually, sometimes excessively.
2. Within the species, not only single individuals but also several populations have been observed thus far, in the field and in the laboratory.
3. These species are well studied. More than 500 papers have been written on tool use in chimpanzees. There are at least dozens of the other selected species.
4. I also refrain from explicit depictions of invertebrates, fish, and reptiles (for example, burrowing wasps, octopuses, archerfish, and saltwater crocodiles) because, apart from anecdotal observations, there are hardly any relevant systematic, experimental studies.
5. For the selected species, there are countless field observations and a large number of experimental studies, not only on tool use itself but also on all relevant issues in the context of physical intelligence, causal understanding, and planning.

Chimpanzees

The use of tools by chimpanzees was already reported by Savage and Wyman (1843–44), but it became known only worldwide through the discovery by Jane Goodall in 1960 and a discussion it triggered about the omission of an important feature that distinguishes humans and separates us from animals. She observed an adult male chimpanzee (named David Greybeard) in the (later) Gombe Reserve in Tanzania not only using tools (bits of straw) to fish for termites out of their burrows but also actually making them by stripping leaves from a stem [201]. To the telegram Jane excitedly wrote to her mentor at the time, the palaeoanthropologist Louis Leakey replied with the following famous words [265, p. 2184]: "Now we must redefine tool, redefine Man, or accept chimpanzees as humans".

Subsequent research has further blurred the line between what we call human and nonhuman because Goodall was able to show that chimpanzees have a much larger tool repertoire than previously thought and can use it very flexibly and purposefully. Large groups of researchers have not only undertaken systematic field observations in all parts of Africa where chimpanzees dwell but also collected very large amounts of data supported by video recordings. They have also conducted sophisticated field experiments to study the individual components of their actions, such as the selection, modification, and hierarchical manipulation of tools [266]. Through these observations, which have already spanned almost six decades, several populations of wild chimpanzees are fully habituated to the presence of humans and can therefore be observed at close range.

As mentioned in the previous chapters, all forms of complex tool use are found in chimpanzees except for the production of compound tools. While not all chimpanzee researchers agree with the latter, my assessment does not detract from the extent and complexity of tool use in these animals. Almost all behavioural scientists agree that chimpanzees are the closest to humans in this respect. Already, in terms of the range of items used by individual animals or groups naturally (not assisted by researchers) and in the wild (not in zoos, laboratories, or sanctuaries), this species stands out among all others—as far as is known. At the population level, 20 types of tools, which are used for various functions of daily life, including food, companionship, reproduction, and self-preservation, were counted [246]. It is interesting to note that populations differ in the primary use of their tools. While chimpanzee populations in the Republic of Congo use them most for extractive foraging, the animals in Ngogo in Uganda use them for hygiene and courtship behaviour.

However, it is not only the number of tools and functions but also the combination of their tools that chimpanzees excel through. In regard to serial tool use, there is an astonishing report of chimpanzees in Gabon that use five different items in a constant sequence to obtain honey. These items were referred to by the authors as the pounder, enlarger, collector, perforator, and swabber [250]. We do not know whether deliberation or simple experience guarantees adherence to the sequence; in any case, this high use of means is unrivaled in the animal kingdom.

Fig. 3.1 Comparison of primate impact techniques. Left: A capuchin monkey in Brazil uses a large stone as a hammer to impact a nut, which was also placed on a large stone as an anvil. Note the size of the hammer and the two-handed striking technique. Middle: A male long-tailed macaque cracking oil palm nuts in Thailand. Right: An adult male chimpanzee cracking a nut in Bossou, Guinea. Note the small stone (called a "wedge stone") that is inserted by the chimpanzee with his foot under the large stone, the anvil, to level its working surface and thus to make the anvil more efficient. In his right hand, the chimpanzee holds a second stone as a reserve

Several examples of ape tool use—the pounding of nuts or other hard objects with stones on stones—have already been mentioned. This fact is accompanied by special public attention not only because it reminds us of the beginnings of human tool culture (Oldowan culture) in East Africa but also because it represents the (thus far) peak of complexity in the animal kingdom. When two small stones are pushed under a large "anvil" stone as wedge stone and another stone delivers the hammering, we are dealing with a tool combination that requires a great understanding of dynamic relationships as well as considerable physical technique (Fig. 3.1). According to the theory of tooling [253], for the analysis of tool actions, this sophisticated technique of nut opening is also quantitatively unique in the animal kingdom given the requirements of this task in terms of frame of reference (allocentric), number (multiple), specificity (orientation of the different stones), dynamics, and temporal sequence of spatial relationships present in each step. It is therefore not surprising that young animals develop the mature technique only after 3–5 years of life, during which they watch the adult animals and repeatedly try to train themselves [267].

Considering how skilfully chimpanzees perform stone-hammering, it is not surprising that a few years ago, a "chimpanzee workshop" was found in Taï National Park on the Ivory Coast [268]. Moreover, 6 years later, approximately 40 kg of nutshells were found in the same area, along with stones approximately 4,300 years old; these nutshells were apparently used by chimpanzees [269]. The stones were apparently brought from rock outcrops and other places to specific destinations to be used as hammers for breaking nuts. Repeated use of these "workshops" led to litter accumulation and site formation. The hammer stones show the same signs of wear as those used by chimpanzees living today. At the same time, they differ from all stone tools known from Stone Age people (Oldovan culture). Interestingly, there are nevertheless functional similarities to human stone cultures, such as the selection and acquisition of raw materials and their targeted use for a specific job in a specific place; furthermore, certain places are repeatedly searched for specific purposes so that residues and debris accumulate there. This Chimpanzee Stone Age is older than the first agricultural settlement of humans in this part of the African rainforest, which is why the chimpanzees could not have copied the technique from them. We must therefore assume convergent stone-knapping evolution. However, we do not know how long chimpanzees and their ancestors have been working with stones. Even palaeoanthropology has not seemed to have found the earliest traces of human stone tool culture; new finds keep coming to light that date its beginning further back [196]. In any case, (non-hominid) "palaeoprimatology" is still in its infancy, but it would be interesting to know whether and why chimpanzees actually began to use their minds to obtain high-energy food so much later.

In addition to stone tool technology, chimpanzees seem to display other cutting-edge skills. They are capable of making several different tools from a single raw material (for example, from leaves a sponge, a pillow, a mouth cloth, a probe) or a single type of tool from different raw materials (a probe made from grass, bark, vines, or twigs). These animals also seem to be the only ones in which tool use varies at multiple levels, from individual to family, from community to population and subspecies levels. This points to the special role of learning and experience, as well as the transmission of knowledge through imitation. Only recently have video traps been used to determine how young chimpanzees learn from their mothers how to obtain honey from bees living deep below the soil surface [270].

The discovery of new tools or tool uses for chimpanzees will not end for a long time, not only because many chimpanzee populations are not yet well studied, as they are not habituated to humans but also because new archaeological tools are being used to find evidence of their technology. For example, in a very dry savannah area in Tanzania (Ugalla), traces of the excavation of plants with storage organs—mainly tuberous geophytes—have been found, suggesting the involvement of chimpanzees [271]. In this case, too, people have long spoken of a human uniqueness that allowed our ancestors to resort to these food sources only in times of food shortage. Digging food out of the ground with different digging tools made from deciduous twigs was also subsequently experimentally demonstrated, with chimpanzees not only being selective in their choice of tools but also using a wide variety of techniques [272]. What is astonishing here is that these were zoo-born animals that taught themselves these techniques.

Chimpanzees in the Nimba Mountains of Guinea take a very different approach, using a hammer and anvil to split fibrous, basketball-sized fruit into manageable, smaller pieces [273]. Finally, an observation in Senegal, where chimpanzees used spears to hunt bush babies, caused a particularly great stir [274]. They made the spears elaborate in five stages, creating a point by shortening it at one end. Interestingly, few adult males were at work here but rather females and juvenile chimpanzees.

Other Apes

Considering all these great reports about the technical abilities of chimpanzees together, one wonders why only they and not the other great apes stand out in this way. The answer became even more difficult when one saw what bonobos, gorillas, and orangutans were actually capable of doing. From zoos, sanctuaries, and reserves, reports of their cognitive and manual abilities increased. In several zoos, researchers presented bonobos with a series of tasks that could be solved only with tools [275]. For example, they showed them food hidden under a rock and left technical aids such as sticks and antler poles nearby. In fact, two out of eight bonobos living in the zoo and four out of seven in the sanctuary made use of them—mostly without hesitation. Not only did they use the available items to dig, but they also used a long stick to lever larger rocks out of the way. Some even used several objects in succession. Three sanctuary apes hit long tubular bones with stones and uncovered food in the bone marrow cavity. Another struck the bone almost exactly along its longitudinal axis, demonstrating a technique previously known only to our human ancestors. One bonobo sharpened a stick with his teeth and made himself a spear. The greater skill of the sanctuary apes compared to the zoo apes may be related to their housing conditions and richer environment. It seems that once the animals have the raw materials at their disposal, they are able and willing to use them in a correct and context-specific way. Why they do not do this in the wild, where they also find stones and branches, requires explanation. However, bonobos have not yet been observed everywhere, especially in their more southern distribution areas, where it is more difficult to obtain food and where the use of tools would pay off more.

Western gorillas in zoos are also very dexterous; despite their large body mass, even silverbacks can very carefully and delicately poke pieces of food out of small maze boxes. In the wild, there are only a few known cases of tool use, apart from the frequently observed targeted throwing of branches and leaves in aggressive confrontations. A major sensation was caused by the observation of a free-ranging adult female using a stick approximately 1 m long to explore the depth of a water hazard. She did this several times while crossing and used the stick as a prop [214].

Finally, orangutans[7] represent an extremely interesting and sometimes puzzling case of tool users. In zoos and sanctuaries, individuals exhibit incredibly dexterous and innovative handling of objects, up to and including lighters and washing rums, partly looking at human usage through exact imitation [276, 277]. In this respect, they are certainly on the cognitive level of chimpanzees. However, tool use in the field has long been poorly known. Orangutans spend most of their time in trees where they are not only less observable but also usually need a hand to hold the tree and are therefore manually handicapped. Not surprisingly, the observations of Suzanne Chevalier-Skolnikoff (University of California), Anne Russon (York

[7]Among the great apes, orangutans are the least related to humans, as the orangutan lineage and that of the other great apes separated approximately 14 million years ago. Today, orangutans are only found on the South Asian islands of Sumatra (*Pongo abelii*) and Borneo (Pongo *pygmaeus*).

University, Toronto), and Birutė Mary F. Galdikas (Orangutan Research and Conservation Project, Indonesia) have been interpreted as indicating that orangutans' high cognitive abilities are most often used for locomotion [278]. In the wild, orangutans are constantly manipulating their three-dimensional environment as they move and forage. However, because the trees, branches, and other vegetation that orangutans use in very sophisticated ways are largely not freely manipulable, these manipulations are not generally called tool use. In captivity, however, they also extensively and skillfully handle loose sticks and other materials.

Over the last two decades, Carel van Schaik and his team at the University of Zürich have systematically studied the use of detached objects in wild orangutans, describing a rich repertoire of tools and their diverse uses [218, 279]. For wild orangutans, a total of 38 (genuine) tool variants (and five nesting variants that could also be considered genuine tool use) have been catalogued. These include seven tool variants used for extractive foraging, 13 for subsistence, 18 for physical comfort, and seven for communication. Some variants are thus used for multiple functions [280].

To date, hardly any complex tool uses, i.e. multiobject linking, have been found in orangutans; these tools represent more than a third of all variants in chimpanzees. Compared to chimpanzees, they also have significantly less foraging and terrestrial tool variants but not fewer cultural tool variants. The contexts in which tool variants were used also tended to differ between the two species. Orangutans have significantly fewer subsistence tool variants than chimpanzees but significantly more physical comfort tool variants and no significant differences in the number of communication tool variants. These new findings contradict the original socioecological theory and instead support the need to use the role of life on the ground ("terrestriality") as an explanation for the differences in tool use between chimpanzees and orangutans and as a driving force for the evolution of complex tool use [209, 281].

Long-Tailed Macaques

Although tool use is relatively widespread in the animal kingdom, only a few animals make routine use of stones. Among (nonhuman) primates, there are only three species: chimpanzees living in western Africa, capuchin monkeys in Brazil, and long-tailed macaques in Thailand and Myanmar.

The long-tailed or crab-eating macaque is a primate species from the macaque genus within the Old World monkeys (Cercopithecidae) family. This species also uses stones to break down energy-rich, trapped food, showing a remarkable similarity to chimpanzees and early Stone Age humans through the selective choice of stones, transport to suitable sites (usually to an anvil), discrete workplaces, modification and collection of tools, and much more. While chimpanzees live inside and outside of tropical rainforests and stone-using capuchin monkeys in dry deciduous forests, long-tailed macaques spend much time on the coasts of Thai islands, where they use stones to crack shells.

The observation that some long-tailed macaques also roam inland caused a sensation. They were found, among other places, on an abandoned oil palm plantation on the island of Yao Noi. And there, stones that were apparently used as an anvil and other stones that were used as hammers were also found (Fig. 3.1). With camera traps, it was found that both males and females visited the sites and did exactly what was suspected: they placed one nut after the other on the flat stone and hammered it with other stones until they cracked open and released the nutritious contents [282].

This discovery is interesting because a very recent behavioural adaptation was found here. Apparently, the monkeys have transferred a technique they developed on the coasts for cracking open mussels, thus opening up a new food source. Humans transplanted oil palms, a food source rich in protein, fat, and vitamins, to this region only over a decade ago. This makes them the first generation of long-tailed macaques that are able to perform this behaviour. The researchers thus testified to a technological response of the monkeys to an anthropogenic change in their environment, and thus also to a genuine behavioural innovation [283]. Now, the researchers want to determine whether, and if so how, this technique spreads, i.e. whether it becomes a tradition. It is possible that other later abandoned places were found where this technique was used, and new places were used by monkeys that developed an improved technique. If one could distinguish earlier use of the stones from current use after years of practice, one would be able to witness an evolution of tool use. As mentioned earlier, chimpanzees probably started using this stone tool technology at least a few thousand years ago. Here, on the other hand, one can study in situ the interplay of behavioural plasticity, innovation, and tradition, which are all essential features of culture. In any case, "long-tailed macaques have potential to exhibit cultural tendencies" [283, p. 872–873].

What also makes this discovery so interesting is that it makes it possible to compare nut cracking between chimpanzees and humans. The macaques could use a similar technique, but just as easily they could use a completely different technique, depending on their perception of the relationship between their movements and their effects on the nuts, as well as their control of the movement. Thus, they might even, in the authors' view, provide a good model for the early evolution of Oldowan stone tool techniques in humans. A group of so-called "(non-hominid) primate archaeologists" thinks that the analysis of primate tool techniques can help us better understand the evolution of tool use in humans and that it is therefore time to overcome anthropocentric narrow-mindedness [284].

Capuchin Monkeys

From a comparative, evolutionary point of view, it is particularly valuable to have New World monkeys as the third in the group of nonhuman stone tappers for the purpose of obtaining food. These are the black-striped capuchins, a primate species from the capuchin subfamily within the New World monkeys. These are medium-sized primates (weighing approximately 2–4 kg) found in the dry deciduous forests

(Caatinga and Cerrado) of northeastern Brazil. They are omnivorous and eat not only fruits but also other plant parts, insects, and small vertebrates. From an early age, they hit the ground with objects, and throughout their geographic distribution, they hit enclosed food (such as hard fruits, seeds, or invertebrates in shells) on hard surfaces to reach the nutritious inside.

It is remarkable that only a few populations have become stone tappers. There are three hypotheses for this, but thus far, none have been sufficiently well confirmed. One is that this tool use is due to an "energy bottle neck", i.e. it occurs only in periods (and places) of extreme food scarcity. Another hypothesis, which is currently quite generally claimed for complex tool use, such as stone tapping, is based on "terrestriality", foraging on the ground, where not only stones and nuts can be found but also good hammering can be performed. The latter is hardly possible in trees. Third, the presence of hard-to-reach food may increase the nutritional benefits of tool use, independent of seasonal availability and terrestriality. However, capuchin monkeys engage in stone hammering in an environment where food is relatively abundant and available year round. In any case, seasonal food scarcity cannot be a reason for this activity in the Cerrado. Rather, a combination of ecological factors, including the spatial distribution of nuts, the absence of ground-hunting predators, and the mechanical properties of the available food types, seem to play a deciding role.

Black-striped capuchins crack open hard nuts with similar skill to chimpanzees (Fig. 3.1) [219]. Even though the hardest types of nuts (especially those of the piassava palm, *Attalea funifera*) are similarly hard to the panda nuts foraged by chimpanzees, and these New World monkeys have significantly poorer physical conditions for hitting them, they are very successful at accessing them. To do this, they use stones weighing half their body weight, stand on their hind legs, and strike with great force and precision at the nuts, which they have previously placed in the slight depressions of their stone anvils. Despite weighing a tenth of the body weight of chimpanzees, they generate a similar amount of force (up to 12 kN) with their technique. Like chimpanzees, they are selective in choosing stones for hammering and transporting them to anvils, which they use as workbenches repeatedly. Selection is usually performed visually, but since size and weight are not always correlated, weights are checked before the stone is carried to the anvil. This behaviour is expedient, considering the ratio of stone weight to body weight. Additionally, at least the females seem to select stones with regard to the hardness of the nut to be opened later—the harder the nut is, the heavier the stone hammer [285].

To obtain a reasonably accurate assessment of the complexity of nut-cracking, it is important to consider what challenges are involved. It has been suggested that monkeys do this through multimodal perceptual learning, primarily haptic perceptual learning achieved through exploration and practice [253]. The animals have to understand not only the multiple static (nut/anvil) and dynamic relationships (hammer/nut) but also various positional options (of the anvil on the ground, of the nut in an indentation on the anvil), choose the right, i.e. less resistant, nut to crack, and, finally, the right stone, both in terms of shape (smooth versus uneven stones) and weight (heavy versus light, as long as they can be lifted) and strength (hard versus brittle stones). Then, the objects are gathered. In a field experiment, it was statistically proven that the monkeys selectively chose the stones before transport. They predominantly took heavy stones and did not change them once they had started hammering [286]. Obviously, the animals in the field have more experience—after all, they need 2–3 years to achieve the dexterity of adults—because this could not be confirmed in comparable experiments in the zoo [287]. In Brazil, if wild animals are not good at visually assessing the quality of the stone, they test the stone by picking it up and striking it. Apparently, they seek the necessary information to avoid carrying unsuitable stones to the workshop. This can be interpreted as planning ahead [285, 286].

The monkeys also proceed very purposefully with the anvil. They bring the nuts into a stable position, mostly in small depressions resulting from previous activities. Before placing the nut on the ground, they tap it several times. In this way, they are probably looking for an ideal, stable position and orienting themselves with their hearing. However, there are differences between the few populations in which nuts are opened with stone hammers on a large scale. These are probably not due to genetics but rather seem to represent behavioural traditions that have been spread through social learning within populations [288].

Some groups of black-striped capuchin monkeys in Serra da Capivara National Park even use stones to dig for roots and tubers in the soil and smash wood to expose insects and larvae [244, 289]. They also knock other stones out of conglomerate blocks, pulverising trapped pieces of quartz that they then lick off (probably to ingest the minerals they contain); alternatively, they reuse the knocked-out stones themselves as tools, which constitutes *secondary* tool use. Stones are not only used as hammers but also sometimes as hooks to remove loosened soil (*multifunctional* tool use!). Some of the monkeys also use sticks to retrieve water or solid food from crevices, holes, and insect nests and partially modify them by removing side branches and leaves or shortening them overall in length. In the few cases observed thus far, they have even thinned the end of the stick, demonstrating the ability of purposeful tool-tinkering. Finally, stones were found to be used multiple times for different purposes, including two stones in combination and even the combined use of stones and sticks. Therefore, we also seem to have examples of *sequential*, *serial*, and *combined* tool use [244]. It is therefore not surprising that these New World monkeys are also major players in experimental laboratory studies, where one tries to systematically track down their cognitive abilities.

Interested to know how long these monkeys have been using stones to crack nuts, and after finding large accumulations of such stones in some places a few years ago at Fazenda Boa Vista (Piaui, Brazil), researchers dug down to 70 cm below the ground in one place and uncovered 69 stones. The size and shape of these stones, as well as their wear marks, were very similar to those currently used by monkeys. Remains of cashew nuts (the kernels of the cashew tree) could be found on the stones [290]. With the help of mass spectrometry, the remains on the stones could be dated to the age of 600–700 years. Apparently, the percussion technique and the choice of stone tools were kept relatively constant over 100 generations.

New Caledonian Crows

The New Caledonian crow is a relative of the common raven, jackdaw, and rook of the genus *Corvus*. Unlike its relatives, the New Caledonian crow is not very widespread, as it is a tropical island crow found exclusively on New Caledonia (Grande Terre) and on the neighbouring island of Maré (Loyalty Islands). It is about the same size as the jackdaw (about 40 cm body length) and has a relatively short bill (42–50 mm). However, it has one of the largest brains of all bird species, relative to its body size [291], particularly with the associative centres standing out in terms of size [292].

As a food generalist, it feeds on a considerable range of food sources, including fruits, seeds and flowers, snails, lizards, adult insects; carrion; and, occasionally, bird eggs and chicks. However, its popularity results from its consumption of wood-boring longhorn beetle larvae (*Agrianome fairmairei*). It finds these in rotten trunks of the candlenut tree (*Aleurites moluccana*), from which it also harvests nutritious nuts by dropping them on hard surfaces. Estimates have shown that where these trees grow,[8] longhorn beetle larvae provide a considerable part of the protein and fat intake of these birds [294]. Only a few larvae can meet the energy requirements of an entire day. This makes them a particularly valuable food source, which seems to explain the emergence of their complex extraction with the help of tools.

How exactly do the birds do this? For more than 20 years, the extraction of beetle larvae has been studied in the field [295, 296], accompanied only a little later by experimental laboratory studies [297, 298]. The techniques used by crows stem from the defensive reaction of the larvae and are similar to those used by chimpanzees to catch ants. Longhorn beetle larvae are fished from the holes in decaying wood stumps with the help of custom-made tools. To do this, the crows insert a specially designed twig or a self-made pointed leaf into the larvae burrow of the rotten tree, irritating ("teasing") the larvae through careful, repeated poking until they bite into the tip of the tool with their powerful mandibles. Once attached, they can then be levered out by the birds.

What is interesting in connection with the question of the origin of tool use in these crows is that they are obviously anatomically and physiologically adapted. These birds possess unusually wide binocular visual fields (up to 60 degrees) [299]. This extreme binocular vision seems to serve less the facilitation of poking from maintaining binocular vision of the tool tips, and more the enabling of each eye to see far enough across the centre line of the visual field [300]. In fact, crows hold poking tools diagonally in their beaks [294], with the side of the handle corresponding to their eye preference [301, 302]. This finding suggested that dexterous tool use in New Caledonian crows has undergone long genetic evolution, but this does not

[8] Interestingly, humans may have influenced the evolution of at least some tool-oriented behaviours of New Caledonian crows through the possible introduction of light nut trees, where longhorn beetle larvae prefer to live [293].

exclude possible individual skills acquired through learning and observation [293, 303].

The scientific community was particularly surprised by the way these poking instruments were created. They are made either from the leaves of the screw palm tree (*Pandanus* sp.) or from the leaf stalks and twigs of the candlenut tree [295]. Initially, many people were surprised when it was discovered that chimpanzees make tools, such as straws and sticks, which they trim to fish termites and ants out of their burrows. The taboo that Jane Goodall had broken by observing tool *use* was magnified with the evidence of tool *making* and *modification*. However, some critics objected that we and chimpanzees have relatively recent common ancestors and therefore must also be relatively smart. Therefore, when it was discovered that New Caledonian crows also make tools, this objection lost traction because our last common ancestor with the birds was a lizard that lived more than 320 million years ago.

As we will see, Woodpecker finches also improve their poking tools, but only with slight modifications. The crows in New Caledonia, on the other hand, make their tools from raw material, leaves, or twigs. They use the twigs of the walnut tree to make hooked fishing rods, i.e. poking tools with barbs that catch the prey when it is pulled out [295]. They do this by first breaking off relatively long branches with leaves and branch forks from low-hanging branches. Then, they proceed in three hierarchical steps [304]. They shorten the detached branches step by step with the so-called snapping technique by first pinching off the side branches and leaves so that a slightly longer stick remains. Then, they detach an end piece just below a side stump, where the side branch was previously detached from the main branch with some distance. This creates a barb at the end. This is now shaped with the beak by chiseling and by sharpening while detaching the smallest parts. Due to the sequence of several different actions (detaching, shaping), one can speak here of genuine *tool crafting* [199]. What is unique is that they reshape the stick even during use until it meets their requirements. This can involve up to a dozen modifications, more than chimpanzees or other tool tinkerers.

Since there are enough twigs on the walnut trees, in which they fish for larvae, the crows could make a living with these. But in fact, they create a second type of fishing tool from leaves of the screw tree. Apparently, these flexible tools can be inserted more deeply into narrow or curved holes. However, since the robust leaves would be much too large for this purpose, they would have to be reduced in size. To do this, the birds bite into a leaf's edge at one point against the direction of the fibres and then carefully pull off a long strip with a pointed end along the fibres [304]. Since the leaves of the screw tree have jagged edges, this in turn gives them a fishing tool with barbs, albeit small ones.

The observation that different varieties of these leaf tools exist has since caused quite a stir [305]. Some are simple, either slightly wider or narrower, while others are more complex in terms of construction. The simple ones are created by tearing the leaf once at the edge, which causes the depth of the initial incision to differ, resulting in different tool widths being created. The other, more complex blade tools are irregularly wide and have a stepped edge of unequal thickness. Thus, they are relatively stiff at the wide end but thin and flexible at the pointed end. Apparently, this makes them easier to poke into different boreholes and pull out larvae. Therefore, these blade tools are also called *multifunctional tools*. Gavin Hunt of the University of Auckland (New Zealand) described the production of stepped, tapered blade tools as 2-dimensional chiseling with high precision that follows a predefined rule, resulting in tools of standardised form [305]. Critics emphasise that there is only one type of action (detachment), therefore one cannot speak of toolmaking in the narrower sense.

Yet, it is unclear how the development of the more complex type of tool came about. It is assumed that there was some kind of cultural evolution through tradition, i.e. the passing on and modification of knowledge [306]. In fact, a pattern of tool-type distribution across the main island of Grande Terre has been found, suggesting a successive spread from simple, broad types in the southern region—a historical origin—to graded, complex types in the northern region [307]. It is possible that different crow populations are responsible for their local traditions or "*tool industries*" [308].

One could imagine a scenario where individual crows with ingenuity produce improved types and then "migrate" to slightly more northerly parts of the island, where the innovation spreads due to the improved functionality through observation and knowledge transfer. If this had happened twice, the described spreading pattern could be explained. One would then have a case of *cumulative* tool evolution, otherwise known only in humans [306, 307]. Since this does not occur independently in single individuals but rather as a successive tradition based on social learning, we would also have tool *cultures*—in the sense of the definition of culture proposed for chimpanzees[9] [309]. As tempting as this hypothesis is, it is hardly supported by hard facts and remains speculative [310].

Too many questions regarding the tool tradition of New Caledonian crows remain open. Some of them could be clarified in laboratory experiments, as we will see, but for other elements of the cultural theory, there are alternative hypotheses. Above all, we do not know the history of the distribution patterns of the individual tool types. Since little wooden branches hardly fossilise, we will never know for sure. We also know little about the exact function of the individual tool types. It could be that simple thin or broad leaf tools are advantageous in particular constellations of burrows, which are restricted to the southern part of the island. This would be an ecological, rather than a social, explanation of the local differences. Against social differences based on local traditions speaks that these birds are not particularly social; rather, they proceed very individually when extracting food and rarely observe each other doing so, and as we will see, the technique can be learned individually [303, 311]. It has now been argued that the transmission of knowledge does not have to be based on direct observation of the actions during tool making, i.e. on *imitation*, but could be explained by *product emulation*. This means that the counterparts of the leaves from which the tools were extracted remain in place and are

[9]This does not mean the kind of culture that is linked to symbol-mediated tradition; this has not been demonstrated in nonhuman animals (i.e., outside the genus Homo).

thus used by other crows as models for their own tools. These tools would consequently have to be closely inspected and then—with a template in mind—formed from an intact leaf [312].

Finally, there are significant gaps in our knowledge about the distribution of crows on the island and the structure of their communities, if any. Are there truly local groups without exchange that have formed and preserved local traditions? Are there truly separate populations on the island or merely one? Is the pattern of tool distribution perhaps only based on the places of their exploration?

Contrary to the opinion of Gavin Hunt and Russel Gray [306, 308], there is no evidence that wild crows show tool sets or combined tool use. Nor is there evidence of the use of objects as meta-tools (think of chimpanzee wedge stones). Even the potential (laboratory) evidence for this [238, 313] is—as we will also see—controversial. Finally, only two modalities of tool use are found in crows: dropping and poking. This number is significantly less than that in chimpanzees or capuchin monkeys. Another difference to these primate species is that poking tool production is consistently observable in New Caledonian crows throughout the species. Nowhere were groups found that did not produce the two poking tools from branches and leaves. As we have already shown, stone tools, as well as other tools, are restricted to specific populations within their range in both chimpanzees and capuchin monkeys, and the differences between populations have been described as characteristics of their tool cultures [309]. Furthermore, the ranges are much larger and more variable than those in the crows of New Caledonia.

Hawaiian Crows

Only recently have it become known that in addition to the New Caledonian crows, which are famous for their use of tools, the Hawaiian crows also have a dexterous beak. Like their relatives from New Caledonia, the Alalā—as they are called in the local language—also use small sticks as tools to obtain food from holes and crevices [314].

The extremely rare Hawaiian crow has been considered extinct in the wild since the beginning of this century, but just more than 100 individuals live at two breeding stations in Hawaii and Maui, from which the first animals were soon released into the wild. Almost all of these birds (104 in total) were tested by researchers from St. Andrews (UK) and San Diego (USA). For this purpose, the researchers drilled small horizontal and vertical holes in wooden trunks, hid popular food morsels in them, and distributed small twigs around them.

Confronted with the task of reaching this out-of-reach food, 78% of the observed animals spontaneously took the twigs in their beaks and fished out the tidbits in usually less than a minute. The percentage is explained by the age of the animals tested: While almost all (93%) adult crows fished with twigs and pieces of wood, only approximately half (47%) of the juveniles did so. There were hardly any differences between male and female individuals.

The adults were especially selective and purposeful. They recognised which branch was suitable and which was not suitable. The material that did not fit was either discarded or shaped by bending and breaking. This method gave the sticks the right shape and length, transforming them into good tools. Some individuals even made their own fishing utensils from the plant material provided.

The authors suggest that the enormous technical abilities of the (2) tropical corvids are due to the unusual ecological living conditions prevailing on tropical islands (such as reduced competition—there are few woodpeckers—for enclosed food and reduced threat from predators) [314]. These conditions also prevail on Galápagos, where the resident Darwin's finches are equally habitual foragers. Although the islands differ in detail, these conditions are nevertheless different from those on the neighbouring mainland. Does this explain the rarity of tool use there?

Since the two crow species have relatively large kin distances within corvids—their common ancestor lived approximately 11 million years ago—and are rather isolated in terms of their abilities within this group, it seems reasonable to assume convergent, independent evolution of their tool use [314]. This is even more true for the less closely related Woodpecker finches.

The discovery of a second corvid species that habitually uses tools has reignited the question of the origin of this rare behaviour. In any case, this opens up a hitherto undreamed possibility of comparison with the New Caledonian crows. These were previously known as the only corvids that use tools innately and habitually and as the only birds that also show special cognitive abilities in the technical field.

With respect to the ontogeny of the highly skilled foraging behaviour, Hawaiian-crow researchers found another commonality with New Caledonian crows. There seems to be a strong, innate disposition, which means that the behaviour starts without social input and is already almost entirely functional in the beginning. The researchers tested this by raising seven young birds and then confronted them with food in tree holes. In fact, all of them spontaneously grabbed the sticks lying around, and four were able to successfully catch the food. The researchers were careful to ensure that the birds had never observed conspecifics using this technique. Nevertheless, most of them were incredibly adept at handling the sticks and even managed to obtain the bait in difficult cases.

The combination of the two facts that young birds acquire poking with sticks without training and without observation and that this ability is not limited to a few individuals suggest that this behaviour has a strong genetic basis. In addition, Hawaiian crows—similar to New Caledonian crows—also have a very straight bill, which enables them to use sticks for poking holes. The two crow species share all these characteristics with a third island species in the South Pacific, the Woodpecker finches.

Woodpecker Finches

Eighteen species of Galápagos finches live on the Galápagos Islands in the South Pacific, approximately 1,000 km off the west coast of South America. All of these species are closely related and descended from a common ancestor. They are known to the public as Darwin's finches because Charles Darwin mentioned them in his diary during his voyage on the *Beagle* in 1835. However, these birds are not finches but belong to the tanager family within the songbirds. They are approximately 10–20 cm in size and differ in the shape and size of their beak as well as in their different ways of life, especially in regard to their diet. Of particular interest to us is the Woodpecker finch, also sometimes called the tool-tree finch. It was named the Woodpecker finch because it occupies the ecological niche that otherwise belongs to woodpeckers. These birds feed, among other things, on insects and their larvae hidden under the bark of trees. They capture them (if they cannot reach them directly with their pointed beak) with poking tools made from cactus spines, leaf stalks, or thin twigs. This is essentially done by appropriately shortening and removing pieces that are obstructive when inserted into the crevices. They also transport their instruments from tree to tree. Like the New Caledonian crows, they hold the tools in their beaks at one end and poke the other end into holes or cracks repeatedly until they hit a larva. Then, they try to spear it, lever it out or poke it until the larvae crawl out on their own and can be picked. To do this, they drop their tool or clamp it under a foot.

Although other Darwin's finches (including those of another tree finch species, the Mangrove finch [315, 316] and the Green warbler-finch [317]) were observed poking at prey with thin objects, these observations were not investigated further. In contrast, my former colleague Sabine Tebbich studied Woodpecker finches with her team on Santa Cruz Island for several years. In addition to her laboratory experiments with animals captured for short periods of time, she also studied the ecological conditions of tool use. In doing so, she was able to determine that the animals only use their tool technique where insects and their larvae cannot be captured freely but remain under the tree bark. This is especially the case in the low-rainfall region of Santa Cruz, which is therefore also called the dry zone. Since the larvae mostly stay under the bark here, the use of the tool is vital for the birds. No less than half of the total food requirement is covered by the instrumental acquisition of prey in the dry season [318]. But even then it is not always dry, and the use of tools varies accordingly. It is particularly rare to find tools for use in the rainy, forested parts of the island, the so-called Scalesia zone, where arthropods crawl out of the trees and can be preyed upon throughout the year without tools.

The differences in food availability based on geographic and climatic conditions could mean that there are two populations of Woodpecker finches that separately employ tools at quite different frequencies in the two zones. However, there is no evidence for this, suggesting that this technique is innate in all animals of the species and can be used when needed. In fact, this assumption was confirmed with the help of a clever laboratory experiment. Animals that grew up without the opportunity to be shown the technique developed into competent tool users, similar to the control group, which was not deprived of this opportunity [319]. Since Woodpecker finches do not live in social groups that promote the learning of complex food acquisition techniques through the observation of competent conspecifics, unlike, for example, the stone hammering of chimpanzees, individuals remain on their own. Only through genetic anchoring could the behaviour be evolutionarily expressed, although a small individual learning component and the accumulation of practical experience are not excluded. Apparently, tool use in Woodpecker finches is an additional form of food acquisition that allows the morphological constraints of the beak to be temporarily extended without affecting the skill or flexibility of the behaviour [320]. The extent to which this extension also entailed cognitive development, or was even made possible by it in the first place, remains an open question. Only laboratory experiments with birds captured in their native habitat and kept in aviaries for a short time were able to demonstrate the technical intelligence of these skilled tool users. For this purpose, a sister species (Small tree finches) was used so that a possible cognitive advantage could be determined from the comparison with the otherwise very similar living and closely related non-tool users, Darwin's finches [321].

Is Tool Use a Sign of Intelligence?

The question of whether the use or production of tools promotes or even requires technical intelligence is vigorously debated. There are studies that have compared brain size or other *proxies* of animal species with and without tool use and found positive correlations [197, 198], but these are methodologically controversial, as mentioned above. Their conclusions are also highly problematic in terms of content because tool use is associated with other behaviours, such as manipulativeness, extractive foraging, or social intelligence, which might require a larger brain volume. There are also environmental factors, such as environmental complexity, climatic variability, and certain food compositions. It could also be that increased intake of high-energy food, such as insect larvae and nuts, made possible by the use of tools can increase the growth of high-energy body parts, such as the brain. We thus have at least two problems in terms of content—that of covariation with other behaviours and that of causality.

In addition, there is a third problem related to the relationship between brain size and intelligence. It is a very general, theoretical one. Why should bigger brains be more intelligent? Because we humans have big brains? Because more cells mean greater computational capacity? However, is not it the case that a large brain would also be beneficial for basal bodily functions and movement control? If there are more muscles, more nerves, etc., the "computing and control centre" must be correspondingly larger. In fact, larger animals also have larger brains. Therefore, it is not absolute but relative brain size that is adopted as an estimate of brain size or capacity.[10] However, even this measure does not seem to be the best approximation [324]. Interestingly, studies of primates and dogs have shown that overall brain size is a better measure for predicting general intelligence [167, 325–327]. The two measures—relative and overall brain size—are closely correlated with each other but also with body size. However, so does the neocortex ratio [328]. Nevertheless, doubts arise, especially if one broadens one's view to other animal species. Even small animals sometimes have to solve complex problems with small brains, whether in navigation or living in large communities. In fact, we find amazing cognitive abilities in animals with tiny brains, such as bees and wasps [329–333].

Therefore, we cannot avoid looking at the matter of intelligence in tool use directly, not via anatomical approximations but rather through good behavioural experiments. We want to simplify the question of whether tool users are more intelligent than animals that do not use tools by looking only at animals that are capable of more complex or flexible tool use. Already, this is a small sample. Many animal species use only one tool for a single purpose in a single situation. As soon as something changes, the behaviour disappears. Tool use is often a specialised behaviour with a very specific, rigid, and context-dependent form. Take the archerfish, for example. It can spit water very sharply and specifically at insects that are just above the water level, usually on mangroves [334]. This allows the fish to be very adept at making prey that it would not be able to reach by jumping. However, this technique is not used for any other purpose, except for "deception" in the laboratory (for which the student Michael Pollirer was able to tell the archerfish to spit on his finger or on a cigarette [335]). We can speak of *tool specialists* in this case. Most likely, many, perhaps even most, tool-using animal species must be placed in this category.

Tool Use as Cognitive Specialisation?

There is no doubt that in many species, tool use is the consequence of evolutionary specialisation with a corresponding adaptive function, similar to other behavioural specialisations, such as certain gestures or facial expressions, or morphological specialisations, such as the beaks of Darwin's finches. In this sense, the use of stones

[10] The most commonly used measure for estimating the cognitive abilities or intelligence of animal species is the encephalisation quotient (EQ) proposed by Jerison [322], which reflects the ratio of a species' actual brain size to its predicted brain size, based on the clade-specific brain–body regression line. This was thought to estimate the proportion of the brain devoted to somatic functions and thus obtain the size of the cognitive portion by subtraction. However, since the brain also controls numerous noncognitive somatic functions, most researchers now agree that it cannot be used for these somatic functions without control. But how else can one control the relationship between the noncognitive functions of the brain and body size? One solution to this problem is to determine the slope of "cognitive equivalence", the line connecting organisms with an identical "Bauplan" but different body sizes. Indeed, it has recently been shown that estimates of cognitive brain size based on cognitive equivalence better fit empirical cognitive studies of general intelligence [323].

by Egyptian vultures to open the eggs of ostriches or the use of stones by digger wasps to block the entrance of their burrows are also examples of behavioural specialisations [199]. Unquestionably, these behaviours are very efficient at fulfilling a specific function but may be very inflexible. It is often said that these behaviours are innate and that the underlying neural architecture is rigid and modular. Rigid would be the "wiring" in the brain because it can hardly be changed by additional experience. The modularity is also understood in a functional sense, not or at least not only in an anatomical sense [336]. Animals would have a special module for tool use. Other actions or cognitive performance are not influenced by this. This *domain-specific* intelligence is fundamentally different from general intelligence [337–339], similar to graduates of universities of applied sciences from graduates of old Humboldtian universities.

However, if tool use is based on general intelligence and does not require cognitive specialisation, this should mean that animals are equally *capable* in terms of tool use but do not *use* this ability equally. Other factors, but not the way animals process information, would be decisive. With this provocative thesis, the psychologist Euan M. Macphail from the University of York (UK) came to the fore roughly 40 years ago [340, 341] and attracted increased amounts of attention. Therefore, is the difference between a pigeon and a New Caledonian crow a virtual rather than a real one—only gradual rather than categorical? If they differ only in terms of input or incoming sensory information, then a crow equipped with the eyes of a pigeon behaves like a pigeon. This is unlikely because it overlooks the enormous power of the brain to process incoming "information"[11] and to create suitable representations of the environment and then still select the appropriate representations for the task at hand. Finally, we must not forget that the existing representations influence and control the processing of excitations coming in from sense organs.

In any case, classifying tool use as a cognitive specialisation must be revised when it becomes apparent that the behaviour occurs spontaneously and without human intervention even outside the natural context. A further step towards greater flexibility would be if the tool could be used for multiple functions, especially if new tasks are solved with it and if the behaviour passes as a genuine innovation [342, 343]. For example, if Egyptian vultures also used the stones to hide (cover) food from the eyes of possible competitors or to throw them at enemies to drive them away. In this case, the classification as a behavioural specialisation would have to be revised.

Cases where multiple tools are used for multiple functions present an even greater challenge for classification as a behavioural specialisation. Particularly exciting are cases where multiple tools are flexibly used for different functions (such as stones used by humans for striking one time and for cutting another and other objects harnessed for striking and cutting when stones are scarce). Animals

[11] I am indebted to Ulrich Kattmann for pointing out that there are strong reservations about the use of the information metaphor for epistemological reasons, since the brain only receives excitations to which meaning is only assigned through processing in the brain and which only then become information.

that show such flexibility in the use of tools are probably beyond the suspicion that their behaviour is based on cognitively narrow, genetically fixed special adaptations. In contrast, one would then have to assume that they have cognitively wide open programmes at their disposal, which they can apply not only to tool use but also to many other manipulative problems. Moreover, the seemingly disparate abilities of spatial navigation (for example, taking detours) and tool use seem to recruit the same cognitive resources in terms of inhibitory control and motor skills [344]. At least in great apes, there also seems to be a link between gestures and tool use, as the same motor control areas in the brain are activated [345]. All of these findings argue against tool use as a module in the brain.

However, how can we empirically test whether tool use is based on rigid or flexible mental structures? Comparisons between closely related animal species with and without tool use are very illuminating. Here, we must make a further distinction: Bonobos rarely or never use tools in the wild, but in the laboratory, when the situation (obviously) calls for it, they do, and they are nearly as adept at it as chimpanzees, their habitual tool-using relatives [275]. I will discuss this in more detail. But the comparison is flawed because bonobos obviously do show great skill with tools, just not in the wild. A better comparison would be with species that never use tools, neither in the wild nor in the laboratory.

This is the case with Darwin's finches. Here, there could be a serious difference in general cognitive terms between sister species with and without tool use because Woodpecker finches habitually use tools, whereas Small tree finches never use tools. To determine how the animals actually differ, Sabine Tebbich (University of Vienna) and Irmgard Teschke (Max Planck Institute for Ornithology in Seewiesen) confronted individuals of both species with a series of different problem-solving tasks [320]. For this purpose, they captured wild birds on Santa Cruz and tested them after habituation to aviaries at the Charles Darwin Research Station.

A first comparison was in a task to connect the position of a hook with its possible function—to fish for food. If the hook was positioned so that it surrounded the piece of food, pulling the hook closer would bring the piece of food closer. This hook was to be pulled in front of a second hook that did not enclose the piece of food. Thus, this test simulated the use of tools with which food could be fished. It should therefore have been very easy for the Woodpecker finches but not the Small tree finches, which had no natural disposition to fish for food with tools. However, this was not the case. While all the Small tree finches were successful at reaching the learning criterion at the given time, only 8 out of the 12 Woodpecker finches were successful.

Another task consisted of combining an action—perching on one of two levers—and, consequently, food becoming available or moving out of reach. This type of food acquisition was also intended to benefit manipulative Woodpecker finches. A hole was located off-centre on a platform. When sitting on one lever, the platform was tilted, and food was rolled. If the bird sat on the right side, the food was rolled out of the apparatus and could be eaten. If the bird sat on the wrong side, the food rolled into the hole and disappeared. However, even in this "seesaw test", there was no significant advantage for the Woodpecker finches. In contrast, more Small tree finches than Woodpecker finches succeeded, as measured against the predefined learning criterion [320].

Flexible behaviour is revealed, among other things, by the fact that an action that *was* successful is no longer so, so that the animal must relearn. If Woodpecker finches require flexible behaviour, they should at least be more successful in this respect than Small tree finches are. The researchers therefore subjected the two kinds of finches to a simple relearning test. Of the two coloured dots, one had to be chosen to obtain food, and the other did not provide food. Once learned, the colour dots changed their meaning, i.e. the positive stimulus became the negative, and vice versa. Surprisingly, the Small tree finches were again more successful, making significantly fewer mistakes in the reversal phase.

The only advantage that Woodpecker finches had was when opening a box. Here, a lid had to be simply levered or pulled up at its protruding end. The Woodpecker finches were quicker than the Small tree finches at opening a box, but this could be due more to their natural way of life and feeding, manipulating objects to extract food, than to their technical intelligence. In fact, they manipulated the lid much more often than the Small tree finches did [346, 347]. Thus, their cousins that forage without tools were equal or even ahead in three out of four cognitive tasks but not in perseverance or intensity in dealing with an object.

This somewhat sobering conclusion is also consistent with the results of other tests with Woodpecker finches. Once they had to choose sticks of different lengths to fish a piece of food out of a Plexiglas tube. Although they rarely chose sticks of exactly the right length, they mostly preferred longer sticks and were successful. Similarly, they learned a second task very quickly, which was close to their natural tool behaviour. Instead of sticks, they were given pieces of wood in an H or S shape. These devices had to be modified essentially by pinching off pieces to fit them into the narrow tube. They also solved this task, but not immediately, after only a few failed attempts. Finally, the animals were challenged with the trap-tube task developed for tests with capuchin monkeys [348]. The trap consisted of a hole at the bottom of the tube. The piece of food lying next to it can therefore be acquired only by pushing it away from the hole to the end of the tube with the stick. As we will see, this is a difficult task for all the animals tested thus far—including chimpanzees—and for children. It can be solved in different ways: by simple means, learning by trial and error and observation, or by causal understanding and logical deduction of the result of the action. Without going into the detail here, the Woodpecker finches again showed only very modest ability, as only one bird out of six solved the task [349]. A second trap-tube task with two traps also showed the narrow cognitive limits of Woodpecker finches. Only a few birds learned the original task, and none succeeded in the extension test, which tested the spontaneous solving of analogous problems [350].

The poor performance of Woodpecker finches could mean that tool use is not a cognitive specialisation but merely a behavioral specialisation, even in animals that improve or even make their tools. This specialisation then relates more to dexterity in object manipulation, along with high motivation and perseverance in food extraction. If anything, cognition would only be at play in a very general form. However, it could also be that the Woodpecker finches of Galápagos simply do not represent a good example of flexible, creative, or even intelligent tool use.

Creative or Flexible Tool Use

The Spanish primatologist Josep Call of the Max Planck Institute for Evolutionary Anthropology in Leipzig (Germany) and the University of St. Andrews (UK) suggests that among animal species with specialised tool behaviour, one should distinguish those that use tools flexibly or even innovatively [351]. Of course, he is referring primarily to the great apes he has studied for decades, but he also concedes this to the other apes and birds I described earlier. All these animals have the ability to detach themselves from the problem space that spans the main use of their tools. I have already mentioned how versatile sticks or stones can be used in these animal species, even to the point of targeted throwing. A very good example of how to contrast specialised and innovative tool use is a study by my former student, Natascha Mendes [352]. She presented great apes with the problem of an out-of-reach, floating peanut. A Plexiglas tube was fixed to the enclosure's bars and half filled with water. The peanut floated clearly visible on the surface. The tested orangutans tried to reach it with a finger but could not. After a few seconds of eyeing it, a female turned away from the tube and went to a water tap. The animal knew the tap; she could drink from it at any time. Now the orangutan did not drink but filled her mouth with water and returned to the tube with the peanut. There, she held her head sideways over the tube and spat most of the water—she still had problems aiming—into it. The water level rose, but since the peanut was still out of reach, the ape repeated the action until she was finally able to grab the nut. Knowing the experience of captive animals, it was certain that the problem was new and that the orangutans had never had to solve one similar problem before. Unlike the archerfish, which has its own morphological adaptation in the oral cavity for spitting, orangutans have no innate scheme of acquiring food by spitting with water.

It is reasonable to assume that the more animal species move away from their natural problems, the more likely a certain intelligence is involved in tool behaviour. Is the use of tools even based on a *behavioural innovation* after a previous cognitive solution in the mind? Although the concept of innovation is also defined very broadly and purely functionally as "the discovery of novel information, the creation of new behavior patterns, or the performance of established behavior patterns in a novel context" [353, p. 788], it intuitively resonates with a higher form of intelligence. Above all, the term is associated with two key characteristics: adaptability and creativity [343]. Individuals who exhibit these characteristics can not only apply old solutions to cope with new problems but also create new solutions when the old no longer work for former problems because circumstances have changed [354]. Thus, in addition to modifications of old solutions, entirely new solutions can also be developed for entirely new challenges [355].

Sometimes, it is no longer sufficient to use a tool, as one has to produce a tool or use a number of different tools to achieve a certain goal. Individuals may even anticipate problems that they could solve with tools, and they may select tools at a time when they have no feedback about the tools' effectiveness, based only on the tools' perceived properties [356–358].

A somewhat sobering assessment by Josep Call [351] is that there has been little progress in the last 50 years in our understanding of how nonhuman animal species generally solve difficult technical problems and manage creative tool use in particular. This may come as something of a surprise, especially in view of the fact that dozens of laboratories around the world have tried to reach the bottom of this question, sometimes at great technical expense. Among them, the Max Planck Institute for Evolutionary Anthropology in Leipzig, Call's former place of work, is at the forefront. However, he did not make his statement lightly but rather aimed at what he considered to be a very important but often ignored distinction between learning and thinking. Comparative animal research could have benefited from this distinction made in the early phase of human cognitive psychology, which goes back at least as far as Norman R. F. Maier and Theodore C. Schneirla [359]. But this did not happen, perhaps because animal-cognitive biology at the time was too cautious about concepts of human-cognitive psychology after the dissolution or fusion of behaviourism and ethology. Tool use was—if at all—only a topic among some primatologists. Despite the groundbreaking discoveries of Jane Goodall, intelligent tool use was still seen as an insurmountable Rubicon between humans and other animals in the 1970s. Of the few known exceptions, such as Wolfgang Köhler's chimpanzee research in Tenerife (1914–20) [100], tool use has not been studied under controlled laboratory conditions and has hardly been considered a cognitive performance, as something that had to be learned individually or through social input and that was based on a technical-physical understanding.

Two Routes to the Tool User

Even species that do not normally use tools can become tool users with proper training. We have mentioned bonobos and will meet other "lab geniuses" later. Some of them can even use several tools in succession and distinguish functional features from nonfunctional features. However, how do they do that?

We now return to the distinction made by Maier and Schneirla [359]—either through learning or through reflection. Through learning, related experiences are linked together. For example, dogs that previously had no preference for one of two strings may learn to pull on the one that is associated with food [360–364]. Success plays a major role in this. The string associated with food will lead to success because the dog will obtain the food it desires. The action preceding this reward is reinforced in the sense of Thorndike's law of effect. Conversely, the action that does not lead to food, in this case, the pulling of the string not associated with food, is not rewarded and is therefore weakened over time. In these actions, the animal also learns to associate certain properties of the objects involved with success. Associations based on mere perception thus also lead to a certain understanding of functionality.

For example, the dog learns that the food forms a (at least visual) unit with the string when it becomes reachable by pulling. This is also the case when pulling boards, a task described by Jean Piaget [89] to test the sensorimotor development of children. However, the dog does not initially learn that the food must be on the board to be pulled with the board (rather than behind or beside it) as long as it touches it [364, 365]. If anything, the dog learns this restriction only in the next step. Although learning by effect is a useful method for solving certain sensorimotor problems, it is limited. There must always be a spatiotemporal link between perceptual properties (the stimuli), actions, and success. In addition, this form of learning is slow, gradual, and error prone—for example, if the dog has not observed the functional properties of the object and pulls anyway.

Another form of problem solving is based on *reasoning*. Maier and Schneirla [359] defined reasoning very generally as the combination of separate experiences. For example, an animal is given the opportunity to explore one part of a maze first and then another the next day. Later, it was tested to determine whether the best way could be found through the whole maze, which requires a complete representation [351]. This coherent representation can emerge without success, merely on the basis of information gathered through *exploration*. Of course, knowledge generated on the basis of success reinforcement can also be incorporated, which means that the two problem-solving strategies—learning and reasoning—need not be mutually exclusive [351].

Another distinction between *reproductive* and *productive* thinking is also useful [366, 367]. Reproductive thinking means that known procedures that have been used before for similar problems are applied again in the same or in a slightly modified way. Productive thinking, on the other hand, involves the use of new procedures, whether for familiar or new problems. Learning and reasoning have a few things in common [351]. For example, both are based on experience. In *trial-and-error learning*, too, the individual does not start from scratch and try everything, but rather, the attempts are channelled, insofar as what has been successful before is given preference. Similarly, reasoning does not happen in a vacuum of experience. In contrast, experiences are even decisive. However, according to the Gestalt psychologist Max Wertheimer [367], the difference lies not in whether experiences are at play but in what kinds of experiences, blind connections or *structural understanding*.

Call [351] argued that this distinction has long been overlooked by researchers of comparative cognition. What is important is the distance between the two experiences to be linked. At the simplest level, the experiences to be linked are connected by spatiotemporal proximity. The closer the connections are, the more likely they are to be linked. For an action to be reinforced, success must occur quickly. In the sense of the reasoning mentioned above, this is not necessary. At an abstract level, the separation of two experiences may be determined by their "symbolic distance" [367]. Here, these are no longer on a perceptual dimension such as colour or size. A pigeon in the Skinner box—an automated learning device developed by the American learning psychologist Burrhus Frederic Skinner [87]—can serve as an example. If the pigeon has learned after a few trials that a green stimulus is rewarded but a red stimulus is not, it will also choose a blue–green stimulus in the so-called generalisation test. But pigeons have problems with *abstraction*, the refraining from the trained stimulus dimension. A pigeon can learn the *concept of* "long versus short" by learning to always choose the longer of two stripes, regardless of absolute length. But when the pigeon is confronted with a completely different stimulus dimension, for example, two tones of different lengths, its ability to generalise fails. The learned concept is therefore stimulus-bound and not abstract.

In tool use, it is important to recognise *functional* features. This is especially difficult when these are (perceptually) very different. For example, a hole on a platform can prevent the displacement of an object from A (in front of the hole) to B (behind the hole). This displacement is also prevented by a barrier. However, holes and barriers have no perceptual similarity, only a functional one. According to Wertheimer, the symbolic distance in recognising functional similarity is small here, while the perceptual distance is large [367]. According to Wertheimer, functional similarity is a structural feature. The distance between two experiences can be reduced by forming a new type of representation. On this structural, functional level, the hole and the barrier become similar, but they still remain distant at the level of visual perception.

Over the years, there have been repeated attempts to reduce these two processes to a single process [368–370], but this runs the risk of overlooking or neglecting essential aspects of animal problem-solving behaviour. If it is all trial-and-error learning, learning via reinforcement, chaining or reawakening of associations [371], then how can one explain that crows solve tool problems but pigeons do not?

Skinner's student Robert Epstein of Harvard University (USA) became famous for his experiment with pigeons, in which he showed behaviour strikingly similar to that of Köhler's chimpanzees in Tenerife [102]. The experiment was also deliberately designed as a copy, not to say a caricature, of Köhler's. To obtain out-of-reach food, pigeons pushed a small box directly under a banana dangling from the lid of the test box, then jumped up on it and pecked at the banana. While the similarity to Köhler's chimpanzees is astonishing, there is a striking difference. In Köhler's case, the chimpanzees were not trained beforehand to perform individual components of the "insightful" final action. Epstein's pigeons, on the other hand, were specifically prepared for the final action with positive and negative reinforcement in step-by-step training sessions. First, they had to learn to push a small paper box forward. Then, they learned to push the box to a specific location on the floor or wall, which varied several times. Finally, they learned to peck at a paper banana that hung far down from the lid of the training box and could thus be reached from the floor. There was no box in the room. Conversely, however, they learned not to peck at a banana that hung down less far and could only be reached by jumping. This behaviour was even discouraged through the absence of reward and a time-out. When, after successful training, they were confronted with the test situation where the banana was hanging high on the ceiling and a box was in a corner of the test box, they were very confused at first. But then they ran to the box, pushed it under the banana, jumped on it, and pecked at the banana. For uninformed viewers, there is hardly any difference between this final action and insightful chimpanzees or humans. However, does this interpretation stand up to detailed analysis?

When the pigeons were first confronted with the banana and the box and looked back and forth in confusion, they expressed the incompatibility of two previously learned behaviours. Because the pigeons were trained not to jump and peck at the unreachable banana, the other behaviours, walking to the box and pushing it, soon dominated. The banana became the target of this pushing because there was no dot on the floor or wall with which they had previously formed a positive association (alongside the banana itself). Finally, by pushing the box under the banana, the pigeons produced a situation where jumping up was rewarded.

The *chaining* of two previously independently learned associations was the secret of the pigeons' success. But Epstein's conclusion [372, 373] that precisely such chaining underlies the creativity of animals and humans was provocative because it lacked something essential: the spontaneity of the final action. It is true that the pigeons displayed flexibility, perhaps even unexpectedly for some, because they immediately linked two previously separately trained actions correctly. But still, these were specifically trained and rewarded. This was not the case for chimpanzees, of course. They do stack boxes in play, which gives them the necessary experience for these actions [374], but they do not do so in the context of food or other problem solving.

Despite all the criticism of Epstein's attempt, one question remained unanswered: Have the pigeons, in addition to the associations formed, possibly understood that the box was an aid to getting to the banana? Köhler's attribution of insight was, after all, based not only on the correct combination of boxes and sticks but also on an understanding of the function of these objects as aids in reaching a goal. Only an experiment could answer whether the pigeons might not have acquired such an understanding in the course of the Epstein experiment. This study was conducted 30 years later by Robert Cook and Catherine Fowler [375] at Tufts University in Boston (USA). After initially obtaining the results of Epstein, Lanza, and Skinner [102] in a faithful replication, they checked in a follow-up experiment whether pigeons, when choosing between two boxes with equivalent reinforcement backgrounds—both of which had previously been rewarded in the same way—selectively chose the one with the required function, namely, the possibility of jumping up. However, the pigeons did not. Apparently, they had not understood the *affordance* of the box and considered it merely as an object that, when pushed in the right way, results in a food reward.

The role of experience has often been misunderstood and used as an argument against spontaneous, insightful behaviour. However, Gestalt psychologists are familiar with the importance of experience solving problems, either as a facilitating or inhibiting agent for potential solutions [376, 377]. Just because experience is an integral part of problem-solving behaviour, it cannot simply be reduced to a concatenation of associations. How similar the experiences made are to the solution to the newly emerging problem is highly important. Many experiences are gained by people in the course of exploration and playfulness in youth. Much later, these previously unrewarded and non-trained experiences are linked to form a new representation of the task at hand. Finally, it should also be considered that the pigeons were placed in a very simplified task because they were confronted with only the relevant stimuli in the learning box; thus, distractions were avoided. When animals are confronted with complex tasks in their natural environment, they are always in a very stimulus-enriched environment. In such an environment, paying attention to the necessary stimuli and controlling the appropriate responses are much more difficult tasks than that of the pigeons. Call [351], therefore, doubts that pigeons are able to cope with the tasks that arise for tool-using chimpanzees.

However, such questions should not be hastily used to impute thinking to chimpanzees or mere memorisation to pigeons or to prematurely classify animal species according to these categories. For example, after decades of operant conditioning in the Skinner box, rats are capable of complex thought processes in spatial tasks of the star maze [378]. Already the British zoologist William Thorpe of the University of Cambridge [101] was inclined to argue that thinking is a widespread phenomenon in the animal kingdom and is not limited to a few individual species, such as chimpanzees. Thorpe, however, also became known for putting Wolfgang Köhler's [100] concept of insight on new explanatory legs.

Insight

In flexible tool use, it is important to recognise *functional* features, which can then be used to spontaneously find new solutions to old problems or old solutions to new problems. Functional features are usually not as easily perceptible as size or shape but are based on abstract, structural representations of physical properties or relationships. A new problem could therefore be visually different but functionally identical at its core. For *insightful* action, however, it must be true that the behaviour has not been performed before in another context, i.e. it is new in the form shown. It is true that experience with individual parts of the problem can and must exist, through the composition of which the new solution results. However, the solution must not exist in its entirety. Thorpe [101] has made a differentiation for insight, which unfortunately subsequently received little attention. He distinguished between insight and insightful learning:

> As a practical and convenient definition, I consider insight to be primarily a matter of the organisation of perceptions and to mean the apprehension of relations. Insight learning, on the other hand, includes as an essential element the appropriate organisation of effector response, and can, I suggest, be defined as the sudden production of a new adaptive response not arrived at by trial behaviour or as the solution of a problem by the sudden adaptive reorganisation of experience [101, p. 110].

Thus, insight learning also has an action aspect. Importantly, insight-based actions lead to a solution, i.e. is adaptive, and this solution is new and occurs spontaneously. Trial and error would therefore be out of the question. Instead, appropriate steps are chosen to achieve a goal.

Cognitive Ingredients of Creative Tool Use

Josep Call [351] suggested that creative (intelligent) tool use requires three cognitive skills: spontaneity, innovation, and flexibility. Only with these can multiple technical problems be solved, especially those problems that are considered innovative in the sense of Reader and Laland [198]. These characteristics also allow animal species that do not tend to use tools in their natural behaviour to skilfully use or even produce and modify tools in the laboratory and under particularly favourable circumstances (see below).

Spontaneity, innovation, and flexibility are certainly important markers for tool-using species. But they are also typical of species that are capable of solving other physical or even social problems. Therefore, what ingredients are specifically important for tool use?

Information Collection

Associations formed through primary or even secondary reinforcement (association with links based on reinforcement) are important for building an experiential memory. However, equally important is the information acquired through exploration, play, or generally latent learning without reinforcement or reward [379]. Information acquired in a context that is not shaped towards a goal can later be crucial for solving goal-directed problems. Such information can be used much more flexibly overall because it is not tied to specific stimuli. Rats that were able to explore a maze earlier without being rewarded for any behaviour could later complete maze tasks much faster than rats that had previously learned a specific path through the maze [101]. Chimpanzees who were able to play beforehand with certain objects later solved tasks that required the use of these objects much faster [374].

The relationship between latent learning and insightful problem solving was addressed long ago [101]. Apparently, it is necessary to collect a sufficient amount of appropriate information to be able to act insightfully. Even actions that initially failed or were not rewarded could later contribute to finding a solution. For example, these actions might reveal relationships between objects or between objects and certain effects that are later retrieved from memory when they matter. These can be relationships that one has perceived oneself or that one has seen from others. In the context of social learning, one speaks of "emulation",[12] when one acquires information about objects, such as physical properties, or about potential functions of objects by watching others.

However, not all information is the same. Two observers can observe the same process but form different representations about it. When attractive things, such as tasty food, are hidden under one of three possible objects (for example, cups), one-year-old children, like chimpanzees, remember the spatial properties of the hiding place, while three-year-old children also notice the visual properties of the hiding place, such as colour and shape [387]. Yet other properties of objects are their functional characteristics, also called "*structural characteristics*" [388]. They denote the relationships between objects, such as connectedness. For tool use, these "dynamic" properties are usually much more important than the "static" ones, such as colour and shape, which determine the appearance of the object. For example, one can remember that object A is on top of, next to, or under object B. If a piece of food can be obtained only with the help of a board, it is much more important to know that the food lies on a board than to know what size or colour it is ("*on-off*" or "*support problem*"). In the *string task,* species also seem to differ in what they perceive or learn about. Pigeons seem to be more attracted to static features, chimpanzees to dynamic features and relationships between objects [389]. When cups are shaken, chimpanzees interpret the sound as being caused by food and prefer this to a second cup where no sound is heard [108]. Dogs do not do this, at least not spontaneously [390]. The crucial question here is whether the chimpanzees understand the sound as a stimulus that merely accompanies food (correlating) or as a causal stimulus—one that also happens to be there when the cup contains food or one that is caused by the food [389, 391].

Information Recombination

In the brain, information is not only stored but also processed. However, how this happens, how it is retrieved afterwards, how it is combined with other information, and how it is itself changed and recombined (restructured) are good questions. Thorpe [101], for example, considered recombination to be the defining characteristic of insight or insightful behaviour. However, how is information recombined, and what processes underlie it? According to Davidson [392, 393], there are three kinds of insight: selective-encoding, selective-comparison, and selective-combination. Others [394] have also proposed a three-stage process that culminates in insight. In the preparatory phase, information is constructed and stored in the

[12] A difficulty with this term is that it is used very differently in the literature. On the one hand, as learning by observation, where the goal or intention of the demonstrator is recognised and adopted. (The more accurate term "goal emulation" is not always used for this.) On the other hand, emulation is used as a synonym for "affordance learning", which is limited to learning by observation, where the function, uses, or other physical aspects of objects are learned [61, 380–386].

brain as a result of dealing with a problem (including mistakes made). In the incubation phase, the subject orients herself or himself away from the problem, and in the process, she or he makes new experiences that are important for the objective problem; these experiences are stored as information that can later be significant. Finally, in the illumination phase, the new information meets the stored error-laden information of the early phase. By combining these two, new insights emerge because old problems are evaluated with the help of new information, and through accommodation and assimilation,[13] the old problem is solved [395]. Sometimes, this is accompanied by a moment of enlightenment, the well-known "Aha" experience. At first glance, these two theories seem very attractive, but they generated little resonance in the subsequent literature on insightful behaviour in animals.

Other authors have addressed the question of what form recombination or restructuring can take [396]. Early theoretical models proposed the following: the perception of stimuli in different ways (for example, reorganisation of visual information); the joining of isolated parts of information or the filling of information gaps (for example, the completion of a schema; [397]); getting over mental blocks caused by old experience (that is, functional fixation; [376]); making an analogy at the level of structural organisation and applying it to the new problem [367]; or reforming the problem, either in terms of the goals[14] [376] or the function of the objects that contain the solution to the problem (such as the "two-strings problem"[15] [398]). Although these models have received some support from empirical findings, they are insufficient for understanding how to gain insight into difficult problems. In perception theory, ambiguous images can be seen differently by changing the focus or viewpoint, which is a well-known phenomenon [400]. Rearranging objects illuminates the context, and a structural problem virtually solves itself. Specific changes are an indication that a certain insight was already present, even if not yet in full form. At any rate, for the Gestalt psychologist Wolfgang Köhler, this was an important difference from the mere trial-and-error learning of the behaviourists. If the problem has already been completely solved in the mind, before execution, that is the prototype of insight for Davidson [393]. Here, there is a certain proximity to reasoning, for example, for reasoning by exclusion.

In any case, it seems important that new representations of the problem emerge even if all the necessary partial information was already present. Either isolated parts are first combined or old representations are fragmented, after which the parts

[13] These two terms are central to the work of Jean Piaget but are hardly used in comparative cognition research. In Piaget's work, the term accommodation means the adaptation of one's cognitive structure, one's schemata, or representations to the external world, when certain perceptions can no longer be classified into the existing schemata, as would be the case with assimilation. See [395].

[14] However, studies of inexperienced problem solvers show that they often have difficulty recognising that one problem can be used to solve another.

[15] Here, two strings hanging from the ceiling are to be pulled simultaneously. However, the distance between the strings is too great. The problem can only be solved with the help of a trick in which an object is tied to a string and thus the string can be made to swing. Again, experience from another context can help. See [398, 399].

are reassembled. Now, one could imagine that animal species have different problems with recombination. Either they can no longer combine more distant pieces of information or old representations can no longer be broken up. Otherwise, the representations are in unfavourable, namely, inflexible formats, because they are bound to very concrete contents. The more abstract the information and the more structured the representation are, the easier it is to find new solutions through recombination.

The term "*functional fixedness*" [376] refers to the problem in which the habitual use of an object impedes its use in a new way. This phenomenon has been suggested for some chimpanzees suffering from the failure of solving the floating peanut problem [401]. Compared to chimpanzees from the Ngamba sanctuary in Uganda, chimpanzees from the Leipzig zoo were much less successful at solving this task. According to the authors, the difference could be due to the novelty of a water dispenser: While the water dispenser was new for the Ngamba chimpanzees, the Leipzig chimpanzees had to use a water dispenser to which they were already used. It seemed as if they were cognitively blocked from using the old dispenser, which they had known for a long time to be exclusively used for drinking, in a new, unusual way, namely, for filling up the tube with the peanut. Obviously, functional fixation through past individual learning occurred here.

Static and perceptual information and the dynamic and functional properties of objects can prevent the solving of new problems. We should understand insight not as a process but as the end product of a process, the final step in a chain of events that begins with the recording and acquisition of information, continues with transformation, and culminates in recombination and problem solving [402, 403]. Additionally, Thorpe [101] distinguished between insightful problem solving and insight.

What remains unanswered in all these models is the question of *how* the pieces of information and representations are combined in the mind. As previously mentioned, the proposed models regarding insight contain a clear demarcation of mental trial-and-error learning. The mere playing out of several scenarios, even if done in a systematic way, seems to be just as insufficient as the simulation of possible consequences. At least in humans, a further distinction is made between the analytical playing through of possibilities until one arrives at the best decision (for oneself) and thus the sudden and involuntary emergence of the solution in the sense of an "Aha experience".[16] The extent to which these distinctions are also meaningful for animals remains to be clarified.

[16] The former head of the University of Vienna's Psychological Institute, Karl Bühler, used this term to describe the sudden and involuntary recognition of a sought-after but previously unknown context of meaning, accompanied by an emotional reaction [404]. The term is often also called the "*Eureka effect*".

Object Manipulation

The third component of creative tool use is object manipulation. Anatomical differences between animal species are often overestimated in relation to object manipulation. This is illustrated in the example of a capuchin monkeys versus a squirrel monkeys. Although both species have similar physical conditions, one is very skilled at manipulating objects often and with dedication, while the other is rarely observed doing so. The much-cited example of New Caledonian crows versus rooks also illustrates how different the propensity for tool use is. However, the difficulty of parrots steering straight sticks in front of their eyes with their strongly curved beaks shows that anatomy may very well be a factor. Nevertheless, the psychological, motivational differences between species are at least as significant for the tendency to manipulate objects as they are for anatomical ones. Similarly, in the case of sound production, anatomical prerequisites of the vocal tract used to be blamed for differences between species, but more recently, neural-cognitive differences have been emphasised more strongly [405].

Both primates and birds provide (convergent) evidence for the strong relationship between object manipulation during exploration and tool use [406, 407]. The difference between ape and monkey species is evident not only in their motivation and propensity to manipulate objects but also in the complexity of manipulative actions, in which different elements and relationships play a role [407]. Chimpanzees and capuchin monkeys stand out in this respect, and it is also those species that use tools to reach otherwise inaccessible food. Is this also the case with birds? This is not easy to answer, as parrots are very skilled and very motivated to manipulate objects, but in the wild, they hardly show any use of tools.[17] The same applies to many corvids, including ravens, various crows, and jays. Unfortunately, unlike primates, there are no large-scale studies on the possible correlation between tool use and object manipulation in birds.

To determine which learning processes are crucial for tool use, Call [351] suggested focusing less on the distinction between learning and reasoning and more on measuring the *amount of information* that is drawn upon in particular actions. This can be done, similarly to social learning tests, with groups of the same species that are provided with different amounts or types of information; or information that differs in spatial, temporal, sensory, or symbolic distance. Furthermore, confronting the animals with various problems can contribute to the quantification or assessment of behavioural flexibility. Finally, learning curves can tell us whether a problem can be solved slowly and only after many repetitions and mistakes or quickly and even spontaneously.

The three components of tool use—(1) information seeking and accumulating, (2) recombination or restructuring of information, and (3) the propensity to manipulate objects—reveal themselves as a mixture of cognitive, motivational, and sensorimotor aspects that appear significant for the evolution and for the individual

[17] I will come back to a recent hint that this could be the case later.

genetic development of tool use. In any case, it is important to distinguish the different abilities and skills involved in tool use. A range from narrow, highly specialised techniques to open, flexible, and creative (innovative) solutions, should be considered. The latter should be understood as a consequence of sensorimotor skills and cognitive abilities. Species without tool use, such as rats, exhibit a number of important creative tool use skills, such as exploration, information gathering and recombination, but do not possess manipulative tendencies. Apparently, all three ingredients must be present or a certain threshold must be exceeded for creative tool use to occur. This could also explain why it only occurs in so few species.

Tool Use as Intentional Action

As we saw at the beginning of the chapter, an important characteristic of tool use is goal-directedness: tool use as a goal-directed action with an appropriate means. In philosophy, this characteristic has been called "*intention*" since Franz Brentano [408].

Thus, the tool is not only an object but also, in a figurative sense, a mental programme. The goal is the intended effect of the action. This includes three abilities: First, the ability to imagine the consequences of one's action before it is carried out, i.e. to represent them mentally. Second, the ability to compare the actual consequences with those that one has anticipated by virtue of one's imagination. Third, the ability to change the next action based on the experience(s) gained from the previous action(s).

These three abilities constitute the dynamic essence of (flexible) tool use. Jean Piaget also emphasised these particular mental accomplishments when he described tool use as one of the crowning achievements of practical intelligence because it required a sophisticated understanding of the world of objects [89].

As human cognition develops, the child learns about the effects of an increasing number of actions, increasingly including actions with objects. Playfully, the child begins to gain an understanding not only of the effects of her or his body but also of the effects that actions with objects produce. The child notices, for example, that hitting objects on hard bodies produces loud noises. Finally, the child begins to recognise relationships between objects; for example, one object may fit well into another or disappear into it. They learn that objects fall down as soon as they are no longer supported or that objects lying on top of another can be moved together with and by the latter without having to be touched themselves. Increasingly, the child develops an understanding of the static and dynamic properties of objects. As acts of adaptation to its environment, they begin to understand which actions produce the desired effect and which do not—what works and what does not. According to Piaget's theory, this occurs in the exchange relationship between humans and their environment through an interplay of "*assimilation*" and "*accommodation*" [395]. The consequences of actions are predicted on the basis of the current understanding; if these "hypotheses" do not apply, they are adapted. On the one hand, this happens through the cognitive integration of sensory perceptions; on the other hand, through the differentiation of these (already integrated) sensory perceptions.

Similarly, exploratory animals could also learn the effects of their actions on the environment in a playful way. In the course of these actions and their changes in objects and their relationships, animals should acquire some understanding of the physical properties of these objects. Thus, young chimpanzees learn that rotten wood, when struck on a hard nut, will break but that a granite stone will not; that termites can be fished with a branch but not with a piece of vine; and that a nut will break on a large stone but remain intact on the forest floor. Thus, young chimpanzees learn properties such as hardness, strength, linkage, and perhaps weight as important, static properties of objects. But they should also learn dynamic properties in the course of their activities. For example, when a round object such as a nut is bumped, it rolls, but a square stone does not; when the object rolls over an edge, it falls; and when the object rolls into a hole, it disappears. At first glance, this seems trivial. However, as we will see in a moment, this kind of basal physical understanding is not self-evident even in tool-making animals.

Chapter 4
Causal Understanding

After these rather theoretical considerations on the cognitive components of tool use, let us now turn to some experiments that have been used to test them. The most studied abilities are causal understanding and planning. Both are considered important not only for creative, flexible tool use but also as forms of understanding and thinking that are most distinct from trial-and-error learning.

What Is Causal Understanding?

A special case of causal understanding involves thinking about physical processes by which a tool achieves a desired result. A purposeful tool user can know that his action caused the result without knowing how and why it works that way. In fact, this is precisely the way in which less technically gifted or physically enlightened contemporaries use sophisticated tools, such as laser pointers and microphones. However, humans are nonetheless capable of understanding something of physics, as is demonstrated, for example, when a coat hanger must be used to fish for a bunch of keys that have fallen down a manhole [409]. This understanding creates a tool with the right physical properties, namely, one that solves the problem. We could also say that we represent certain physical properties in an *abstract* or *principled* way because properties such as hardness, weight, stiffness, and connection cannot be grasped by direct sensory perception and because they contain certain *principles of action.* A simpler alternative would be that our thinking is initially or perhaps generally guided in everyday life through superficial, perceptible features such as shape, haptics, or spatial orientation. It is only in school classrooms, laboratories, and workshops that we penetrate deeper into the physical nature of objects and their related principles of action. In any case, it is undeniable that humans have much potential for this. But what about animals? Let us review some laboratory experiments that have become classic of cognitive biology.

L. Huber, *The Rational Animal*, https://doi.org/10.1007/978-3-031-60803-2_4

The Trap-Tube Task

In the 1990s, Elisabetta Visalberghi and her team from the Istituto di Scienze e Tecnologie della Cognizione in Rome (IT) confronted capuchin monkeys with a technical problem that not only became a classic but also subsequently became a standard test for causal understanding. The monkeys were offered a treat, for example, a nut, in a horizontally arranged Plexiglas tube. The tube was too narrow to be able to grasp the nut with bare hands, but a stick was available as an aid. With this, the nut could be pushed out of the tube. Inserting the stick and pushing the nut out were easy. But when bundles of sticks or other objects that did not fit into the tube were offered, the first difficulties appeared. Only after a few failed attempts were the right sticks used or the unsuitable sticks modified. It was already evident here that the animals first had to learn to pay attention to the essential characteristics of the auxiliary objects.

The experiment now specifically tested anticipatory action [348]. The task was made more difficult by the fact that a "trap" was installed in the middle of the tube bottom. It consisted of a cup-shaped depression. If the nut was pushed or pulled over this space, the nut fell into it and was lost to the animal. With the necessary understanding of this connection, the monkeys were supposed to avoid this trap and push the nut away from the trap to the correct end of the tube. This required pushing the stick into the tube through the opening from which the nut was further away than the trap. To avoid the simple solution of holding on to the side where you just succeeded, the nut was randomly positioned either to the left or to the right of the trap in successive tests.

Four capuchin monkeys were confronted with this task. They could not solve the task immediately or spontaneously, so the "test" was continued, with a total of 14 blocks of ten tests each. Even after this, three animals were still at chance, and only one animal was able to complete the task after almost 50 failed attempts. This animal, named Roberta, was almost always able to get the nut out in the last six units (95% of the time). Apparently, this animal had learned a rule that enabled her to (almost) always insert the stick from the correct side, depending on where the nut was in relation to the trap. This assumption was confirmed in subsequent tests with this animal. She had learned to always insert the stick from the side from which the nut was further away. Since the trap was in the middle of the tube, this was the working strategy. However, when, in further tests, the trap was made ineffective by turning the tube and thus pointing the trap upwards or when a tube without a trap was offered, the animal stuck to this rule of thumb. The authors soberly summarised that the capuchin monkeys lacked the necessary understanding of the effects of their actions on the nut and the causes of success or failure. The only successful animal in the original task did not rely on the functional properties but merely on the perceptual features of the tube trap.

The surprising poor results observed in capuchin monkeys could be explained by several factors, including general difficulty with Plexiglas tubes. Therefore, other tool-using primates were tested. Chimpanzees were the obvious choice. After being confronted with the task [410], only two of the five animals came up with the solution, and again, not immediately but only after 70 and 80 individual tests. Was the solution of these two older chimpanzees (a 13-year-old male named Darrell and an 11-year-old female named Sheba) similar to that of Roberta? This was to be shown by further tests with two successful chimpanzees, in which a second, novel tube was used. To check whether the animals also acted on the basis of the distance to the nut, the trap in this tube was moved away from the centre so that the nut could be positioned exactly in the middle of the tube. Both chimpanzees (but not the capuchin monkey, which was also tested with this new apparatus for comparison) were able to master the new challenge. Apparently, they had not only used the simple distance rule but also chosen a strategy that took into account not only the position of the nut but also the position of the trap. The authors of the study even concluded that the two chimpanzees acquired an understanding of the causal relationship between the action with the stick and its effect [410].

With respect to capuchin monkeys' credit, it must be said that their suboptimal behaviour does not harm them as long as they can always obtain another chance. They lose nothing but only need slightly more time and effort to reach their goals. In the case of deception with an ineffective trap, the application of the simple rule of thumb was harmless. Therefore, it may not be surprising that Woodpecker finches [349], New Caledonian crows [411], and chimpanzees [412] also fall for deception and avoid ineffective traps. In the present case, even humans acted similarly "irrationally" to Roberta. Most of the human participants continued as before, even with the tube turned and the trap no longer working [413]. Apparently, they also chose a safety variant.

In general, the use of simple heuristics does not, of course, preclude the use of more advanced strategies in other situations. Furthermore, such imperfect strategies are even advantageous in situations that require quick action when the smarter strategy is too costly. They provide evolutionary advantages, especially for species in harsh and competitive environments. Humans also often use "fast and frugal" heuristics rather than slow and costly thought processes to solve everyday problems [414, 415]. Who has not tried to put a key in the lock without first looking to see which way the prongs are facing? So-called *ratiomorphic* decisions (discussed in the Chap. 2) also usually lead to useful solutions and are preferred to lengthy, rational deliberations [85, 86].

However, there are two alternative explanations for the seemingly suboptimal behaviour when dealing with a tube trap. One is the distance of the nut from the trap, which may not be as easily perceived by the tube entrance. In fact, if chimpanzees can move the treat directly, they have much less trouble [416]. On the other hand, poking with sticks is difficult for animals that do not have a natural disposition to do so. Therefore, researchers at the University of Cambridge (UK) have devised a different arrangement in which the treat can be fished for with a pusher already placed in the tube [417]. The experimental animals were rooks, which have no natural tool behaviour. These animals subsequently proved to be true masters of both general technical problems and tool-based tasks in particular.

The rook is one of four European species of the genus *Corvus* from the family of corvids. This species is related to both common ravens, which there is much to report about, and the often-mentioned New Caledonian crows. It is a rather large crow—usually the genus *Corvus* consists of smaller species—in comparison to the slightly larger ravens with a prominent bill, high head, and shiny metallic black plumage. Their diet is extremely varied, with a large proportion being animal food, which they access with their pointed beak that they use as a universal tool for digging, chopping, probing, and poking. Tool use therefore does not seem to be necessary for them and has never been observed in the wild.

To avoid the trap, the subjects had to learn to pull the piece of food over an opening in the tube on the floor, which looked very similar to the trap in the experiments before but did not function like a trap [417]. In one experimental condition, the false "trap" was closed with a lid (Fig. 4.1a); in another condition, the false "trap" was open at the bottom, releasing the food (Fig. 4.1b). A total of seven out of eight animals solved these tasks after a few trials. Next, the animals were subjected to a critical test. Either the formerly correct side was closed with plugs at the end of the tube (Fig. 4.1c), or the tube was lowered onto the table so that the previously opened "trap" was now closed (Fig. 4.1d). The experimenter's trick was thus that the visual stimulus, which could have been used as an arbitrary cue to solve these tasks in training, namely, the position of the black disc on top of one of the traps, was the same in both tasks, but each task required the opposite response (pulling the food away from the black disc in Fig. 4.1c and towards it in Fig. 4.1d). Therefore, the

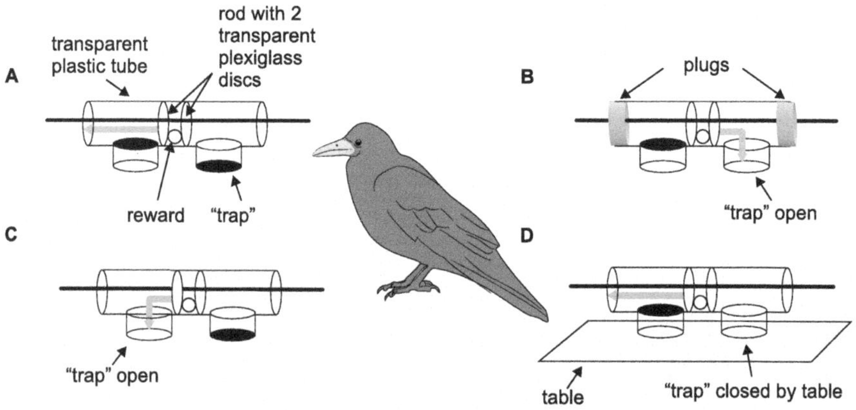

Fig. 4.1 The four tasks of the trap-tube problem that rooks were confronted with at the University of Cambridge (UK). First, the animals were confronted with tasks (**a**) and (**b**), and then they were tested with tasks (**c**) and (**d**). At the beginning of the experiment, a rod was already inserted into the tube, to which two transparent Plexiglas discs were fixed, enclosing the piece of food so that it moved with the rod no matter in which direction it was pulled. There were two openings at the bottom of the tube. When these were closed at the bottom (with a black disc), they acted as "traps" for the food piece (**a, c**); when the black disc was fitted at the top, the "trap" became ineffective, and the food could be pulled across it to the end of the tube (**a**) and (**d**). In (**b**), this path was closed by plugs. The arrow shows the path of the food to success. (From [417])

tests could not be solved by applying only a single, simple procedural rule based on the position of the black disc in the tube (left or right).

Six of the seven rooks failed the critical test, but one, Guillem, solved the two difficult test tasks with flying colours (nine out of ten and ten out of ten correct attempts, respectively). Therefore, this animal clearly acquired a more fundamental understanding of certain physical properties—in this case, blockage and permeability. Later, chimpanzees were also tested with this novel arrangement, and here, too, one animal passed the test [416].

The related New Caledonian crows are sometimes capable of transferring success from a kind of trap-tube problem to one that is visually entirely different, suggesting understanding at an abstract level [418]. Nevertheless, there is a considerable degree of individual difference among these attempts, the reasons for which still await clarification.

The String-Pulling Task

If a goal can be achieved only via several steps, it is advantageous to think of these steps ahead and then integrate them into a suitable scheme of action. This *means-to-an-end thinking* is, in my opinion, inherent in insight learning in the sense of Thorpe [101]. It is the basic prerequisite for action to begin spontaneously and to be carried out fluently. A good example of this is string pulling, also referred to as "scooping" in the older literature [419, 420]. The task, which Pliny already describes, is to reach a piece of food that cannot be reached directly but is tied to a string. The food must therefore be drawn in by the string. This is easy if the arrangement is horizontal, i.e. the string is on the ground. With each pulling movement, the food comes closer. It is true that the principle of connection must also be understood here, which dogs, for example, often fail to do [360]. But the motor part is simple. The action part becomes much more difficult when the string is attached to a branch, hangs down, and must be pulled up. Here, repeated pulling alone is not enough. The string must be fixed after each pull before the next scooping action starts. The longer the string is, the more often this combination of (1) grasping down, (2) pulling, (3) fixing, and (4) releasing has to be repeated. The main problem here is fixation. For animals with two hands, this action sequence is relatively easy; one hand is always on the string while the other is reaching. The marmosets at the Institute of Zoology at the University of Vienna got the tidbits in no time. But for birds, this task is much more difficult. If they pull on the string with their beak, they first have to put the string on the branch at their feet and then get on the string with one foot; only then can they open their beak again and pull up another piece of string.

Since the experiment is easy to carry out—practically every bird breeder could do so at home with pet birds—and of great knowledge value, there are many studies on it. Margaret A. Vince [421] concluded after testing great tits, greenfinches, canaries, and chaffinches that birds follow the principle of trial-and-error learning. Budgerigars, Indian starlings, and jackdaws also had major problems [422]. At the

Max Planck Institute of Comparative Ethology in Seewiesen, Bavaria, a large number of finches—goldfinches and siskins—were tested by Uta Seibt and Wolfgang Wickler, and large individual differences were observed [423]. Approximately half of the birds learned to solve the problem; others learned by watching successful birds, and some never learned.

We confronted kea with this task because we knew from previous studies that these birds acquire an understanding of certain functions of new objects only by watching conspecifics. They also demonstrated highly developed technical intelligence in other manipulative tasks [424]. The kea is a winterproof, large parrot from the family Strigopidae, an ancient parrot family consisting of only five extant species, including the kaka and the kakapo. Unusually for parrots, the kea live in the rugged mountains and glaciers of New Zealand's southern island. They became known for their irrepressible curiosity and playfulness, with which they damage cars, campervans, tents, and mountain huts with enormous endurance. They also made themselves unpopular with local sheep farmers after they one day began not only eating dead sheep but also forcibly causing their deaths. As explorative opportunists, they have an enormously large and expandable plant food spectrum. Although they are born vegetarians, their enormous technical skills and perseverance enable them to tap food sources that remain hidden from other animals, which allows them to survive in snowy winters. In doing so, they are not afraid to steal eggs from migratory birds nesting in rocky cliffs. With their long, curved upper beak and short, strong lower beak, they can dig, poke, drill, pull, crack, and much more. Pre-equipped with this kind of Swiss army knife, they master their lives in the harsh conditions of New Zealand's mountains. However, tool use or nest building is not known among these birds.

In Vienna, the animals had been kept at the Konrad Lorenz Institute of Ethology for decades and studied for their social and cognitive abilities. Despite evidence of their intelligence in dealing with conspecifics to solve a seesaw problem [425], we were amazed at the dexterity with which they solved the string problem [426]. Without any experience with strings—the birds had no strings or string-like objects with free ends in their aviaries—all birds (except one) were able to solve the problem in the first attempt. The exception was a young bird that still lacked the necessary motor coordination. The detailed analysis showed that the individual actions of the other animals were performed in a fast and almost error-free sequence. Although we used a very long string to force a greater number of scooping acts, four birds showed a rapid and flawless sequence of actions that lacked any signs of unsuspecting trial and error. In less than 15 seconds, they took control of the attached food can. These birds were unable to increase the efficiency of the scooping in another nine attempts, suggesting that the first attempt had already succeeded perfectly. The most remarkable result of this study, however, was that some birds kept developing new solutions to be successful. Either they used a foot instead of the beak for scooping, or they threw the string to the other end of the branch, or they moved the string so strongly that it wrapped itself around the branch and, finally, chose the most efficient method: they let the string slide loosely through the beak while they moved to the side, whereby the end of the string with the food landed all by itself in their beak. While there is no doubt that the mountain parrots displayed a fair amount of exploration and joy of movement here, the spontaneity and speed of execution, as well as the almost 100% success rate, attest to a remarkable degree of motor skill and technical intelligence. After the rooks, we are already dealing with a second bird species that does not use tools in the wild and yet displays enormous sensory-motor skills.

Solving the string problem requires at least two steps. First, the animal must have a high degree of movement coordination. Not only do two parts of the body—the beak and the foot—have to work together, but the efforts must also have precise timing. I have observed many birds that put the string towards the foot but already opened the beak before the foot fixed the string. These examples are somewhat reminiscent of Sisyphus. In addition to this motor part, there is also a cognitive element, an understanding of *connection*. In subsequent tests, we wanted to investigate this understanding and confronted the kea with a number of variations of the string problem. Among these were set-ups with two strings, one with the food attached to it and the other with a piece of wood. The kea were not swayed by this. They did not pull blindly but only after checking where the food was hanging. Likewise, they remained successful when the two strings were pulled to the side so that the end of the string with the food was exactly below the place on the branch where the string with the piece of wood was attached. Only when the two strings were crossed did the errors creep in. Apparently, the perception of the continuity of the string from the branch to the piece of wood was important for establishing the connection [426]. This continuity makes the string the mediating element and thus the means to the end. Many "means-to-an-end" solutions are of a similar nature because they involve connecting the actor and the target via a mediating object.

How important is the unrestricted and unbroken view of the target? Is it enough that this is captured once, or must the contact between actor and target always be maintained? An experiment with New Caledonian crows provided a surprising answer. At least during string pulling, immediate feedback seems to be important for these birds [427]. Although even naive animals solved the problem in a single, flowing action on the first try, they failed when a plate with a hole deprived them of visual feedback at the end of the string. Then, they only tugged on the string but did not pull it up. Even the performance of experienced animals was strongly affected. The importance of visual feedback has also been shown in chimpanzees [428]. However, once they had solved the problem, they were able to succeed without direct feedback. Apparently, they had stored the consequence of their action in a permanently retrievable form. Thus, visual feedback may be a necessary element of goal-directed action in New Caledonian crows but not in chimpanzees. However, this is not the "end of insight", as some have proclaimed [429], even if quick and spontaneous solutions do not always and entirely result from a previously conceived plan [399, 409]. On the one hand, experimental manipulations of the event, such as restricting the view of the end of the string, produce a rather unnatural effect that can throw animals off track. And on the other hand, the importance of feedback points to another feature of goal-directed action. Clearly, scooping is not a simple, preformed instinctual action (such as the egg retrieval movement of the greylag goose described by Lorenz, which continues even after the egg has already been taken away [430]). For even if feedback is necessary and the solution of even the most successful animals is not based on total insight or even prior planning, important components of insightful action are still present in kea and crows.

The Towel Test

It is certainly the case that actions that are only carried out and continued if there is a repeated effect on the reward have the characteristics of *trial and error*. They then correspond to Thorndike's *law of effect* [431] because, as he wrote, responses that are accompanied or followed by satisfaction for the animal become more strongly associated with the situation, so that when the situation occurs again, those responses also occur again. This is why the experiments were scheduled for multiple times, during which the animals were presented with a spontaneous choice, i.e. they could not learn by trial and error. The correct choice, if it is not random but (statistically) significant from the beginning, requires the correct conception of the consequence of action. In the question of "means to an end", the means open up the possibility of achieving the end through the means. In this way, the understanding of physical properties, which also play a role in the use of tools, can be tested. Tools are also means to an end in this sense. When fishing for objects out of reach, this is the principle of *connection*. We have already learned about this phenomenon during string pulling. A variation of this is the principle of *support*. Here, the connection is created by gravity. An object lying on top of another object is moved with it. Even if gravity, which is a physical force that creates a connection, is not understood, the principle can be learned as a rule with the help of repeated experiences.

We have mentioned the *support* or *on-off problem* in the chapter on information collection. This task has its origins in Jean Piaget's theory of cognitive development [89]. Like string pulling, this task can be studied with minor adaptations in many species. At its core, it consists of a forced choice between two options: a functional or effective option and a nonfunctional or ineffective option. In the version with children, a toy lies either on or next to a towel [432]. With animals, a piece of food is used as a target instead; this lies either on or next to a small board. The animal can reach the board but not the food at the back. Therefore, again, we are dealing with a means-to-an-end task. If the animal understands that the board, when pulled, brings the food on it closer, but not the food next to it, the core task is solved.

An average of 160 repetitions were required for the pigeons to make statistically significant correct choices [433]. They orientated themselves exclusively on simple visual features, which they learned to associate with success or failure through trial and error and were unable to develop an understanding of the "support" functional feature. In contrast, great apes showed a statistically significant difference after only a few trials [434]. However, they did not choose correctly from the beginning and remained at a low-hit rate of less than 70%. Most of our kea, on the other hand, were sure of themselves from the start [435]. They already managed a percentage of more than 90% as a group in the first experimental run with ten trials, and six out of eight animals were already correct the first time. They maintained this elevated performance in subsequent more difficult tests. In a test with a "splitboard", there was food on it, but it was separated from the part of the board that the animal could reach by a visible gap. The "right" tool was an undivided board. Here, we tested whether the birds understood the principle of "connection"—as they did when pulling the string. Even more difficult were the tests on the principle of "enclosure". Here, in the "right" choice, the piece of food did not lie directly on the board but in a recess of the board at the rear end, so that it was also pulled along. The recess on the "wrong" board, on the other hand, was open at the back. Again, most of the kea chose correctly on the first try and achieved outstanding results in the following trials. Even the four ape species studied (chimpanzee, bonobo, gorilla and orangutan) lagged behind the kea in their performance [434]. While caution is always needed in a direct comparison because the apes may have had visual problems and did not see the necessary details precisely enough, this performance is nevertheless astonishing for a bird that does not actually use tools. Unlike pigeons, the mountain parrots were able to immediately and consistently recognise and use features that are crucial for ensuring the effectiveness of the tool.

When Will the Platform Collapse?

Another example of understanding the "means-to-an-end" principle was provided by rooks in an experiment with the collapsing platform. Here, the physical principle of *displacement,* which involves weight and pressure, was tested. For example, an object can be displaced when it is pressed down. This pressure also occurs when other objects with a corresponding weight are on it. For example, some animals can learn over time that branches can not only be pushed down but also give way when you sit on them. In the experiment, one piece of food at a time was placed on a drop-down platform—it had a hinge on one side—at the bottom of a tube [436]. This piece of food could therefore not be reached directly; a way had to be found to push the platform down. At first, the animals could not solve the problem. But during training, they learned to push small stones lying at the edge of the tube into it, which pushed the platform down and released the food piece. The authors were aware of the possibility that the first successful act of pushing into the tube was random. However, what did the animals learn from the subsequent successful actions, and what was their effect in common? Did they understand anything about the functional cause of the platform collapsing, namely, weight and pressure? In fact, in a subsequent test, they were able to generalise the solution they had learned earlier to achieve the goal with other objects and other actions. They fetched stones when there were none near the apparatus, chose them in such a way that the heaviest stone was preferred for the respective size of the opening, and avoided throwing in objects that were too light. Even more interesting was the result of the test where there were no suitable heavy objects, only a long, light stick. Instead of throwing it in, they pushed it downwards onto the platform and released the food by applying pressure. This is an indication that they were performing a functional generalisation, from the weight of an object to pressure with their own bodies [436].

Aesop's Crow and the Pitcher

A physical principle related to *displacement by pressure* is *fluid displacement*. This is visible in liquids that are displaced by denser objects. In a vessel, this causes the liquid to rise to the top. Animals could learn the principle of displacement when they climb into puddles of water. Our kea like to bathe in the small pools of water in their aviaries. When an animal climbs in in the morning (after the pools have been filled to the brim by our keepers), the water spills out over the edge. This principle is also the subject of the famous fable attributed to Aesop. In "The Crow and the Pitcher", a thirsty crow finds a half-filled pitcher of water, but despite its stretched neck, its beak is not long enough to drink from the vessel. By virtue of her intelligence, however, she finds a solution. She throws pebbles into the jar, which raises the water level and allows her to quench her thirst.

This fable was the template for several experiments. I have already reported on the clever solution of the orangutans who wanted to reach the floating peanut [352]. As with Aesop, however, crows were also used in this task, namely, the rooks, which were already known from earlier experiments to be particularly clever [436]. A Plexiglas cylinder half-filled with water was placed in front of the birds, in which a worm floated on the water surface. Stones were placed next to it. In fact, the rooks threw the stones in one by one until the worm came within reach. The problem was thus solved. But what did the animals understand during the process? Did they actually acquire an understanding of displacement, or, in the manner of Thorndike, learn to produce the positive effect of their actions more and more often?

Again, further tests and control experiments should clarify this phenomenon. Instead of water, a tube was filled with sand. The animals did not spontaneously avoid throwing the stones in, but they learned very quickly. They apparently paid specific attention to the effect of their actions on the water surface. If the worm was approached by their action, they continued; otherwise, they did not. Later, experiments were also performed with opaque and interconnected ("communicating") tubes. Again, the birds learned quickly. Action based on an understanding of the underlying physical principles could be ruled out, but unlike many other animals, they do not have to make hundreds of mistakes to learn to pay attention to certain perceptible properties; rather, they seem to utilise cause–effect relationships very quickly, sometimes after only a single attempt.[1]

[1] A word of caution must now follow at this point. The quick filling up with stones can be interpreted as an understanding of the principle of displacement. But what experience had the animals brought with them? Were they individuals who had already been tested in the experiments with the collapsing platform? Had these animals thrown stones into Plexiglas cylinders before? After all,

Recently, my collaborators, led by Raoul Schwing, conducted a similar experiment on water displacement with kea [437]. These animals had no experience with throwing stones into glass cylinders, but they did have experience with water, as discussed earlier, and likewise with tubular objects into which they occasionally inserted or threw other objects. Confronted with the Aesop task, within only five trials, they learned to throw stones into that of three outwardly identical cylinders, which were half filled with water instead of sand or stones. In a second experiment, they generalised the behaviour to differently coloured liquids, in each case in the very first trial. Therefore, they obviously did not use the colour or transparency of the water as a clue because some cylinders now had liquids of the same colour as sand or stones or became opaque. Finally, they were given a third, even more difficult task. One of two cylinders filled with water had a hole on the side so that water leaked out as soon as it rose from the thrown-in stones. Would the kea learn to avoid this broken cylinder? They did not, which means that they solved the first two tasks based on certain characteristics of the water or the stones in the water but did not pay attention to the change in the water level. Like the rooks, they learned extremely quickly, apparently because they already had a lot of experience with the different carrier materials, but they remained far from a causal understanding.

This conclusion must probably also apply to all the other Aesop experiments that have been conducted with various animal species in recent years. They not only examined what animals understand about (1) the displacement potential of carrier materials but also what they understand about (2) the effectiveness of objects as a means of displacement. For the first question, the classic distinction between liquids and solid materials was tested. Some animal species, including jays [438], New Caledonian crows [439–442], and raccoons [443], learned this relatively quickly, while others did not at all (Western Scrub Jay [444]). To answer the second question, effective items that displaced the liquid and ineffective items that floated on the liquid were used. However, even the habitual tool users from New Caledonia learned only through success and failure. When, in a highly sophisticated experiment, the physical nature of objects was pitted against food rewards in terms of their effectiveness as a means of displacement, the crows acted on previous success [442]. First, they were rewarded for choosing an object that was not suitable for the aesop task. Then, when confronted with the Aesop task, the birds performed poorly: They dropped the unsuitable objects down the tube on their first attempts. This suggests that they do not have an a priori understanding of water displacement but rather act according to pre-existing object preferences. However, this may not be surprising when one considers that even children under the age of seven have the same weakness. The children learned quickly but did not fully understand the problem. Only 8-year-old children could no longer be fooled because they seemed to have developed a real understanding of the effectiveness of the various materials [442, 445, 446].

---

the tests were conducted in the same laboratory, by the same experimenters and published in the same year [235, 436].

Hook Tinkering

From all the examples thus far on the question of understanding the physical properties of objects in terms of their effectiveness in helping individuals acquire out-of-reach food, one can draw at least three general conclusions. First, all animal species (and infants) tested extensively and systematically thus far have only a very rudimentary innate understanding of certain physical properties, such as connection, support, displacement, and their effects. Second, some species remain oriented to certain superficial features even after long sequences of learning, while other species improve on their rudimentary understanding of effects. Third, these skills are not limited to natural, habitual tool users and tinkerers. In birds, Woodpecker finches probably belong to the first group, and New Caledonian crows certainly belong to the second. In addition, some species of birds from the corvid families, such as the rooks, and the parrots, such as the kea, can maintain the latter, although they manage entirely without tools for natural foraging.

However, there is a report that has astonished the experts and delighted many a bird lover because it seemed to most strongly refute these three conclusions. In addition, it made Betty a celebrity [298]. This New Caledonian crow, living in Alex Kacelnik's laboratory at the University of Oxford (UK), was offered a treat (a piece of lamb's heart) in a planned experiment but out of reach in a small basket at the bottom of a Plexiglas tube. As an aid, she was given two pieces of wire, one of which was straight, and the other bent into a hook. With the hook, the food basket could be pulled upwards out of the tube by the handle. But another crow, Abel, was quicker and grabbed the hook. Betty stared at the meat, then in distress took the straight wire, and tried to spear the piece of the lamb heart. But this failed, forcing her to take another action. She took the straight wire at one end in her beak, stuck the distant end into a crack on the floor, and then moved to the side, thereby bending this end into a hook. With the hook first, she guided the modified wire into the tube and positioned it under the handle of the basket; when the hook gripped, she pulled the basket out [298]. Even though this was the first time she had ever seen a wire, she quickly realised that she could form a hook from it and where and how to bend it to make it fit. The researchers concluded that she had a highly developed cognitive ability, especially since Betty later also made hooks from a flat piece of aluminium and used other bending techniques [447]. The behaviour was described as new, intentional, and specific because Betty had seen a hooked piece of wire before but had never made one herself.

A few years ago, the story took an unexpected turn. A team of researchers led by Christian Rutz at the University of St Andrews (UK) discovered that New Caledonian crows have natural potential for bending objects into hooks. First, the researchers were able to observe the production of a series of 85 hook-shaped twigs on 18 briefly captured crows. Unexpectedly, ten animals began to produce these hooks not normally, by breaking off side branches but by bending the shanks at the end [448]. This could be confirmed in further field studies. The bending is performed quickly and skilfully; the branches[2] are bent very much beyond the final shape, similar to what Betty did with the piece of wire [449]. A total of three different bending techniques were observed, one of which was shown by Betty; the end of the shaft was put into a hole, and the shaft was bent back. Betty was also born from New Caledonia, which suggests that she already knew the bending technique. Her "invention" in Oxford may have been the result of natural, species-specific behaviour or an innovation that once spread in the population of New Caledonia. Generally speaking, we can say today that the tool behaviour of New Caledonian crows is shaped by genetic predispositions, individual and social learning, and ecological influences [448]. Concerning the latter, raw-material properties have a strong influence on tool morphology, and together with age and manufacture method, are drivers of tool shape variation [450]. Concerning the spread of the behaviour it is important to note that the crows exhibit strong population-specific material preferences for the manufacture of hooked stick tools [451].

Hooks were a key invention of the Stone Age people approximately 90,000 years ago because they allowed them to fish much more efficiently and thus expand their range of prey considerably [452]. Apparently, New Caledonian crows also developed hook technology a long time ago and integrated it into their behavioural repertoire. The experimentally assessed finding that crows acquire food several times faster when using hooked rather than non-hooked tools, implies that small improvements in tool shape, and the resulting efficiency through higher profit at constant cost, are strong selective forces of technological progress [453]. Does this therefore speak more in favour of a genetic cause for technological progress and less for cultural accumulation of innovations?

It seems that we have to put Betty's insight and innovation to rest with this. But is this true in general? Could birds manage hook bending even without any disposition of the species for this behaviour? In fact, Christopher Bird and Nathan Emery of the University of Cambridge (UK) found a positive answer [235]. Four rooks, which we already encountered in the collapsing platform test, were first confronted with the same problem as Betty, fishing a small basket of food out of a 15 cm high tube. To do this, they were given a 16 cm long stick with a hook on the end. In a total of 30 individual trials, the animals learned to use the tool correctly, namely, to push the stick into the tube with the hook first, to hook it in the handle of the basket and then to pull it out carefully. Then they were given two tools to choose from, one with an effective hook and another with an ineffective hook; the sides were reversed. The animals mastered this task spontaneously. Three out of four animals chose the working tool at the first attempt.

Last but not least, the animals were also confronted with the crowning task of hook making. For this purpose, they were given a 17 cm long, straight piece of wire. Amazingly, all four crows bent a hook and succeeded. Three animals managed to do so in the first attempt, the fourth in the fourth. In a video attached to the publication,

---

[2]The crows use the flexible side branches of the perennial shrub *Desmanthus virgatus*, which is known by many common names, including wild tantan, prostrate bundleflower, dwarf koa, and desmanto.

a crow named Cook can be seen pushing the straight wire into the tube, then, using the leverage, bending it over the edge of the tube at the end, taking out the bent wire, pushing it back in hook first, and finally pulling out the basket by the hooked handle.

Since the previous history of the rooks in Cambridge was known and they were therefore proven to have no experience with hooks or to show signs of hook bending in their natural behaviour, this achievement was hailed as a real innovation. It is also worth noting that this unexpected behaviour was shown by all the subjects and thus does not represent a singular feat. Although they had previously used hook tools made of wood in the experiment—often successfully in a total of 60 individual trials—these tools were only functionally, but not materially comparable to the ones they had formed themselves. Like Betty, however, the rooks might have memorised the hooks they had used successfully before and, with this image in mind, recreated their hook tools out of wire.

The beak dexterity of New Caledonian Crows, Hawaiian crows, and rooks suggests that we are dealing with a special form of tool intelligence in corvids. The use of poking tools in Woodpecker finches has been debunked as an innate special adaptation without any particular higher-cognitive development. Could we therefore assume the singular emergence of flexible tool use in one bird group and one mammalian group, here corvids, there primates? Although the conclusion seems obvious, it would be unlikely because one would have to assume a simultaneous frequent disappearance of this feature in the two vertebrate classes. In both cases, the trait has been indisputably demonstrated for only a small number of species. Presumably, this is a case of multiple convergent evolutions of special intelligence [454]. This matter becomes even more complicated, however, when one takes into account recent sensational reports on a bird species that neither belongs to the corvids nor uses tools habitually in nature but is in no way inferior to the previously mentioned corvids in terms of the corresponding abilities.

Goffin's Cockatoo: A New Model Species for Technical Intelligence

The Goffin's cockatoo, also known as Tanimbar corella, is a snow-white parrot species from the cockatoo family and one of the smallest species in the genus of true cockatoos. Again, we are dealing with an island bird south of the equator because these animals are native to the two islands of Yamdena and Larat, which belong to the Indonesian Tanimbar Islands in the southeastern Molucca region. There, they live in the forests, breed in tree hollows, and fly in swarms to the fields of the local farmers to feast on their crops, such as maize [455]. With their short, strongly curved beak and side-set eyes, there are no visible anatomical adaptations for (stick) tool use. These parrots also lack the typical behavioural predispositions. Neither are there certain heritable dispositions for tool use, such as in the New Caledonian crows [411] or the Woodpecker finches [320], nor do they build complex nests

[456], nor do they have ancestors that gather and hide food [303]. One can therefore conclude without hesitation that habitual tool use is not present in this bird species (at least it has not yet been observed in the wild). However, their foraging behaviour exhibits two features that are characteristic of tool use and innovation. On the one hand, they are extremely opportunistic in their foraging as generalists, taking what they can get. Their strong exploration instinct makes them constantly seek out new food sources favoured by plants introduced or cultivated by humans. On the other hand, food acquisition is predominantly "extractive" because food can be uncovered only after persistent and sometimes elaborate manipulations. Precisely these features have been associated with the onset of tool use in great apes [457–460]. In addition, it may be their playful urge to combine objects in different ways [461], which is also considered to be an important psychological component of tool use and innovation in primates [462, 463].

Cockatoos are popular pets because they can also be kept and bred in larger groups, provided that the aviaries are suitably large. However, it is important that the animals not only be able to develop a natural group life but also be cognitively challenged in diversely equipped aviaries so that they can live out their exploration instinct and their joy of playing with objects. My colleague, Alice Auersperg, who was already able to gain experience with the technically adept kea during her doctoral thesis, built a research station with large and well-equipped indoor and outdoor areas in Lower Austria in 2011, which not only does justice to the animals' natural behaviour but also allows interesting behavioural studies under controlled conditions.

Not everything that birds are capable of reveals itself in meticulously planned experiments; sometimes chance helps out a lot. This was the case one day when a student, without any experimental intentions, observed a bird playing with a small stone on the grid wall of the aviary, and it fell out through the grid. This male cockatoo not only maintained his interest in the little stone that had fallen out of reach but also wanted it back. Figaro did not linger long at the grid wall but returned after a short while with a small bamboo stick, which he used to fish for the longed-for toy through a mesh in the grid. The student excitedly reported her observation, whereupon the surprised Alice took a closer look at the matter herself—or rather in front of the video camera. Now, Figaro was given cashew nuts instead of the little stone on a support beam a few centimetres outside the aviary. After a short search in the aviary, he made for the inside of the wooden beam on which the aviary grid was mounted, through which he could see the nut. With a few powerful bites, Figaro broke off an approximately 7 cm long splinter on the inside edge of the beam and used it to fish for the coveted food. He repeated this behaviour in nine more attempts, with the chips hardly varying in length. Only once was the splinter too long, and he had to shorten it afterwards. Once he fetched a piece of bamboo from the aviary, he otherwise made his own fishing tools. In the tenth attempt, he chose another alternative; he picked up a branch from the floor of the aviary and shortened it in four steps to the usual length. The cockatoo managed the angling of the nut with similar skill, despite its unsuitable short and curved beak. Either it placed the end of the splinter behind the nut and pulled it towards itself or it wiped the nut sideways towards the grid wall.

Unsurprisingly, Alice Auersperg and her team have generated much interest in the scientific community due to their reports on the tool-making of a bird species that is completely unknown for this issue [464]. Unlike the laboratory studies on rooks, it was not literally "suggested" to the cockatoo to use a technical tool because Figaro found the solution spontaneously and on his own. Although cockatoos, like many other birds, nibble at wood, a deliberate breaking-off of a splinter had never been observed before. Since the construction of his first tool took about four times as long as that of all the others and since the aviary beams were still intact before that day, both experience and innateness can be ruled out with great certainty. Thus, we agree with the title of the report by Alice and her team that Figaro showed spontaneous innovation in tool making [464]. His inventiveness was confirmed by his aviary mates who, like Pipin, never tried or, like Figaro's mate Heidi, failed miserably. Obviously spurred on by Figaro, Heidi broke off small splinters of wood, but they were completely useless as fishing tools.

All the animals exhibited enormous beak skills in a series of further aviary experiments. They easily managed to open a small box with five latches [465]. They show great stamina and an irrepressible curiosity, using not only their beaks but also their tongues and feet. Interestingly, Pipin, who had previously shown no interest in nut fishing, overcame five whole obstacles in just a few attempts, including a screw, a turning wheel, a splint pin, a bolt and a latch. The animals demonstrated great exploratory and manipulative skills, alongside functional generalisation ability for modified locking mechanisms, and an understanding of the effects of the closures on each other (such as blockages). Their learning was also characterised by ratcheting up, as they hardly ever made mistakes on tasks they had already solved.

Not all animals learned the sequential manipulations individually; some acquired the skill by copying from successful conspecifics, which was shown in a subsequent experiment [466]. Although their actions with the fishing tool differed from those of Figaro, they were also able to replicate his success with their own actions.

As we have seen in chimpanzees and New Caledonian crows and, to some extent, in Woodpecker finches, tool hobbyists can generate tools from different materials. This suggests a certain abstraction from the specific action and flexibility in achieving the goal. Alice and her team have tested whether Goffin's cockatoos are also capable of doing this [467]. The task was to retrieve a cashew nut from a transparent box. The nut was placed on a small platform in the middle of the box, and at the front, there was a small round opening into which an elongated object could be inserted. When the nut was pushed off the platform, it rolled out of the box over a slope. To make the necessary tool, the cockatoos were each given one of four different materials in individual tests. Three plates (5 mm thick and 5 × 15 cm in size) were made of larch wood, cardboard or beeswax with chambers. The fourth raw material they were offered was a long (20–30 cm) beech branch with leaves. All of these raw materials require a different production technique. The cockatoos also passed this test impressively. All the animals were able to form the required stick from the larch slab and from the beech branch; in the first case by biting off, the whole length in an elongated manner, and in the second case by breaking off the side branches. A little more difficult was the production from the cardboard, from which they bit off an elongated strip along the edge, which was then used as a stick. Interestingly, they did not nibble the whole length of the cardboard but only just exceeded the minimum length with which the nut could be reached. This suggested that the birds were estimating the distance to the unreachable piece of food and then transferring this to the length of the tool to be made. Only the attempts to shape the wax failed because the pieces of wax stuck to their beaks, which soon made them lose the motivation to continue with this material.

For the effective use of a tool to reach a spatially distant target, correct length estimation is important. New Caledonian crows, which make poking tools from the leaves of the screw tree, seem to take this into account in the elaborate manufacturing process [305]. A laboratory experiment confirmed this [297]. After Goffin's cockatoos first showed signs of this phenomenon when working with different basic materials [467], it was obvious that this ability could be tested in more detail. For this purpose, a transparent box with a hole and an inclined base plate was again chosen, but the nut could be positioned at different distances from the opening—4, 10, or 16 cm [468].

To test whether the animals could also adjust the thickness of their fishing tool, the diameter of the hole was varied according to the second condition (10, 15, or 20 mm). This time, the experimental animals—six adult cockatoos—two of whom were previously reported (Dolittle and Figaro)—were provided only cardboard plates as raw material (Fig. 4.2).

The animals used two different strategies. Either they broke strips over the full length of the plate or stopped breaking early, usually at a length corresponding to the current distance of the nut from the entrance hole. Sometimes they discarded strips before using them (as the experimenters' check revealed, they were indeed too short). The width of the tools, on the other hand, was not adjusted to the hole diameter. One reason for this could be a morphological limitation because the width of the cardboard strip corresponded exactly to the length of the upper beak. The precise length adaptation, however, was remarkable and once again confirmed that these birds exhibited high technical creativity.

The prime example of creative toolmaking, according to Betty's sensational report, was bending wire into a hook [298]. Although the creativity of this activity was later clearly put into perspective because New Caledonian crows, in their natural foraging behaviour, shape bendable twigs into hooks [448, 449], the paradigm was maintained for at least two reasons. On the one hand, rooks were able to form hooks despite a lack of natural disposition to use tools [235]. On the other hand, children without experience fail miserably with this up to the age of five and do not reliably show the ability until 8 years of age [469]. The task seems so difficult because only the initial and final conditions are available, but the path in between is completely in the dark. Only when one has had certain relevant experience can one solve the problem. There is no doubt that by the age of 8, children not only have seen but also use many hooks, such as clothes hooks or pokers. Seen in this light,

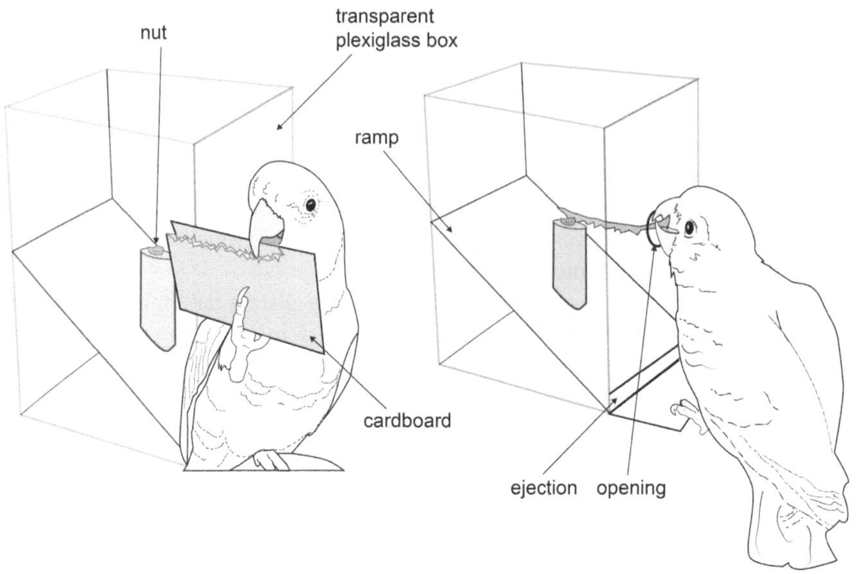

Fig. 4.2 Length estimation during the production of a fishing tool by Goffin's cockatoos. On the left, a cockatoo breaks off an elongated strip from a cardboard box with repeated bites; on the right, the strip is inserted through the hole in the transparent Plexiglas box to push the nut off the small wooden column. The nut then rolls out over the ramp through a gap and can be eaten. (From [468])

and after all that Goffin's cockatoos have already achieved, it would have been a great omission not to confront these animals with the hook problem as well.

A total of 13 members of Alice Auersperg's Goffin colony were tested with the "Betty task" [470]. All had experience with straight poking or fishing tools but had never previously seen, let alone used, flexible material or hooks. This also distinguished the task from that of the rooks, as they were already allowed to work with hooks before the hooks were produced [235]. Before the actual experiment, all the animals were individually familiarised with the problem of getting food out of Plexiglas tubes. In this habituation phase, the tubes were either arranged vertically, and the cashew nut was in a small feeding cup with a handle (Fig. 4.3a); alternatively, the tubes were arranged horizontally, and the cashew nut was lying open inside (Fig. 4.3b). The Plexiglas tubes were very short (3 cm); therefore, the food could be grasped and removed with the beak. In addition, they could observe an experimenter wrapping material around a pen to inform the cockatoos that wire, but not string, maintains the shape after bending.

In the test, the tubes were clearly elongated, and the cashew nut became out of reach. Without aids, the animals therefore had no chance of getting to the coveted nut. Therefore, they were offered two materials, a piece of string and a pipe cleaner (with a flexible wire inside). To check how innovative and flexible the cockatoos were, they were given two tasks: In the bending task, the tubes were arranged

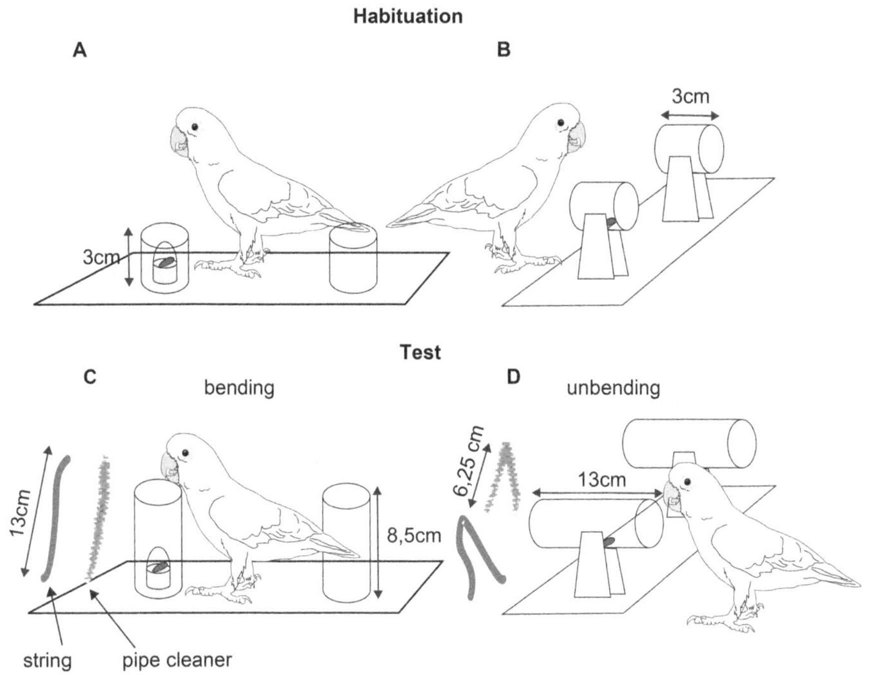

Fig. 4.3 The experimental set-ups for testing the tool modification of Goffin's cockatoos. In the habituation phase, the food could be reached from a vertical (**a**) or horizontal tube (**b**) without a tool. In the test phase, either a wire had to be bent to serve as a hook tool (**c**) or bent open to be used as a fishing tool (**d**). (From [471])

vertically, and the two aids were in a straight form (Fig. 4.3c); in the bending-open task, the tubes were arranged horizontally, and the two aids were bent in the middle (Fig. 4.3d).

To determine the influence of previous experience, the animals were divided into two groups. While the animals in the control group had to approach these two tasks completely without prior experience throughout the entire test phase, the animals in the other group, let us call them the experience group, were able to gain relevant experience later through appropriate assistance after the first test phase, in which they were also given the two tasks without prior experience. However, both groups were confronted with the two tasks shown in Fig. 4.3c and d the same number of times.

First, all animals were given the opportunity to solve the task ten times each in the first test phase to check whether a bird would spontaneously be successful. For this purpose, the string had to be ignored, and the pipe cleaner had to be bent into shape for later use. To reach the nut in the vertical tube, the straight pipe cleaner had to be bent into a hook at the end; to reach the nut in the horizontal tube, the hook-shaped pipe cleaner would have to be bent open. Although all the animals failed both tasks in the first test phase, three animals were able to complete the bending task, and four animals were able to complete the bending-open task in the subsequent test phases. Two birds learned to solve the bending and bending-open problem without any assistance.

The large individual differences in this experiment made it clear that this is a task that is close to the limit of the cognitive possibilities of Goffin's cockatoos. Even the successful animals were not immediately error free. The hooks were not bent correctly at the first attempt, their shape gradually improved, and errors occurred in the use of the new tool, especially in terms of orientation, suggesting imperfect solution finding. Nevertheless, this is remarkable considering that two control animals solved the task even though the cockatoos had no experience with bendable materials, hooks, or bending activities. All species that have been confronted with the hook problem thus far have previously seen and used ready-made hook tools. As far as a possible natural disposition to bend thin, elongated objects is concerned, it is not only the New Caledonian crows that are "predisposed to bending" [449] but also the rooks [235]. They build nests of thin twigs, which they pull out and twist again and again, and in the middle of the nest, there are many bent twigs [472]. The hooks of the three Goffin's cockatoos can thus be regarded with greater justification as genuine behavioural innovations. Incidentally, something similar also applies to two orangutans who were successful at solving the bending problem at Leipzig Zoo [471].

The remarkable results of these laboratory studies of Goffin's cockatoos motivated us to research the tool-making abilities of the animals in the wild. Mark O'Hara, my former PhD student in the Kea laboratory, and his partner Berenika Mioduszewska, PhD student at the Max Planck Institute for Ornithology in Seewiesen, set up a small research station near the village of Lorulun on the largest island of the Tanimbar Group, Yamdena, with the energetic help of local farmers. From there, they observed the cockatoos for almost 6 years using state-of-the-art tools such as drones and video traps. They were looking for tool use, but this did not come to their attention (since the animals breed in tree hollows, manipulation of twigs and branches is not applied for nest building [473]).

Although lacking relevant sightings, this did not mean that individual animals were not innovative enough to facilitate their extractive food acquisition with technical aids. Evidence that the cockatoos manipulate small twigs or pieces of branches was found when the two could occasionally observe the birds throwing smaller twigs from the trees. However, the function of this behaviour could not be clarified, nor could it be linked to food acquisition, especially because it was hardly possible to observe the animals at a height of 20 m or more. As a way out, the short-term keeping of wild animals in a forested experimental aviary was devised, in which a small group of animals could be observed and tested at close proximity. After a few weeks, the animals were released back into the wild.

In the summer of 2020, the researchers finally made a sensational discovery [82, 474]. The aviary group of 15 animals was given Wawai fruit as part of their daily feeding.[3] The Wawai fruit is a widespread local fruit of the sea mango (also called the suicide tree) that contains seeds that are poisonous to humans. The fruits are ovoid, 5–10 cm long and turn bright red when ripe. They consist of a thin exocarp and a fibrous mesocarp. Within the hard, woody endocarp, up to three seeds are each enclosed by a central woody plate and a peripheral leathery coat. For cockatoos, the seeds are not poisonous but are a tasty treat.

When these fruits were offered for the first time, one animal immediately set about penetrating the inside of the fruit with deliberate, strong manipulations of its beak. The cockatoo was observed drilling into the fruit with a piece of wood in its beak. In two subsequent feedings, another animal showed similar actions on the fruit. This prompted Mark and Berenika to observe the two animals more closely in a separate aviary area. Owing to their great astonishment, it was not only possible to capture on film how the birds used pieces of twigs to purposefully and efficiently expose the seeds from the fruit's hard interior but also how they made their own tools beforehand by breaking off twigs and then modifying them. The two birds were fed Wawai fruits in a controlled manner several times in a separate area of the aviary so that the sequence and number of pieces of wood made could be precisely determined. Each time, the birds began by splitting a fragment from a suitable branch. Even before applying them to the fruit, the two birds modified and refined the pieces of wood by twisting them with their tongues inside their beaks and subjecting them to selective modifications until an object with characteristic properties was produced.

Because the human eye cannot capture the details of the actions with the tools, the researchers used an elaborate frame-by-frame analysis of the video recordings. The resulting data could thus be quantified and meticulously documented. On average, seven tools were used to break open one fruit, and each tool was applied to the fruit an average of seven times before it was dropped. A precise video analysis produced an unexpected result because the piece of wood from which the seed was ultimately extracted was used in a way previously undescribed in nonhuman animals. The authors called this activity "spooning" [474], an addition to the 22 types of tool use of Shumaker et al. [199].

No less spectacular was the finding that proved the production of three different tools used in a specific order. For each of 50 collected pieces of wood, a 3D model was created on the basis of a total of 108 macro-images, with which physical characteristics such as size, volume, and shape could be precisely determined. A cluster analysis of these measurements yielded three well-defined types of tools; called "fine" (thin, sharp, low volume), "medium" (medium thickness, medium-sharp, medium volume), and "sturdy" (thick, blunt, high volume). A combination of qualitative observations of how the two birds manipulated the different tool

[3] The genus name *Cerbera* is derived from Cerberus, the hellhound from Greek mythology, because it refers to the deadly toxicity of the seeds. In Madagascar, these seeds were used to pass judgements on gods and to poison kings and queens in punishment rituals. In previous centuries, they were responsible for the deaths of over 3000 people per year and were used for murder and suicide far more often than any other plant.

types—including in particular, the directions of movement and angles of insertion, as well as precise determinations of the order in which the three different types were used—suggest that at least two of the three different tools serve different purposes. While fine tools were used for spearing and poking to pierce the leathery coating encasing the seed, medium tools were used afterwards and more often as spoon-like tools with vertical and horizontal movements to scoop embryonic material through the open slit. Sturdy tools, ultimately, were used as powerful chisels to widen (and wedge) the slit.

If one relates the three techniques to the tooling theory of Fragaszy and Mangalam [253], it will be easy to see that we are not dealing with a simple, spontaneous, or random chain of action. Indeed, it requires not only the production of a functional and specific type of tool but also the consideration of two types of spatial relationships (the simultaneous holding of the fruit with the claw and that of the tool with the beak with different forces and positions), as well as the sequential application of the tool to the fruit in several successive steps. The actions of the tool on the fruit are dynamic and, above all, allocentric, that is, oriented according to objective standards and very specific in placement. We believe that the "spooning" of the two Goffin cockatoos in particular represents one of the most complex types of tool use yet is recorded in animal species without hands [474].

These findings shed new light on the key question of tool use in animals raised at the beginning of the chapter. Until now, we knew the use of tool sets with different functions and precise sequences of use only from chimpanzees. Wild chimpanzees in the northern Congo use toolkits to access honey in stingless bee nests or to reach termites in their burrows [246, 475, 476]. However, unlike chimpanzees, we do not have—as far as we know—any population-wide, regularly occurring tool traditions. Rather, in the two cockatoos, we seem to have witnessed a genuine behavioural innovation that is the result of a rare combination of several conducive behavioural dispositions. This was caused by the interplay of creativity and exploration, playfulness and high motivation for object manipulation, considerable perseverance, precise sensorimotor control, and the ability to immediately memorise and repeat complex but efficient behavioural sequences [351, 461–463]. However, it is highly unlikely that these two animals took the innovative step in the forest aviary. It is much more likely that they had already developed techniques for opening the Wawai fruit beforehand. The scale and complexity of the production and precise application of tool sets are too great to be generated in a single flash of inspiration. Rather, we seem to be dealing with the result of a series of creative acts that have been improved, refined and completed over time. Since only 2 out of 15 animals displayed these enormous feats, it seems reasonable to assume that these are rather rare exceptions, perhaps even developed by only one of the two animals and copied from the other.

Although no tools are involved, another species of the genus has been found to exhibit innovative behavior. Sulphur-crested cockatoos have been observed opening of rubbish bins in Sydney (Australia). It is not only the opening itself that is remarkable—we found a similar behaviour in kea [477]—but also the geographic variation and spread through social learning. This was interpreted as putative case of the development of a cultural adaptation to urban environments [478], which, together with bin-protection innovations by human residents, has resulted in a kind of inter-species "innovation arms-race" [479]. In conclusion, it seems that the combination of large brain size, high sociality, and slow life history, features that parrots share with primates, cetaceans, and corvids, are the necessary ingredients for behavioural innovation and (animal) culture [309, 480].

Chapter 5
Looking into the Future

The use of tools requires not only special sensory and motor skills, but also increased action control and flexibility. Simple stimulus–response chains are not sufficient to use tools efficiently. The use of a tool is particularly efficient when it is embedded in a larger context of action. Often, several actions must be combined and executed in the right order. Thus, such behaviour extends over longer periods of time than individual actions.

Chains of action do not necessarily have to be represented as a whole. They do not have to be thought out from beginning to end, but could also arise from chains of associations. For example, when animals walk a long, complex path back to their burrow, they do not have to have the whole path in mind when they start. They can also orient themselves by way markers that indicate the next part of the path. In this case, they only need to memorise the individual markers and the corresponding directional decisions.

In general, one can imagine chains of action as natural sequences of individual cause–effect chains, namely, when the result of one action gives the indication for the next action. It is then only necessary to make the right decision at the beginning. This can be based on mere chance or depend on the prevailing situation at the time. An external stimulus as a cue, for example, a just perceived prey, or an inner drive such as a strong feeling of hunger, are strong effective factors in the decision to perform the first action. Sometimes, however, it is advantageous to anticipate future events, situations, or actions and use them for the current decision. In other words, it would make sense to prepare for future events and adapt current actions to them.

In the animal kingdom, there are many good examples of preparatory or future-oriented behaviour. Gathering food in autumn for winter, hibernating, lowering metabolism or limiting energy-consuming activities, and even building nests fall into this category. These are all future-oriented activities. However, again, it is important to note that these do not require foresight or representation of future events. They are species-specific, innate behaviours with corresponding trigger mechanisms and matching instinctual actions.

L. Huber, *The Rational Animal*, https://doi.org/10.1007/978-3-031-60803-2_5

Can some animals nevertheless see into the future? In other words, can they see, in their mind's eye, future events or situations in which they will 1 day find themselves? From the point of view of cognitive research, such abilities are problematic in that the future does not affect the present since there is no causality acting back from the future. Words such as foresight and future-oriented action are merely metaphors from the realm of visual, spatial cognition. An animal can see something at a distance and orient itself to spatially distant things, but it sees nothing in the future. Its behaviour is determined by the present and the past. Current influences, sensory stimuli, memories, and experiences determine its actions.

The prevailing opinion among philosophers, anthropologists, and human psychologists is that only humans can anticipate the future and make appropriate preparations. Only they can imagine future needs that differ from their current needs. The view that animals do not have corresponding cognitive abilities, making it impossible for them to think ahead and plan chains of action, is referred to in the literature as the Bischof-Köhler hypothesis. It goes back to the German developmental psychologist Doris Bischof-Köhler [481] and her husband, the Zürich psychologist Norbert Bischof [482]. Later, this hypothesis was advocated by the German psychologist Thomas Suddendorf (University of Queensland, Australia) and the New Zealand-Canadian psychologist Michael Corballis (University of Auckland, New Zealand), who argued that nonhuman animals, including great apes, are confined to a "present" that is limited by their current drive states [483].

An (nonhuman) animal seems unable to imagine what it will be like to be hungry the next day when it has just eaten. Many researchers have added to this admonition by stating that we will not and cannot experience the animal's subjective experience or even its imaginings [484, 485]. We can determine functional similarities or differences only. Neurobiologists must also maintain this functional level when comparing animals and humans. Moreover, the search for a yes-or-no answer to a question about human uniqueness is not a productive path for comparative cognitive research. This path leads to the search for proof of existence—a single animal with the ability would suffice, followed by endless arguments about whether the ability has actually been proven. In line with evolutionary thinking, it is far more productive to proceed step by step, first examining simpler but more easily ascertainable components. In particular, complex cognitive phenomena such as planning or future thinking should be deconstructed and resolved by asking the following questions: which components of process X occur in which animal species and why, what functions do they fulfil, how might they have arisen phylogenetically and individually, and how might they function?

When planning actions, one can basically distinguish between two areas: planning sequential chains of actions for current needs and planning for future needs. In the first case, the animal performs actions that satisfy its current need. If two or more successive actions have to be carried out in the right order, the animal must also act purposefully but without looking into the future. The monkey fetches a stone, puts the nut on the anvil, then hammers to get a tasty meal. But what about planning actions for needs that do not yet exist but are expected later? Before we can answer this question, we must turn to an intermediate stage. In this stage, the future need is identical to the current need, but actions must be taken that do not satisfy the current need but only the latter need.

Store Tools

If one assumes that the use of tools is an elaborate behaviour that requires a great deal of physical effort on the part of the actor in addition to a cognitive effort, sparing their use should be advantageous. Just think of the capuchin monkeys, who use stones of half their body weight. Now, they should not only choose the stones carefully, as they often have to be carried over long distances to the "workshop", but also store them after use for later use. In fact, as I have already discussed, such accumulations are found in long-tailed macaques and chimpanzees as well. However, not much thought needs to be put into this. The stones are simply left lying around, as in an unorganised workshop.

If the tool and its procurement are valuable and worth reusing but storage is unguaranteed because, for example, you are working in a tree, you have to find ways to actively preserve the tool until the next use. This is exactly what they found with the New Caledonian crows. First, using mini-cameras attached to some free-ranging birds, blades of grass, which were shown to be successfully used as tools for finding insects, were carried in the beak when the birds flew to another feeding site [486]. These tools are valuable because crows use them to obtain a significant proportion of their daily protein and fat diet. If the tool is to be reused in the same place, they have different strategies for doing so. In the simpler case, they leave their poking tool in the current branch hole. Here, however, it is not known whether there is truly an intention behind this approach, or if it is simply convenient. It is different when they stick the tool in another hole and retrieve it from there whenever they continue working. The problem with this approach is that the branch or screw tree leaf can easily fall to the ground. Therefore, the birds also have to stable the tool with their feet and thus actively store it [487]. This tool "safe-keeping" behaviour is not as safe as storing the tool in branch holes but is apparently good enough when things have to be done quickly. But, if a bird has trouble processing and eating food, which increases bodily involvement, the risk of losing the food is high. In this case, it would be useful to choose the somewhat more elaborate variant of stroing the stick in a branch hole. This context-dependent tool-handling behaviour is exactly what has been demonstrated in aviary experiments with wild-caught crows [487]. Moreover, the crows seem to modify their tool "safe-keeping" behaviour in response to tool type. They prefer hooked stick tools over non-hooked stick tools, suggesting that they "value" the costlier but also more efficient hooked stick tools more than the cheaper and less efficient non-hooked stick tools [488].

I have already pointed out several times that the New Caledonian crows can draw on a long evolutionary experience in their almost excessive use of tools and that this apparently expedient behaviour has therefore developed over long periods of time, anchoring it in their behavioural repertoire. Again, comparisons with birds that do not use tools are useful for making more general conclusions. Goffin's cockatoos were just right for us to do this. So, we chose an experiment in which the animals had to acquire food with a poking tool several times in succession at the same location [489]. Five boxes were placed side by side on a table top with a hole at the front and the food inside. In front of them, a wooden pole served as a perch. When eating food after successful fishing, the tool could fall to the ground where it would have to be picked up again for the next box. To simulate the difference in effort, the boxes were on a table 1 m high in one experimental condition and on the ground in another experimental condition. To vary the risk of losing the tool, the animals were given 1/6 of a cashew nut either loose or enclosed in a gelatin capsule from which it had to be laboriously extracted.

Eight animals were tested in this experiment. To secure the tool (a small stick), they either held it with 1 foot or pushed it into a hole and stabilised it with 1 foot. Since both methods were very difficult and probably also strenuous, they kept losing their stick. However, the effort to retrieve a lost stick clearly showed an effect. When the perch was close to the ground, they held the stick in their foot more often, but retrieved it less often than when the perch was at a height of 1 m. Obviously, the animals wanted to avoid fetching the stick if it meant more effort. On the other hand, they secured the tool more often when the food was easy to eat than when it was difficult. This sounds illogical at first but can be explained by the fact that it is very difficult to obtain food from the gelatin capsule while balancing on 1 foot. This was also the case with New Caledonian crows [448].

In sum, the results of Goffin's cockatoos show that even without special adaptations for tool use, animals are capable of choosing their actions with auxiliary objects on the basis of a cost-benefit trade-off, comparing the relative effort of different options in relation to the reward to be expected in each case. This situational weighing-up of options, which precedes the decision, enables a high degree of flexibility and situation-appropriate behaviour. The scope for action comes from a number of options: (1) immediately or later?; (2) for which goal (different foods)?; (3) which action (with or without tool)?; (4) if with a tool, which one is most effective (different effectiveness)?; (5) with or without modification?. As we shall see, the deliberation process is limited by this large amount of options. Limitations may be due to those of both working memory and executive control. Finally, how far into the future one can go to decide on their next steps will be limited. For longer chains of action it is useful to see further ahead. Are there indications of this in nonhuman creatures?

Two Steps Ahead

Although we are already dealing with foresight when the next step depends on a goal to be reached—for which the best means must be chosen—we are still moving very short distances in terms of time. The effect intensifies when further steps are added. For example, sometimes intermediate steps must be carried out before the actual means can be used to achieve the goal. As we have already shown, capuchin monkeys use a small stone to knock out an enclosed large stone, which they subsequently use to crack open nuts. Gathering the small stones is an intermediate step that does not yet lead directly to the goal. Are the monkeys aware of this when they start? Did they first have the goal, the coveted nut in mind, and then think back to hitting it with a stone, and, in order to get the stone, to hammering the stone out of the conglomerate rock [244]? Analogously, one may ask whether the monkey has the nut and the hammering in mind when it searches for good tools [285]. In any case, chimpanzees on the Ivory Coast have been observed to bring their hammering tools to places far out of sight [490], and in the Republic of Congo, they take their tool sets to the places where they collect honey [251]. But even so, it is difficult to determine by pure observation what motives the animals act with and what decision-making basis they had for doing so. Therefore, tests were again carried out in the laboratory.

New Caledonian crows also take a step-by-step approach to foraging, especially when they perform several actions in succession to make their poking tools. In Alex Taylor's laboratory at the University of Auckland (New Zealand), the question of whether these birds can use tools sequentially was investigated using seven crows [238]. The task was to use a freely accessible object to anchor an out-of-reach object, which in turn made the out-of-reach food accessible. To do this, a small piece of meat was placed in a small, narrow box (the "meat box") 15 cm from the front of the aviary, inaccessible but clearly visible (Fig. 5.1).

In the wild, animals solve problems by using a branch or a stick. In the experiment, an adequate, sufficiently long stick (18 cm) was also offered to the animals. However, it was not openly accessible. The stick was placed in a second box approximately 2 m from the first box, the front of which was blocked with vertical bars.

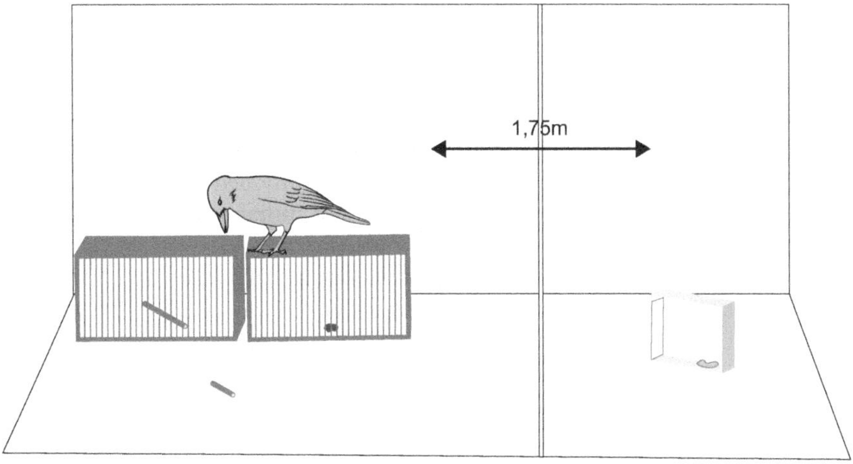

Fig. 5.1 Experimental setup for sequential tool use in New Caledonian crows. (From [238])

The stick was 4 cm from the front, just far enough that it could not be reached with the beak. Near the box was a short, 5 cm long stick, long enough to catch the longer stick in the box but clearly too short to reach the piece of meat. Next to the box with the long stick, there was also a second, identical-looking box with a small stone. This was used to check whether the crows would randomly use the small stick to fish for an out-of-reach object or whether they would specifically take the large stick.

Of the seven crows, three showed the correct behavioural sequence on the first attempt: they first picked up the short stick, fished out the longer stick with it and went to the meat box to fish for the anticipated piece of food. Only one animal went to the meat box with the small stick. Of the remaining three animals, one was not motivated to pick up the small stick, and the other two tried to fish the long stick with the short stick but failed. However, the initially unsuccessful animals took further opportunity to solve the problem, so that by the end of the experiment, all seven animals exhibited the correct behavioural sequence.

According to detailed performance records, two of the animals performed absolutely flawlessly, and two others performed with only one mistake in ten attempts. With regard to the question of the birds' competence in the gradual use of several tools, the result is even clearer. Overall, six out of seven animals used the small stick as soon as they had it in their beak for the first time to fish for the long stick, instead of flying with it directly to the food box or grabbing the small stone. Thus, the crows seemed to have passed with flying colours in a task that even apes face with major problems. Five out of six gorillas, three out of five orangutans, and three out of five chimpanzees were successful in similar experimental set-ups [239]. However, the three successful chimpanzees initially took the wrong, direct path to the food before taking the correct path to the target tool in further tests. In monkeys, task performance is even worse. Only two of a total of eight capuchin monkeys solved the problem in the first attempt [491], and Japanese macaques did so only after many (unsuccessful) attempts [492].

On closer inspection, however, the task successes of the New Caledonian crows also have to be given certain limitations. Why, for example, was the meat box placed almost 2 m away from the small stick and the toolbox [238]? Isn't it literally much easier to bring the small stick to the toolbox instead of the meat box? And did the animals even know about the different functionality of the two sticks? It is not clear from the publication whether the crows first looked at the meat box, then at the two sticks, and only then, after gathering all the relevant information, took action. The execution of the correct chain of action alone is not sufficient to identify the animals' mindset.

The control for trial-and-error learning must also be taken with a grain of salt. The fact that all animals—with only one exception—always went to the toolbox with the long stick instead of the one with the stone after picking up the small stick can, but need not, be interpreted as deliberate action. A strong preference for sticks over stones can be assumed in New Caledonian crows, as well as a preference for a long stick over a short stick as a fishing tool [297]. We also do not know how often the birds lingered by the box with the long stick and squinted at it before picking up the short one. After all, they had several opportunities to retrieve the long stick from the toolbox with their beaks before the experiment. Even if they picked up the short stick right away, it was obviously to use it to get the preferred, attractive tool right away—even more so when they had already gained experience with the long stick before the actual experiment, and therefore they knew how well it worked for fishing the piece of meat [234]. We do not know how the animals proceeded in detail, but a control with reversed conditions, where a long stick must be used to acquire a short stick as a target tool, would be informative.

The researchers added a second experiment that strongly resembled such a control experiment. They switched the positions of the two sticks. Now, the longer stick was in fact freely accessible in front of the boxes, and the short stick was inaccessible in a box. However, the short stick was not necessary to obtain the food, and the long stick remained the target tool. This arrangement was designed to test whether the crows automatically or unthinkingly use a freely available tool to acquire a locked tool or whether a stick in the beak entices them to poke immediately at the next opportunity. Interestingly, two thirds of the animals tested made this very mistake. Only two animals picked up the long stick and immediately went to the meat box. The other animals quickly corrected themselves, but the temptation to poke the toolbox with the long stick remained for some time, even in an animal that had shown the correct behaviour at the first attempt.

In summary, how should we assess this behaviour of crows? Is it justified to follow the researchers' preferred interpretation and attribute both analogical thinking and causal understanding to the animals? The former, since they used their experience with tools to solve a structurally similar yet novel problem, instead of fishing for food with another stick? Is this not merely a generalisation? After all, conclusion by analogy is a logical process of inference in which the correspondence of two things in some respects leads to the conclusion that they are the same in other respects as well. This is not only a matter of similarity but also of equality on the basis of a logical operation along the lines of A being similar to B, and B having the property of C; therefore, A also has the property of C. We do not know which characteristics or actions fill the units A, B, and C in the experiment in question. Thus, this question must be left open for the time being.

In the case of the crows, understanding causality refers to understanding the function of the stick as a means of collecting out-of-reach objects. A strong term (causality) is preferred for this (again). The leaner alternative would be the term functionality, i.e. the property of the object to fulfil a certain function or group of functions. So, the function of the stick would be to collect distant objects. There is no doubt that the animals were familiar with this property of sticks and used it in this test. I do not think it is necessary to have an understanding of how this function is made possible, i.e. what the exact relationship is between cause and effect.

It must be acknowledged that the majority of crows use short stick not as a target tool but as a tool for acquiring a better tool. But, apart from the problem of the great difference in distance between food and other tools, the great attractiveness of a long stick also calls into question an interpretation in the direction of "What do I need for what and when? Researchers have attempted to determine how to address this issue by using the indirect argument that causality involves the quick correction of mistakes. This argument was used to explain the erroneous behaviour in the second experiment, as the crows very quickly corrected their mistake of poking the box with the long stick into the box with the short stick. However, this argument stands on very shaky ground. First, the animals that made initial mistakes have shown that they do not act in an exclusively deliberate and forward-thinking way, or that they have difficulties abandoning the behaviour often repeated in the first experiment—first poking in the toolbox then at the meat box—for better action. Second, the correction in further experiments could simply be because they were again encouraged to use the long stick for fishing in the meat box, while poking into the tool box was unsuccessful.

Chains of Action

The crucial question is, as Alex Taylor and coauthors suggested, what the attribution of meta-tool use actually means.[1] Which cognitive processes best explain the behaviour of the crows in their experiment? Tool use for other tools is the key event here. During hominid evolution, our ancestors improved the functionality of stones by hitting them with other stones, thus sharpening them [493, 494]. As mentioned earlier, chimpanzees appear to use small stones to stabilise large blocks of stone that they use as an anvil for pounding on hard nuts [495]. In both cases, the use of one tool alters the functionality of the other tool. This is not the case in the crow experiment. The short stick does not change or improve the functionality of the long stick but rather makes it available. What remains is the peculiarity that a tool habitually used to acquire food is now used to acquire nonfood. This certainly requires a kind of flexibility and breaking out of habitual courses of action—although it must be said that the action itself, namely, poking and fishing, remains the same. However, if we now know that longer sticks are extremely attractive objects for New Caledonian crows—in nature, they make their fishing tools from the leaves of screw trees pretty much exactly the length of the long stick used here [306]—and that they already had positive experience with the long stick before the experiment, acting as a secondary reinforcer [234], it can be argued that this long stick itself became the target object. With respect to the application of a tool—here: the short stick—to another object—here: the long stick—instead of food, it is now known that these birds use sticks not only to fish for food but also, more generally, to explore their environment [236]. The argument that they must inhibit the target action leading directly to food in favour of the action directed at the other tool is weakened by the large difference in distance (see above) and again by the attractiveness of the long stick. Finally, it is questionable whether the crows must integrate a new, innovative action (tool on tool) into usual action chains hierarchically as a secondary goal

---

[1] As discussed in the section on tool use, this is strictly speaking *sequential* tool use. In true meta-tool use, one tool used at the same time acts on another, like the wedge stone on the anvil.

under the main goal (tool on tool on food). This would mean that the animals subsequently form recursive loops of thought, as is necessary and common in human language use.

It is undisputed that the animals had to perform a two-part action chain to reach their goal. However, the classic counterargument since the experiment by Epstein, Lanza, and Skinner [102] with pigeons is that it is a matter of learned chain formation (*chaining*). It has nothing to do with insight, planning, or innovation. Indeed, another study showed that after successfully solving the meta-tool problem, New Caledonian crows continued to use the short stick to grab the long stick, even when there was no food. Apparently, the long stick was indeed attractive enough (see above), or the action chain, once learned, continued to be executed, with or without the target object.

In a follow-up experiment, Alex Taylor's crows even had to use three different tools in the correct order to reach the target [237]. The last link in the chain was a task they were familiar with: using a small stick to fish a piece of meat out of a hole. Again, this was not lying around freely but in the barred toolbox. A short stick was again supposed to solve this problem. But this time, the short stick was not lying freely in front of the box with the long stick—it had to be pulled by a string and then released before it could be used to fish for the long stick. In contrast to the earlier experiments [238], the meta-tool was also made unattractive in prior training sessions. It was offered as a tool for fishing the out-of-reach food but was unsuitable due to its shortness. Therefore, the animals were not supposed to start by pulling the string that the short stick was attached to. The concatenation of positive experiences made earlier, of associations between stimulus, reaction, and positive consequence, should no longer be possible, rather, planning with backwards steps away from the goal should be: for the food I need a long stick; for the long stick I need a short stick; I get this by pulling the string.

Since the researchers primarily wanted to test this plan action, they first trained four freshly caught crows on each of the three parts of the action chain before presenting them with the entire task. (We recall how pigeons were similarly trained to push a small box and peck at a paper banana [102]). Not surprisingly, the crows very quickly found the solution as a concatenation of the individual tasks. The only difference was that here, three instead of two components had to be put in the right order. Although, this task would be a mere quantitative improvement of the pigeons' performance, and would justify neither planning nor insight. For this, animals would have to succeed spontaneously without having practised the subtasks beforehand.

In fact, spurred by the performance of the trained animals, the researchers tested three more crows in this way. Without error, two animals succeeded in executing the correct chain of actions, and the third animal had to perform some trial and error before doing so. Interestingly, all of them looked at the objects for a while before taking action.

Another research group from the University of Oxford was skeptical about these experiments and conducted its own on the same question. Here, New Caledonian crows had to master a sequence of two or three steps with objects to reach food [236] (these three steps had a significantly lower random probability of eliciting the correct sequence of actions than did the other two steps). Again, several items were offered, which the animals could select or manipulate. Items differed in their functionality, which was intended to test whether the animals selected items merely because of past rewards or actually because of their usefulness in achieving the goal. To test for goal-directed, planned action, the so-called controls included conditions without food, conditions in which unattainable items were replaced with attainable ones, and conditions in which action sequences were unnecessary. Finally, the amount of pretraining was varied so that spontaneous, deliberate action could be distinguished from learned stimulus–response sequences.

In their first experiment, in addition to a tube of dislocated food where the distance from the entrance varied (from 7 to 25 cm), the researchers offered four boxes of rake-like tools, one long (20 cm) and three of medium size (10 cm), all out of reach, and finally a so-called tabletop tool, a small rake (6 cm) lying free on the table. The individual tests, which were randomly intermixed, differed in their difficulty and control conditions mentioned above. While three animals had prior experience with the tools, three others were tested without any prior experience.

Prior experience did indeed play a major role. Only the animals with previous training of the individual action components were able to master sequential tool use. When only an unreachable rake had to be fished with the small table top tool, the animals were successful from the first attempt. However, only one individual (Betty) could master the triple action sequence, which had not been demonstrated by any nonhuman animal before. For one of the three middle rakes offered, Betty had to fish with the small, exposed tabletop rake and with this rake for the long rake and, finally, for the food lying at the back (25 cm) of the food box.

The other two experienced animals only managed the double sequence. The inexperienced animals fell far short of this performance. Only one animal out of five fetched an out-of-reach rake to grab the food with it, and even this success came only after many failed attempts. This large difference between trained and untrained animals not only weakened the conclusion of the authors from New Zealand [238] that New Caledonian crows are capable of spontaneous, goal-directed action based on analogical reasoning and planning but also highlighted the importance of prior experience and learned action sequences. It does not seem to be easy for birds to transfer fishing for food to fishing for other objects—or in other words, to understand fishing in its full functionality—as a means to different ends. Additionally, the findings from Oxford [236] do not allow any conclusions to be drawn about detailed planned actions. Even in experienced animals, it could only be shown that they understood when they needed another rake but not which one. The recurrent errors of poking in the feeder with the short, exposed tabletop tool when there was no food at all or fishing for the wrong tool because it was too short, show that deliberatly planning for tool use in New Caledonian crows is a rather implausible explanation. Rather, the errors show that the animals did not pay attention to all the necessary details of the tasks, that they could not always overcome or suppress habitual actions and that they did not make full use of their prior experience.

However, it would be premature to deny the animals any deliberation when they are acting in tool-use sequences because of failure. The Oxford researchers were aware that sensory and motor difficulties may have facilitated the occurrence of errors. For example, it might have been difficult for the birds to estimate the length of objects through the reflective acrylic. In addition, their actions might have been affected by limited attention given to the toolboxes. Therefore, the Oxford researchers conducted a second experiment with three birds, taking these problems into account. In fact, the success rate was much greater than that of the first experiment. This time, all of the animals were able to complete the sequential task with two tools, already in the first experiment and one even without any previous experience at all (Corbeau, the other two animals were already tested in the first experiment). The success with three tools was particularly convincing, but only Betty, who we already knew to be particularly clever in hook-making, succeeded. In this task, called the "tertiary" by the authors, the short (6 cm), exposed tabletop tool had to be used to fish out a slightly longer tool (10 cm) from a toolbox, and in turn use this tool to fetch an even longer one (20 cm) from another toolbox, ultimately using the longest tool to fetch the food from the food box, which was placed 25 cm at the back. The animals considered the respective distance of the food in the food box, which proves that the fishing of further tools was not based on frustration or the inaccessibility of the food.

In summary, New Caledonian crows can master sequential tool use that is strongly influenced by prior experience but not necessarily enabled and that they pay attention to some, though not all, relevant aspects. Again, there is considerable inter-individual variance, with birds failing despite long training and others completing tasks with no training at all. For the cognitive processes involved in these performances, we should consider the presence of analogous thought processes as very unlikely but not impossible. With all due caution, we can grant individuals of this species (in this case, Betty and Corbeau) the ability to weigh options for action, including imagining the consequences of the next steps. What is decisive here is what information was available to them in this process, whether they selectively obtained missing information before taking action, or whether they rather went with the simpler two-way strategy, namely, to first try what was available and to choose another option if this was no good. Therefore, it seems that at least some animals act purposefully and with foresight, while others are more explorative and rely on familiar courses of action.

Soon or Later?

Can the behaviour of Betty and Corbeau already be understood as genuine planning? When people plan, they use two cognitive abilities in the broadest sense. They are able to suppress current needs to satisfy later needs. We speak of self-control or inhibition control. And they can imagine possible or future events. For the question of animal consciousness, the second faculty is more important, but we also want to look at the first, especially since there are already many experiments and interesting theories on this topic. These take their start in the investigations of Walter

Mischel—keywords: "Marshmallow test"—as already presented in the opening chapter on the subject of self-control. In long-term studies, Mischel and his colleagues were able to show that the ability to defer rewards is related to success later in life [496]. These successes seemed to come from the fact that these individuals were better able to cope with frustration and stress, as well as to resist temptation. In any case, as far as the issue of time travel is concerned, although there is great individual variation in this topic, humans have the ability to wait not only minutes or hours but even days, weeks, and years to reap a greater reward.

This decision depends on how large the difference between the two rewards is and how long the waiting time is. Some monkey species, dogs, corvids, and parrots have already shown such waiting decisions in tests [149, 164]. They have suppressed quick-drive satisfaction in favor of a better option for a few seconds to a few minutes. One could assume that they made a flexible assessment of relative costs and benefits before making the decision.

Although experiments on delayed gratification [497] are the first approach to test underlying capabilities, the complexity of decisions involving the mere contrast of two alternatives falls short of the reality found in nature. To maximise profit, the agent may have to consider several factors. These include the predictability or expectability of the higher reward, as well as the attainability and, ultimately, the availability or effectiveness of the means needed to achieve it. If any of these conditions are not met or are met only to a limited extent, choosing the immediate reward may be the better alternative: "Better a sparrow in the hand than the pigeon on the roof".

Although there are many experiments on this topic, only a few have been embedded in the context of tool use. The use of tools not only increases the time needed to reach a goal but also brings about a better end goal itself. Instead of picking less nutritious fruit from the tree, one could invest in cracking nuts. Although this alternative is not immediately available, the time delay is compensated for by the higher quality of the nut. In fact, capuchin monkeys have confirmed this ability in experiments [157]. They could either immediately eat an elongated celery or pretzel stick or use it to reach much higher prized peanut butter out of a tube. They did the latter, especially in contrast to a control situation in which the tube was empty and they immediately consumed the less popular food. Great apes were able to increase this behaviour even more. Chimpanzees and orangutans preferred a long plastic tube to an immediately accessible grape, only to use the tube 70 minutes later to suck a much more popular fruit soup from a bottle [498].

All of the primates mentioned are also known for their ability to use or even make tools. It is therefore conceivable that there is a connection between this ability and the ability to self-control and weigh alternatives with temporal difference. However, instead of a direct causal link with habitual tool use, there could also be an indirect link if general cognitive mechanisms enable these two abilities. Goffin's cockatoos are an ideal test group for this purpose. They have already shown strong indications of self-control, as they can wait for better food and are more likely to do so the better it is [161]. Thus, motivational and cognitive abilities play together in their choice behaviour.

To more closely examine the decision-making processes in the face of different food qualities under varying waiting conditions, the cockatoos were confronted with a range of tool choices and associated uses [162]. Can they weigh the qualities of food, the accessibility of food with and without tools, and the effectiveness of their means before taking action? To this end, individual binary decisions of increasing complexity were tested.

First, similar to the previously mentioned monkeys, the birds had to decide between two differently valued food rewards in the context of using tools that made the included reward available. This pre-test revealed which of two foods was more valuable. They then had to factor in the effectiveness of the tool into this decision. The reasoning was as follows: To act in an economically optimal way, they would prefer a tool to a directly available reward only if the more valuable food could only be obtained with that tool. This approach makes it possible to check whether the tool was chosen partly out of manipulative pleasure, or whether more work was accepted only if there was both a better goal and one that could be achieved with the available means.

The researchers developed five different tests for this purpose [162]. In the tool test, the birds were asked to choose between two tools, a small stick and a small ball—in the face of either a food box where the stick was useful (it could be used to reach the reward through a hole) or a food box where only the ball was useful (inserted through a tube to push the reward out of the box). The cockatoos chose the stick for the box with a hole and the ball for the box with a tube with approximately 80% certainty. In the motivation test, either the box with the hole or the box with the tube and the corresponding tools were presented in successive trials. In addition, equally valuable food was available both inside and outside the box. As one might expect, they preferred the immediately accessible reward and refrained from using the tool for the same reward. However, they did not always do this, but only in about three quarters of all cases. The third test checked the trade-off between reward qualities under four different conditions. Either the more valuable food was freely available or was in the box, which had to be combined with either the stick or the ball. In the tool functionality test, the less valuable food was always freely available, and the more valuable food was always locked in a box. Again, the animals chose the tool according to an economic decision, namely, in approximately 80% of the cases, if the more valuable food could be reached with it. However, the effective tool was available in only half of the tests; the ineffective tool was available in the remaining tests. Did the birds also resort to the ineffective tool in this situation? When deciding economically, birds should choose the less valuable food if the wrong tool is in front of the box, regardless of whether it is a ball or a stick; however, they should use the tool if it gives them access to the valuable cashew nut. Again, the animals chose correctly in the previously described economic sense in nearly 80% of the cases and in all four conditions. Finally, the animals were still confronted with a combination of all three factors, as both boxes, both tools and both reward qualities were present. However, unlike in all previous tests, their choices here were no different than random choice. The birds did not choose the ball (statistically) significantly more often than the stick when the valuable food was in the box with the tube and the less valuable food was in the box with the hole. The same was true in reverse. It can be assumed that the amount of information to be taken into account here for an optimal decision exceeded the cockatoos' capacity to weigh things up. Possibly the birds were mentally overloaded here.

Their performance in the four other tasks was amazing. When the right tool was available, they chose it to obtain more valuable food and ignored freely available but less valuable food. But, if the available tool was not good enough to get the preferred food out of the box, they chose the less valuable but still attainable reward. This showed that the cockatoos considered the consequences of their actions—can I open the box with the available tool or not?—and the goals that could be achieved later—will I get my preferred food or the less valuable food?—in their heads before they took action. An alternative explanation would be that the animals slowly learned the right decision through success and failure of the individual tests. But, this was not the case. Although they did not choose optimally, i.e. always correctly in any test, their probability of success did not increase within the tests.

If such "economic" or even "rational" decisions are found in animals in the wild, one can never be sure whether the animals have learned over time to decide correctly, i.e. reward effectively, in individual situations. After hundreds of trials, a trial-and-error approach moves increasingly more towards the decisions that yield greater psychological or physiological gain. Over many generations, genetic dispositions to certain actions, such as poking and fishing with twigs in knotholes, might also evolve. But all of these limitations could be ruled out in laboratory experiments, even more so for animals that have not integrated tools into their natural foraging behaviour.

The Power of Imagination

Now that we have discussed in detail the first important ability in connection with planning, namely, the suppression of current need in favor of future one, let us turn to the second ability. It comes, as already said, much closer to the question of consciousness in animals because it concerns the question of whether animals can imagine future events. There are, as we shall see, prominent contemporaries who deny this ability to animals and make it a sole characteristic of humans [499–501].

In any case, thinking and imagining the future seems to have great adaptive value because it enables us to act now to secure future benefits and avoid future difficulties. Indeed, the ability to imagine future events appears to be an essential component of human cognition, as much of our behaviour is guided by foresight. We direct our actions towards distant goals and make appropriate plans for them.

But what about animals? Can at least some animal species remove themselves from the here and now and travel in time to the future? Can they think of nonexistent things? Again, the already often strained logic of starting with simpler questions applies here. Instead of talking about the ability to imagine, we could content ourselves for the time being with looking at the ability to extrapolate. After all, it may be enough to project things that one knows from the past into the future.

When dealing with experience, one has to bring the capacity of memory into play. Here, one must distinguish between two forms [502]. *Semantic* memory, or memory of knowledge, concerns things that one knows without having to make a personal reference to them. For example, one can infer from films that it often rains in London, at least more often than in Cairo. However, this knowledge does not have to be linked to personal experience. Personal experience, in turn, is the characteristic property of *episodic* memory. You may know that it often rains in London because you have been there yourself several times and it usually rained. I well remember being surprised by a downpour during a stay in London because the sun was still shining in the morning. This memory includes the place (London), the event (a meeting of the Royal Society on social intelligence), and the approximate time (the year 2006). Above all, however, the subjective impression is attached to this event, which is where self-awareness comes into play. One undertakes a journey into one's personal past and associates an emotional component with it. The aforementioned event in London was particularly unpleasant because I had to listen to the first lectures heavily drenched. Episodic memory has been repeatedly considered as a human uniqueness by its discoverer, the Estonian-born Canadian psychologist Endel Tulving of the University of Toronto, in contrast to semantic memory that birds also possess, for example, because of their self-conscious, "autonomous" character [502].

When remembering events experienced in one's own body, the person embarks on a journey through time into his or her own past. The projection into the future was accordingly seen as a journey with a reversed sign. It is based on the ability to vividly experience events in the personal future in advance. The journey through time is thus undertaken from the so-called first-person perspective. Therefore, cognitive processes similar to those involved in episodic memory should be involved—at least this seems to be confirmed by neurological studies in humans [503]. On the one hand, there is a network in the frontal and medial parietal lobes that is active in both processes, episodic memory and foresight, but there is also another network (in the left lateral premotor cortex, left precuneus, and right posterior cerebellum) that is more likely to be active in simulating movements and imagining a future event [504]. Similar activations of the two networks for past- and future-related tasks seem to be because they both require reactivation of previously experienced spatial–visual situations. Thus, it seems that future events are embedded in well-known spatial–visual situations.

We will return to the question of whether animals have episodic memory in the next chapter. Notwithstanding this, however, prospecting into the future could also be based on a counterpart to semantic memory, the ability to extrapolate "cold", impersonal knowledge [501]. This knowledge is rule-based, abstract, and neutral; i.e. it does not contain any personal special cases or peculiarities and refers only to repetitions, regularities, or constant processes. Certainly, it can be used to see into the future, but the question remains how this semantic knowledge is stored and how it becomes available. Tulving's theory of semantic memory assumes that knowledge about the world requires a linguistic structure [499]. He has called this a "*mental thesaurus*", an organised collection of symbols, their meanings, reference points, relationships to each other, plus rules of usage, formulae, and algorithms. These things are facts; they are abstract because the specific perceptions on which these memory contents are based are not included. In contrast, the contents of episodic memory are concrete because they store temporally structured episodes (events). They are bound to perceptions as well as to the subjective.

Although the two forms of memory overlap and interact well in the mental life of humans, probably even having arisen in parallel during evolution, it has been suggested that they are of different kinds [499]. On the one hand, a so-called *noetic* consciousness enables us to become aware of things that cannot currently be sensed. We can take such things into use as facts, to establish relationships and derive rules. It thus conveys a sense of knowing. In contrast, episodic memory requires an *autonoetic* consciousness, one that is autobiographical because it refers to personal experiences. In this process, memories are awakened and virtually re-experienced. The person can position herself in time.

The existence of these two forms of consciousness in conjunction with the two forms of memory became apparent in the case of a patient (K. C.) who suffered a severe brain injury in a motorbike accident and was left with high-grade amnesia. He could not remember any personal experiences, even those that occurred only minutes ago, while other mental activities, including factual knowledge and even perspective taking, remained almost undisturbed. His time horizon was, as he himself stated, empty [505]. But he was apparently able to use his still intact semantic memory to make decisions about the future. This dissociation of memory could later be confirmed in other patients with similar clinical pictures and was also verified in test subjects who were questioned in a special way about things in the past and in the future.

A possible connection between foresight and episodic memory was celebrated as the scientific discovery of the year in the journal *Science* in 2007. The findings of the previously mentioned brain studies, which revealed similar activation patterns in both thought processes, suggest not only this connection but also the parallel developmental course in children. Both abilities occur at about the same time, at approximately 3 years of age, which led the authors of this study to call it "episodic future thinking" [506]. In the past decade, research on the role of memory in imagination and future thinking has produced further insights. The similarities between remembering the past and imagining or simulating the future have been confirmed, and the supposed "*default mode network*" that underlies both memory and imagination has been discovered, such as in daydreams [507]. Nowadays, this phenomenon has been demonstrated not only in humans but also in monkeys, rats, and dogs [508–511].

Together, these findings led to the "Janus hypothesis", referring to Janus, the Roman god of the New Year, who looks both into the past and into the future [501]. Overall, these two processes of time travel seem to be related, as are other processes by which simulations of phenomenal consciousness are accomplished, such as perspective taking and spatial imagination. The network described above can be flexibly coupled with other networks to support complex goal-directed simulations. The close connection between remembering the past and imagining the future suggests that a key function of memory is to provide a basis for predicting the future through imaginary scenarios. In any case, from an evolutionary perspective, the ability to flexibly recombine elements of past experiences into simulations of new future events appears to be an adaptive process.

However, what is the fitness-maintaining or even fitness-enhancing value of memory-based simulations? It has been suggested that the ability to travel in time provides a motivational brake on current drives in favor of future rewards [512]. Sensory experience in conscious episodic memory evokes the emotion associated with that episode. This competes with other emotions that occurred further back in time. One can equate the competition between immediate and delayed rewards with this struggle between emotions. This is how the future comes into the present. The memory of a reward is projected into the future and can compete with a present reward. If the memory is strong enough, it can be used to make a decision for the future. One could also say that cortical structures thus trick subcortical structures because they are deceived to react emotionally and motivationally to a simulation as if it were a real experience. However, the question remains as to what influence consciousness has in this process and possibly even had in this process before the emergence of man. Could future-related decisions be based on "primitive" systems of sensory input and emotion, or does it require reflective self-consciousness?[2] Some believe that simple phenomenal consciousness without self-reference might suffice [513] and could be a simple, unreflective act based on the strongest motivation. The reasons why the emotion arose remain undiscovered to consciousness along with the mere fact that an emotion arose at all. Even the generation of an episodic construct may be an unconscious process. A decision based on an autonoetic simulation can be blind to consciousness to the same extent as one made "in the moment" and without simulation [514]. In any case, the involvement of consciousness in episodic processes is a nontrivial matter and will occupy research for a long time to come.

Planned Actions

At the beginning of this chapter, we asked whether animals could move away from the here and now and take a journey into the future. After some theoretical considerations that revolved around the question of imagination of non-existent things and the involvement of consciousness, we now want to ask how we can deduce from the behaviour of animals whether they can make decisions with a future perspective. From an experimental point of view, it should be easier to exclude influences that

[2] I will discuss the highly controversial question of self-consciousness in animals in detail in the penultimate chapter.

come from nonexistent things than influences that come from the past. Therefore, journeys into the future should be easier to determine than specific memories, which can be overlaid by many past and present factors. Indeed, the preview can be determined only by the present situation; thus, an experimenter can exercise much better control over it. However, he still needs clear behavioural criteria, which in turn must be based on clear definitions.

All the previous criteria were created for the act of *planning*. If decisions are made in the process, which are followed by actions, one can speak of *planned actions* [515]. While this excludes mere imaginings and leaps of thought from empirical investigation, evidence of planning at least proves some form of mental journey into the future. Abstract planning is a cognitive ability that has a long tradition in human psychology and cognitive science; after all, following a plan is one of the essential criteria for rationality and intelligence. Planning is also an important subject of research in neuropsychology, especially because this ability is attributed to "executive functions".

Planning can be defined simply as making decisions in the mind that are future-oriented. More precisely, these decisions are based on a mental anticipation of action steps that seem necessary to achieve a goal. Importantly, several steps and their respective consequences are considered in advance, as are the necessary measures, the required means, and the ways to achieve the goal in the future. In doing so, external sources of information and experience are used. In any case, a decision must be made among several possibilities to become recognizable as a decision. Only then can one assume that a mental anticipation of future action has been made by weighing up various alternative courses of action. One could also say that "Plan B" must be mentally prepared in case the original plan cannot be implemented or expected events do not occur or occur differently. But if plan A is executed, it becomes difficult to determine whether plan B has been mentally prepared. Only if plan A does not lead to success, then plan B can emerge. So, all this is rather difficult to verify.

In principle, planning can also involve a single step, but there is also danger of blurring the difference from goal-directed action. Unlike mere goal-directed action, planning involves the goal and consequence of actions and a future situation to be considered mentally. This is especially evident when behaviour is shown that does not lead directly to the goal (we remember the means-to-purpose actions, for example, pulling up a plank or pulling up a string). Here, all actions are directly directed towards the goal. With planning, however, it is different. If one anticipates what might happen in the future or what situation might occur, one can begin with "preliminary work". These activities are clearly different from other activities that would make sense in the current situation, and ideally, they are activities that do not even reveal the goal to an outside observer.

In his essay "Zur Psychologie des Schimpansen" (On the Psychology of the Chimpanzee) [516], Wolfgang Köhler, described such preparatory action as an indication of genuine planning. Cautiously enough, Köhler speaks of recognisable approaches and admits the revision of his earlier pessimistic stance. But he goes even further and sketches out a scenario that could be used to study planned action in chimpanzees. He describes a situation in which a monkey is forced to plan ahead if he or she wants to reach the goal, a coveted food item, on a later occasion.

Since Köhler was no longer on Tenerife at the time of writing and thus could no longer carry out the experiment himself, his theory remained so for a long time. In his Master's thesis, Thomas Suddendorf [517] finally expanded the theory to a two-room scenario in which the decision between different motivations also played a role. Tulving [499] designed such a scenario called the "spoon test". In it, a girl dreams of attending a birthday party. There is chocolate pudding, which all the children eat with a spoon they have brought along. The girl herself has no spoon with her and therefore leaves empty-handed. The next evening, she takes a spoon to bed with her to be prepared for a similar case. Tulving defines three criteria for the test. First, behaviour should not arise from a momentary need or be accompanied by current physiological needs. The child does not take the spoon to bed because it is hungry in the moment. Second, in the planning phase, no stimuli from the environment should cause or guide the planning: There should not be a cup of chocolate pudding by the bed. Third, the behaviour should satisfy a future need. The spoon should be used exclusively for eating from a cup and should have no other function when getting into bed. The spoon could be a gift or a souvenir, something the child likes to take to bed.

Authors who are particularly critical of the possible capacity for planning in animals later added further criteria to these studies. In animals studied, planning should not be based on either associative learning or innate responses [518]. Therefore, animals should be confronted with problems in which different solution strategies are required so that innate behaviour can be excluded. Furthermore, the planned action should become visible in a single act to exclude learning by trial and error. In addition, the action directed into the future and its consequences should be clearly separated in time and space to avoid solution cues or long-term memory strategies. Finally, the ability to plan should be possible in different behavioural categories—not only foraging—so that a certain flexibility and generalisation can be recognised.

The Swedish author couple Mathias and Helena Osvath from Lund University (Sweden) also insisted that the motivation that prevails during planning should be different from that during goal attainment [498]. It is not enough that the drive, let us take hunger for example, only occurs in a reduced form. For example, one could demand that the animal not be hungry during planning (a demand, however, that is not always fulfilled even in the case of people who go shopping). However, the Osvaths go one step further and demand that a conflict of motives should occur during planning because it is precisely this that underlines the functional value of the ability to think ahead [498]. If future rewards are greater than those currently available, the current rewards should be suppressed in favor of future ones. This, like the Marshmallow test, is about self or inhibition control. However, it would be better if two strong motives were present during the time of planning and if the animal chose the action that involved the future reward. We have already discussed an experiment with Goffin's cockatoos in which they were confronted with a range of tool choices and associated uses [162]. For the economically optimal decision, they had to weigh the qualities of food, the accessibility of food with and without tools, and the effectiveness of their means. We learned that they preferred a tool over a directly available reward, only if the preferred food was in an apparatus that could be opened with that tool. The competition of motives was thus between immediate food intake and tool use, perhaps also motivated by manipulative pleasure. In any case, a greater amount of work was accepted only if the later goal was more rewarding and could be achieved with the available means. The Osvaths argue that in the case of anticipating a future motivational state, episodic elements come into play. Thinking about what it will be like to be hungry the next day, they argue, requires that one has oneself experienced what it is like to be hungry in the past. These memories of previously experienced states made it possible to project the state into the future. Semantic knowledge is not enough to suppress a current drive.

The literature is dominated by experiments on the ability to plan, in which a need must be anticipated in the future but which is not present in the act of action-planning because it has just been satisfied. These so-called specific-satiety feeding tests are based on the consideration that the actions performed cannot then be motivated by existing drives but solely by the idea of the need at a later date [501, 519]. But general, semantic memory is sufficient for this purpose. The "cold" knowledge that one always gets thirsty on hikes and that it is good to prepare oneself accordingly is enough motivation to fill the water bottle. One does not have to put oneself in the state of being thirsty. Therefore, the Osvaths suggest a self-control procedure. Instead of suppressing and waiting for an urge triggered by a stimulus, this requires two stimuli to be present simultaneously. These evoke two different drives or needs between which the individual must choose. Unlike in satiety tests, the need states are not reduced or absent but induced and present. The competition between the two existing need states opens up the possibility that the future state will be linked to an episodic experience. If prior experience of some aspects of the future state induces the motivation to work towards its attainment, this would be an example of linking a first-person experience to the future event. Information about the emotion from an event retrieved from memory that corresponds to the future event defines a key aspect of an episodic foresight system [500].

The first reports on planning experiments in animals appeared in the mid-2000s. On the one hand, owing to the still widespread opinion that only humans are capable of planning and imagining future events and, on the other hand, challenged by Tulving [499] to use his spoon test as a paradigm for demonstrating planning, Call and a collaborator conducted an experiment with five bonobos and five orangutans [520]. All the apes first learned to open a box with a specific tool to reach food inside. In the first experiment, three individuals of each species (the others were tested later) entered the test room where the box was visible but not accessible. There were also several tools there, six unfit ones and two fit ones that resembled the familiar tool from the pretraining. The subjects had 5 minutes to choose a tool and were then let into another room. In this waiting room, they had to wait for an hour before they were allowed to return to the first room. There, the box was now accessible, but there were no more tools. Like the girl with the spoon, they were supposed to have taken the right tool into the waiting room before bringing it back to the test room, so that they could now open the box with it. The apes showed this apparently planned behaviour with disproportionate frequency (70%). In a second experiment, the waiting time was drastically extended. The two most successful apes in the first experiment, a bonobo (Kuno) and an orangutan (Dokana), were allowed to return to the test room only in the morning of the next day, after they had spent the night in their sleeping room. The two apes also successfully completed this task. Although they did not take any tools with them in the first experiment, they almost always did afterwards. Another experiment checked whether the animals would take tools with them even if the box did not contain any food or was not present at all during the selection of the tools. This was to check whether the animals only took the tools with them in the earlier tests to satisfy their momentary hunger. However, the animals also solved this task, albeit not as impressively. Finally, naive individuals were used to check whether the apes in the earlier tests actually made a connection between a tool and its subsequent function in solving the task. These animals were rewarded with a treat for taking and returning the correct tool, much like a pigeon in the Skinner box was rewarded for randomly picking a slice, but after returning, there was no box at all, so the tool had no functional value at all. Had the animals perhaps only learned that taking the tool that matched the available apparatus when selected would be rewarded? In other words, had they learned this behaviour only associatively, as a connection that occurs, as it were, blindly through reinforcement ("Take tool A when you see box A!"), but not anticipating and knowing the functional value of the object as a tool for performing a later action? In fact, these animals could not keep up with those in the third experiment, confirming the assumption that the apes had selected and taken the tools for later use rather than merely for an arbitrary reward.

A similar experiment with chimpanzees and orangutans, conducted by Mr. and Mrs. Osvath, produced stronger data in relation to action-planning of great apes because it also introduced controls for the motivational state of the animals [498]. First, the animals were given the opportunity to suck fruit juice from an apparatus using a straw. Then, in another room, they had to choose one of four familiar tools, one of which was the straw. The subjects could then take the chosen item to an area where they were with the other animals in the group. After 70 minutes, they were allowed to return to the test room where the apparatus was now installed. Only with the straw could they feast on the fruit juice. Unlike in the study by Mulcahy and Call [520], these apes succeeded almost flawlessly in choosing the right object (the straw), and the rate of retention and return was also high. A second experiment was conducted to determine whether the animals would suppress their current motivation in favor of later tool use, which would lead to a greater reward. The procedure of the first experiment was identical, but in addition to the objects, a valuable fruit, a grape, was now offered. However, the animals were not very impressed by this, and the straw was chosen much more often. A third experiment tested whether the planned actions were not based on learned associations. The straw could have acquired a positive value through its frequent previous successful use, which would explain its preferred choice over the grape. Therefore, when the animals had chosen the straw from the range of tools, the experimenters offered them the same items and the grape again, afterwards. The animals now invariably chose the grape and not the same tool again, which they already possessed to solve the future task. The fourth experiment tested behaviour towards a completely new object that could be used as a tool. This was offered at the same time as other objects, some of which resembled the familiar straw from earlier experiments but were unsuitable. This experiment was repeated several times, always using new combinations of new-suitable and familiar-yet-unsuitable objects. Once again, the animals were not fooled, and the suitable tool was chosen significantly more often. Thus, the authors concluded that apes have the capacity for episodic foresight.

As already mentioned, one objection to these experiments is that action-planning, if at all, is shown only in the narrow context of tool use. In fact, experiments with chimpanzees on planning social episodes were much less successful [521]. In these experiments, the animals had to choose and store objects to be exchanged for treats later. These animals not only performed very modestly overall but also performed significantly better when tested in a tool trial along the lines of Mulcahy and Call's study. This study, however, must be taken with a grain of salt since, on the one hand, it had certain methodological problems, especially a strong dependence on previous training, and on the other hand, a similar study produced significantly better results [159]. Finally, chimpanzees were even able to show in further studies that they were capable of planning in the context of confrontations and even of deceptive manoeuvres [522, 523].

Another often made objection relates to the artificiality of laboratory experiments. Is there any evidence that animals plan actions when they are not encouraged to do so by experimenters? Do they do so spontaneously? The answer to this question is of enormous importance for the evolution of the ability to foresight and planning, because only then can we understand, or at least guess, why this ability arose, and what it was useful for. Until around a decade ago, there were only very few indications of this phenomenon, and they were only mentioned with reservations. It took one chimpanzee in a Swedish zoo changed the picture.

Three years after this male, named Santino, took the dominant position in his group, he started throwing stones at visitors [522]. This hostile behaviour is not unusual in itself; dominant animals are repeatedly observed to exhibit it, apparently to consolidate or even increase their position in the group. However, chimpanzees in zoos usually do not exhibit this behaviour towards visitors to assert their rank over them but because this behaviour consolidates their position within the chimpanzee group. It goes without saying that zoo keepers do not enjoy this. Therefore, the keepers tried to remove all the stones from the outdoor area, whereupon Santino started to collect stones from the moat surrounding the outdoor area. He collected them early in the morning, before the zoo opened, and stacked them in small piles at the edge of the moat that separated his enclosure from the visitors. Hours later, he would use them as projectiles. When he ran out of these stones, he broke off splinters from an artificial rock in the outdoor area. The concrete that had been used to build the artificial rock started to loosen due to the freezing of water that had penetrated it. The chimpanzee took advantage of this by first finding hollow spots by careful tapping and then breaking off the detached surface section in discoidal pieces with strong, well-aimed blows.

Three features of this behaviour are indicators of a planning approach [522]. One is that the gathering of stones and the chipping of concrete slices occurred with a calm mind, clearly different from the aggressive and emotionally charged state during throwing. This fulfils an important criterion of planning, namely, that the prevailing motivation during planning differs from that during goal achievement. The second is that a clear temporal and spatial separation between the preparatory action and the target action was fulfilled. The throwing of the missiles happened hours after collection, and at least the production happened at a different location. Finally, neither the collection nor the knocking off of the concrete splinters was influenced or even induced by external stimuli because at this point, the spectators were not yet even in the zoo. The entire sequence from creating the missiles, collecting and preparing them, to throwing them corresponds to an ideal and typical planned action, which impressively refutes the Bischof–Köhler hypothesis. Since the collection of the missiles was observed and documented more than 50 times and the concrete discs were broken off 18 times, the phenomenon can be considered certain and removed from the realm of the merely anecdotal. Neither was this behaviour observed in any context other than in the course of throwing, nor did it occur outside the visiting hours and season. The position of the collection points also suggested that these actions were carried out only and precisely for the purpose of later throwing.

Certainly, the observations and analyses of Santino's behaviour cannot explain how rock collecting originally arose. However, in 2010, previously unobserved behaviours involving both deception and novelty were documented in terms of missile planning in Santino [523]: The chimpanzee began concealing his collections of missiles and deceiving keepers. He made haystacks that resembled rock piles and he hid the rocks under such haystacks. Moreover, the chimpanzee seemed to consistently combine two misleading strategies: hiding missiles and suppressing dominance behaviour. The creation of forward-looking concealments suggests that chimpanzee can imagine the behaviour of others—in this case, zookeepers—while they are not (yet) present. In doing so, the chimpanzee takes action in relation to keepers' potential future behaviour. These results suggest that the chimpanzee recombined episodic memories into perceptual simulations of the future.

The act of gathering for later use is not unknown in animals. Gathering and hiding food in autumn for consumption in the food-poor winter is a hallmark of many food-hoarding animals, such as jays or squirrels. However, this behaviour, which is essential for survival and is exhibited by all individuals of the respective species, represents an innate behaviour that has evolved over millions of years, in which intention and arbitrariness—as far as we currently know—are absent. Collecting nuts in trees, on the other hand, to later reach the nutritious inside by hitting them with stones on the ground, can almost certainly be categorised as a planned action. Stones and nuts are collected not only by chimpanzees but also by capuchin monkeys, and are subsequently brought to suitable places where the target action takes place [524].

In all likelihood, the apes and monkeys are already hungry while collecting the nuts in the trees, which means that the motivations of the preparatory and the final actions are not separated. However, would animals also collect and store items for food acquisition when they are already satiated? To this end, an orangutan in a zoo was tested after ingesting an amount of juice previously determined to be sufficient to satiate him [513]. A cardboard box was filled with juice and placed outside the grid in another enclosure. A piece of tubing was then hidden in the orangutan's enclosure, which he found by accident. An hour and a half later, he was let into the other enclosure where he could reach the juice box through its mesh barrier with the tube. By this time, however, he had already disassembled his tube and made it unfit for extracting the juice. On a second attempt the next day, he was more deliberate in his approach. After he had found some tubing again, he waited until the juice box was accessible and then used his tool. He then kept this with him for a full 3 days, including during the night. Unfortunately, he did this only after he had destroyed it once and then painfully discovered its lack. The tube thus became a sought-after object. In this respect, it would have been convenient to offer the tube in a controlled situation and then food sources that could be tapped with other tools. Precisely with these controls, a mere associative learning of successful action chains was excluded in the laboratory experiments with chimpanzees and orangutans [498, 520].

We previously wondered whether there was good evidence that animals plan actions when they are not encouraged or even forced to do so by experimenters. Examples of this all came from zoos, i.e. from animals that usually have intensive contact with humans over many years and have solved many tasks set or influenced by humans. An additional difficulty for animals is the pressure to make decisions, especially in binary tasks. If by chance they succeed on average in half of the cases, this could well be satisfying and increasing the success rate through obsolete reflection. This could be a possible explanation for the "failure" of chimpanzees in the "forked tube" task. Along with children aged two to four, chimpanzees were presented with a task in which a reward emerged unpredictably from one of two tubes [525–527]. The 4-year-old children immediately and repeatedly held each of their hands under both openings. They were thus able to prepare for two mutually exclusive versions of an impending event. However, 2-year-old children, 3-year-old children, and all the chimpanzees held their hands under only one tube and were therefore successful in half of the cases. While some of the children learned the correct strategy over time and achieved maximum success with both hands, the chimpanzees never did. Does this prove that only humans can develop higher forms of foresight, especially about one of the most fundamental aspects of future thinking, namely, that it is uncertain? Do only we have the ability to prepare for different possibilities and make alternative plans?

Are there no examples from the wild of animals solving natural problems important to them and their survival and making forward-looking decisions in the process? Yes, there are. A good example was again provided by chimpanzees. On the Ivory Coast of West Africa, five adult females were observed over 2 years during the food-poor season to determine what pattern they followed in seeking out trees with faster or slower perishing fruit [528]. Some fruits, such as figs, only ripen for a very short time and are also sought after by many other forest dwellers. Therefore, one must be there early enough in the day, while one can arrive later at those trees that bear fruit with a long ripening period. In fact, females left their roosts earlier—often before sunrise, when it was still dark in the forest—moving to trees that bore quickly perishable fruit, especially if these were further away. In addition, females tended to build roosts closer to these trees. Other explanations for this difference, such as morning temperature, could be ruled out. Apparently, wild chimpanzees flexibly plan the time, food, and place for breakfast, taking into account various pieces of information. The function and benefit of these anticipatory actions are obvious. Early access to nutritious food can simultaneously mitigate the effects of seasonal food shortages and increased interspecific competition.

Semantic Prospecting

The previous section on action-planning in animals focused exclusively on great apes, which dominate research on this topic. This primatocentricity was in turn first broken by corvids. Western scrub jays have already been shown to be capable of episodic-like memory performance [529–532]. If it is true that future planning is closely related to this episodic system, these animals should also be capable of planning into the future. With this in mind, Nicola Clayton and her team at the University of Cambridge (UK) created the so-called "*planning for breakfast*" experiment [533]. The scrub jays were given the opportunity to plan for food hiding. Like many other jays, they hide food as a stockpile when food is scarce. However, how much thought and forethought are involved?

First, the animals were familiarised with two cage compartments. In one of them, in the morning after a night without food, there was always powdered pine nuts as food, which could be eaten but not hidden. This was the breakfast room. In the other compartment, they never got anything after a night without food. After training, in a third, middle compartment, the animals were offered pine nuts in the evening that they could hide. They could choose one of the two compartments from training because, in both, there were containers filled with sand, which they had already used in earlier experiments for burying seeds. The hypothesis was that if they could act with foresight, the jays would hide the food in the compartment where there was no breakfast. This is exactly what they did, not exclusively, but statistically more often than hiding the food in the breakfast room.

Now, however, the objection was that the animals simply had a preference for laying out hiding places where they were hungry. As a rule or preference, this would be semantic knowledge or custom; they would not have to bring in a specific, subjective memory for this and would not have to plan ahead and consider what would await them in which places the next morning. They only associated the spaces with hunger. To counter this objection and prove that the birds not only made arrangements for a future need by preferentially depositing food in a place where they have learned that they will not find anything there the next morning—and thus remain hungry—but also store a specific food in a place where it will not be available the next morning, the researchers conducted a second experiment [533]. The logic of the experiment was based on the assumption that the jays would prefer to have a choice between different types of food at breakfast rather than having to settle for just one. This assumption wa based on the finding that when given a choice between two foods to eat (or store), the jays preferred to eat (and store) the food that was different from the one they had just been presented with [534]. Therefore, the jays were now given breakfast in both rooms but different in each case: dog food in one room and peanuts in the other. They should only be able to deal economically with this new experience if they can consider the future. If they want to maintain their freedom of choice for breakfast, they should rather hide peanuts in the compartment with dog food the night before and vice versa. Again, the jays supported this hypothesis through their actions. The authors concluded that the birds have the ability to anticipate and take action for future needs, thereby challenging the Bischof-Köhler hypothesis.

Another study by Clayton's team aimed to clarify whether the animals also fulfilled another criterion for planned action [519]. Could they suppress the motivation or need at the time of the preparatory action—hiding food—in favour of the future need that occurs at the time of the target action—digging hiding places? Jays mainly hide food that they have not found for a long time; that is particularly desirable. To test the hypothesis that they base their actions on future needs, the jays were given enough food of variety A before they were allowed to hide food of two varieties; A and B. At this point, they were satiated with food A and should therefore be more likely to hide food B, in line with our freedom of choice hypothesis described above. (Since they were not hungry at this point, they would not consume it immediately. This was checked and confirmed in a pre-test.) Before the actual test, the jays were given training: before they were allowed to dig out of their previously created food hiding place, they were always fed with food of the other type, type B. With this knowledge, the jays were supposed to hide food A in the test because they would have both varieties available after digging up food B. With this ingenious arrangement, the researchers put the jays in a conflicting situation, or more precisely, in a competition of motives. Which would the jays hide: the other food that would be economically better from the point of view of their current satiation, or the same food that would be more desirable at the time of future excavation of the hiding place? We would not have learned anything from this experiment if the jays had not this time also shown the behaviour that corresponds to the hypothesis of anticipatory action. Although they had just been given food A, the majority of them hid it—completely against their natural preference. In a sophisticated experimental design, the researchers varied the types of food used for hiding and retrieval and whether the same food was used for these two actions or—as just described—different types of food. With the various controls, the results and, consequently, the hypothesis of anticipatory action could be confirmed.

As impressive as the scrub jays' results were, they in turn elicited the standard criticism that this behaviour was limited to the context of food hiding and that it was therefore impossible to know whether the birds could also plan and act strategically away from it. Human planning, they argue, is flexible and context-independent, ranging from planning a barbecue to making career plans. But the behaviour of scrub jays, remarkable as it is, is more like a laser beam, with a very narrow focus on hiding and retrieving food [535]. This very justified objection was countered by the Swedish researchers Can Kabadayi and Mathias Osvath in a recently published experiment with common ravens [536]. They were able to show that the ravens planned events that had nothing to do with the use of food hiding places but with tool use and bartering. How did the two researchers go about it?

In a total of four experiments, five ravens were tested for their ability to plan in different areas. Each experiment contained two tasks for which the ravens had no innate behavioural dispositions: one technical (tool use) and one social (bartering with humans). Ravens are not habitual tool users, and bartering has never been observed in the wild. In the first experiment, the researchers tested whether ravens would choose either a tool for immediate use or an exchange item for later use. The second item, a bottle cap, became effective only as an exchange item 15 minutes after selection and in a different location.

In the tool task, the ravens were given five training sessions to learn the functionality of a tool (a stone of a certain size) on a box-shaped apparatus, using rewards. They then had the opportunity to learn that other objects, which later served as distractions, did not open the apparatus. The next day, they were allowed to spend some time with the food-filled box but without tools at their disposal. This training served to create a possible incentive for later planning. Afterwards, the box was removed in the presence of the raven. One hour later, the ravens, who were always tested individually, were offered a tray with the suitable tool and three unsuitable items at another location. After they had chosen, they had to wait 15 minutes before the box was installed in front of them. If they chose the (correct) stone, they could open the box and receive the reward. This test was repeated 14 times, but only in the first test were the ravens shown the food-filled box in advance so that they could see what the stone was useful for.

In the exchange task, the birds were given 35 training sessions so that they could learn that an item could be exchanged for a treat. The barter partner was one of the two researchers who asked for the item and, upon receipt, handed the treat in return. To create a planning incentive, the ravens were exposed to another researcher who wanted the exchange item when they did not currently possess it. (For the ravens were to experience that there are times when something is needed but is not in possession). Therefore, the ravens were to conclude that they should take appropriate precautions, namely, to keep things that might be needed in the future. As with the tool test, 1 hour later in a different location, the ravens were allowed to choose the exchange item or one of three other distracting items from a tray offered by a new person. But unlike in the tool test, the ravens were allowed to choose from three trays in a row (to make this study comparable to some studies on great apes), allowing them to collect and later use up to three exchange objects. Again, the test—there were 12 in total—took place 15 minutes later in a different location.

The results all showed impressive performance by the ravens. They had the right tool available 86% of the time on average, and an exchange object available 91% of the time on average. Moreover, they chose and collected 143 out of the 144 barter objects offered. What is particularly impressive and important for the interpretation of these results is that all the animals were successful in the first experiment. In the second experiment, the researchers extended the time between choosing and using the object in both conditions to 17 hours. In each of the six tests, the animals were successful at least five times and again achieved success rates of 90% and more.

The researchers then tested the ability to plan with self-control. Would the animals ignore an immediate, valuable reward in favor of an item that would give them access to an even more valuable reward later? In ten control trials, when the animals were presented with only one piece of food on the tray, they consumed it without hesitation. When they were presented with the food piece in another 14 trials per condition—either together with a functioning tool or an exchange object and some distractions—they chose the object for later use in three quarters of the trials and the food piece in only one quarter. In the fourth and final experiment, the researchers tested the hypothesis that the frequency with which the ravens prefer the functional object to the immediate food reward can further increase if they have to wait less time for it. This is because the closer the larger reward comes to the time of receiving the smaller reward, the more willing they should be to accept the delay. If, on the other hand, the item, be it the tool or the exchange item, had intrinsic value, for example, because it made a large reward possible often enough, then it would be chosen regardless of these delay variations, and there would be no difference from the third experiment. The ravens behaved as the researchers had predicted for they preferred the objects to the immediate reward in all cases where the delay was shortened. You could say the ravens took into account the temporal distance to future events.

Not surprisingly, these results provoked opposition. The birds had been trained in advance with rewards to prefer a proper tool and an exchange item over distraction items [537]. In the test, they chose these items on the first attempt but without knowing whether they would later receive a reward for them. Thus, this choice seems to be merely attributable to learned associations from training. As a control experiment, one would therefore either have to reward the distraction objects as well, or destroy the apparatus in front of the birds so that the tools would be unusable. Flexibility could be tested by making the same objects usable in one future context and unusable in another. Additionally, the fact that the ravens chose the functional tool even more often in the fourth experiment, in which the future event occurred earlier than in the third experiment, could be attributed to a consequence effect—the animals simply learned from the previous experiment. It would be better to either have a control group in which the shorter time interval of the fourth experiment is tested before the longer time interval of the third experiment, or to let the ravens experience both time intervals alternately. Critics believe that the animal studies to date have not met the conservative criteria for demonstrating competence in children [527, 538]. They cite a total of four criteria for this. First, animals should be tested only once at a time to prevent a continuous strand of experience with the same stimulus–reward relationships and to prevent memory of a past problem that can be used on a particular future occasion. Second, one should pose several problems from different domains to test flexibility and cross-domain competence. Third, one should test animals with different temporal and spatial distances between the time and place of decision and future action to prevent solution clues and demonstrate the ability to link past and future events. Finally, one should create novel problems to eliminate the effect of learning and to require the use of future-directed cognitive processes [538].

In their response, the authors of the raven study emphasised the importance of the following distinction. One can plan from long-term knowledge and make preparations with indefinite expectations without planning for a specific, predicted time in the future. Similarly, one packs an umbrella when one sees dark clouds coming, but one cannot know exactly when it will rain. The ravens likewise had taken preparatory actions, with the learned knowledge of what works and what does not. In addition, they were not immediately brought from training into the test situation but were confronted with a situation an hour before the test, in which they could have solved a problem with the tool or with the exchange object but they did not have either with them. The ravens learned from this shortcoming and thus chose the tool or the exchange item in the following test to be prepared for further occasions.

Overall, the objections are coherent only at first glance because, upon closer inspection, future-oriented behaviour requires both experiences (whether specifically trained or made by chance) and indications of the future event. Without experience with tools, planning the use of tools is not possible. It would then be only a matter of imagination or speculation, not planning. These experiences or memories may have been acquired in different ways and thus may also be of different qualities or intensities. Incidental, one-time or accidental events are remembered less strongly than memories of events or activities that happened intentionally and repeatedly. In both children and several great-ape species, the second form of experience has been shown to be of greater value for later tool activities and planning [539].

As far as the temporal distance to past events is concerned, it may well be years in the past and may have happened only once to be retrieved from memory and be effective in finding a solution to a problem, including a planned action. As we will see in more detail during the chapter on episodic memory, chimpanzees and orangutans were able to remember a tool-finding task that happened 3 years ago [540]. As with autobiographical memory in humans, an associative retrieval process set in motion by certain stimuli could trigger the apes' memories. The requirement of perfect context exemption as a criterion of proof of future-oriented action must accordingly be dropped. The action must be "triggered", as they say, and thus must evoke experiences, knowledge, and possibly episodic memories to prepare solutions for future events in the here and now.

Not only must there be stimuli for the evocation of memories, but stimuli are also necessary for planning. Unlike dreaming, which actually takes place without sensory influence, the act of planning is determined by the current situation and present influences. There must be a reason why one begins to plan. For example, a chimpanzee finds a stone suitable for hammering nuts, which leads him to consider whether he should now look for nuts, whether he should save the stone for later use, or whether he should even take it to an anvil to be prepared for later use there [541]. Conversely, the ape could find nuts and plan further actions in the absence of a stone.

The objection that the act of planning should take place in a motivational state different from that of the future target action is justified but not compelling. It is certainly favorable for proving future-oriented decisions but not necessary for their existence. The decision depends very much on the actions that are performed as a result of planning considerations. If a chimpanzee finds a kola nut and thinks about what it has to do to get it open and thus sets a chain of actions in motion, this does not have to begin in a different motivational state than in the future target action of hammering.

Preparatory actions are not the same. These are typically motivated differently than the final actions. When people go shopping for the weekend on Friday, they do not necessarily have to be hungry. In fact, it is usually better for more economical shopping behaviour if they are not. The chimpanzee in the Swedish zoo does not have to be aggressive in the early morning when he hoards stones for the afternoon. In his case, the suppression of motivation is concomitant with planning; it allows economic action and makes time travel even more successful [512]. However, it is neither necessary nor causative for patients with episodic amnesia, as intact self-control has been proven [542]. Thus, even animals, which often act impulsively and are only partially capable of self-control, are not necessarily "trapped in time", as one critical author remarked [543].

In summary, foresight, future-oriented decision-making, and planning require a variety of cognitive abilities, including good (executive) control systems to suppress current sensory input in favor of an unobservable goal and means-to-an-end thinking to decide which actions would eventually lead to the desired goal. Finally, decisions must be based on a well-developed memory. To avoid being stuck in the present, the animal must be able to recall experiences and memories to project them into the future. Experiences are based on semantic knowledge, which is acquired through learning. The greater this type of knowledge is, the more extensive and long-term the plan can be. John F. Kennedy could not have announced grandly that he would send a man to the moon within a few years if extensive preliminary work had not already been done for this purpose. Individual career plans can also be made on the basis of knowledge about training paths, job opportunities, one's own abilities, the experience of others, and so on. It therefore seems quite possible that the planning ability of animals is limited more by the lack of semantic knowledge than by the lack of memory of subjective experiences.

Despite the importance of semantic memory, research on planning and foresight has been overshadowed by the question of the role of episodic memory, self-referential consciousness, and subjective experience. Today, both subjective memories and ideas about future events tend to be based on inaccurate constructs of past impressions rather than faithful reconstructions of experiences at the time [544]. Episodic memories and semantic experiences merge. These new constructs become part of planning and thus blur the distinction between memory and foresight. The more extensive the store of experience is, the more easily episodic memories are retrieved and the better they can be recombined; additionally, the more far-sighted and creative an animal will be able to act. Let us now turn to episodic memory.

Chapter 6
Episodic Memory

A paradigm shift began in memory research when Endel Tulving [502] contrasted episodic memory with general, fact-based memory, which he called "*semantic*". For him, only episodic memory refers to the past, while the other refers to the present. What was revolutionary here was the idea of "*mental time travel*" through subjective time. For Tulving, episodic memory is a recently evolved, late-developing, and early-deteriorating past-oriented memory system that makes mental time travel through subjective time possible [545].

As we have already discussed, mental time travel enables the possessor of episodic memory not only to remember one's past "thoughts about" experiences through the medium of autonoetic consciousness but also to "think about" one's possible future experiences. The adaptive function of the episodic memory system lies less in keeping accurate records of the past and more in what these memories can offer to present and future fitness [499, 501, 503, 546]. Almost all subsequent work on memory in humans and animals has adopted Tulving's separation of it into semantic—effectively our database of knowledge about the world—and episodic—a collection of temporally dated episodes or events and their temporal–spatial relationships—memory. However, how autonomously and independently these memory systems work, how great the functional overlap is, and what influence they have on prospective thinking are still disputed today. In particular, the involvement of autonoetic consciousness and the mental conception of time are highly controversial. Some define episodic memory as only the kind of memory that allows people to reflect on their personal experiences [547]. It is clear that this makes comparison with animals difficult, if not impossible.

We therefore want to put aside the question of mental-time travel and the associated subjective phenomena for the time being to look at current research on episodic-like memories in animals. In the long list of studies on memory in animals, one is quickly struck by the great variance in memory performance, even within the same species. In some tasks, animals seem to remember relevant events in detail and for a long time, while in others, they seem to forget surprisingly quickly. For example,

L. Huber, *The Rational Animal*, https://doi.org/10.1007/978-3-031-60803-2_6

jays can remember the location of hidden food for many months [548] but have difficulty remembering simple colour cues for more than 40 seconds at most [549]. How can we reconcile these large differences in memory performance?

For some, the large differences are evidence of the problem of laboratory experiments involving overly artificial stimuli and tasks. For others, they are an indication of a very limited, domain-specific memory capacity. Instead of a single system, animals could therefore possess two types of memory systems. In addition to having a *general* working memory for all types of stimuli and events, some animal species would also have a *specialised* memory system that reflects adaptations for processing specific types of information [550]. These might be adapted to different tasks in different ecological niches and to the times and needs required for these tasks [549].

What, When, and Where?

Some animals became famous for their achievements in hiding food. Both the duration of several months and the number of hiding places, which can be several hundreds, testify to highly specialised adaptations [548]. In rodents, the memory of having consumed food, which lasts for hours depending on the need, is impressive. Apparently, it is necessary to associate possible toxic effects of novel foods with the memory of eating them, because such effects often only appear several hours after consumption. The famous experiments by John Garcia (University of California, Los Angeles, USA), which collapsed the behaviourist learning theory that had been valid until then, showed that rats remember specific features of taste without having to go through long learning sequences [551–553]. Birds can use visual cues, instead of olfactory and gustatory ones, for memory formation [554]. Finally, we know several types of imprinting that can also be considered as products of a specific memory system. In sexual imprinting, for example, recognition features of potential sexual partners can lead to lifelong memories [555, 556].

In contrast to specialised systems, the general memory system seems to be very similar across species. It can store information about many types of events, but can store them only for a very short time, from a few seconds to a few minutes. Standardised test procedures have been developed for use in the laboratory to demonstrate and determine general memory capacity in many animal species; the best known is the "*delayed-matching-to-sample*" test. A single stimulus is presented for a short time, followed by a stimulus-free pause, and subsequently followed by the presentation of two stimuli, one of which is identical to the sample stimulus [557]. The task is therefore to remember the first stimulus, the sample, over the stimulus-free pause and then to recognise it again in the presence of a distracter stimulus. By manipulating the stimulus-free period, one can determine the recognition rate as a function of memory duration. It is assumed that the general memory system plays the main role here since the animals must remember artificial stimuli such as abstract pictures or coloured lights. However, the task can also be solved with the specialised memory system if the recognition feature is taken from the natural stimulus area.

For example, rats, with their distinctive spatial orientation, can draw well on positional features and even, on average, remember them longer than most primates (e.g., [558, 559]).

The question now arises whether the specialised memory system is able to store more than just individual sensory features of an object or place that needs to be found again after a certain time. In addition, are animals, for example, still able to remember an event of hiding itself? Quite independently of the subjective, personal experience of this, quite a few think that animals lack a system for explicit memories of past events [543, 560]. Only humans are said to be able to remember virtually arbitrary events for very long periods of time without a noticeable decline in performance, especially during tasks in which most other species perform poorly [561]. According to other researchers, this is far from a foregone conclusion, as there is now sufficient evidence that (some) animal species also have "episodic" or "episodic-like" memories that are not very different from those of humans [532, 562, 563]. What is clear is that demonstrating consciousness is an enormous obstacle to demonstrating episodic memory in animals because, in the absence of a well-developed language system, it is not clear what such evidence should even look like (but more on this topic later in Chap. 9). Therefore, what is the evidence for "*episodic memory*", i.e. memory processes with analogous, functionally similar properties in animals?

What functional properties would the memories have to have to be "episodic", in other words, what exactly has to be remembered and how would the remembering be expressed in ananimal's behaviour? First of all, episodic memories must refer to past events (episodes). Experiences, on the other hand, refer to current knowledge, even if the act of experiencing or acquiring it happened long ago. I "know" where my house keys are, but I do not have to remember the act of putting my keys down the night before. I could put the keys in a different place the next night and know where they are the following day. My knowledge is updated, but the past events disappear from my memory. This is what Tulving and Lepage called [564] "*proscopic*" (forward-looking) memory and distinguished it from "*palinscopic*" (backward-looking) memory, where the past event is remembered (I remember that yesterday, when I came home, I first took off my shoes and then hung my keys on the key board).

In humans, this distinction was tested experimentally by asking participants, after reading a list of words, to indicate in a subsequent test which words were not on the list [565]. They were also asked whether they actually remembered the words they perceived as not being new or whether they only seemed familiar. In fact, the participants were not only very accurate, but the majority also said they remembered reading the words. This correlates with episodic memory variables (such as word frequency and divided-versus-undivided attention) [547] as has it been confirmed in studies of amnesia patients [566]. This distinction between familiarity and memory will concern us in several of the animal studies that follow. However, before that, it is necessary to consider the other features of episodic memory.

On the basis of Tulving's description of episodic memory, it became apparent that the memory of episodes contains at least three types of information: the content of what happened, the time at which it happened, and the location of the event [502, 567]. *What*, *when*, and *where* are the three central components of the memory of a personal experience, thus, they should be used as behavioural criteria for episodic or episodic-like memories [529]. As I have already exemplified in the chapter on the power of imagination, I can recall some of the circumstances of my visit to the Royal Society in London for a discussion on animal social intelligence. The what and where are very much in my mind, but the when is only vague; it must have been in the spring of 2006. Semantic memory helps me here because, first, these discussions usually take place in the spring, and second, I was busy in 2006 with the very research on imitation that I was discussing in London.

While adult humans can report their memories, animals and infants cannot. We have to use indirect methods to extract evidence of memories. For example, if infants are shown certain bodily movements and they later copy some of them faithfully, we might assume that they must recall certain episodic aspects of their experience [568]. The key point here is that the imitation occurred without explicit training that would generate semantic or associative knowledge.

This process is quite different when training comes into play. An animal that has learned in a matching-to-sample experiment to later choose a red comparison stimulus, when the sample is a vertical-line stimulus, and to later choose the green comparison stimulus, when the sample is a horizontal-line stimulus, does not have to use episodic memory but learns these rules associatively. Any learning that could be described as the acquisition of a rule—in this case, "If vertical, then red, if horizontal, then green"—cannot exclude the possibility that semantic knowledge has been acquired, even if the memory of the pattern stimulus contains information about what, where, and when.

Tom Zentall and his colleagues at the University of Kentucky (USA) developed an elegant experiment using pigeons to test the ability to retrieve episodic-like memory content without training, i.e. when prompted suddenly [569]. Since the animals cannot be instructed verbally, one has to use indirect methods, namely, training the animal to respond to one question with a certain response and to another question with a different response. In this case, the pigeons had to learn to peck at the screen when a vertical line was presented as a sample stimulus (Fig. 6.1). Then, two comparison stimuli, red and green, were presented, and only the response to the red stimulus was rewarded with food. In the presence of a horizontal line, the pigeon had to learn to refrain from pecking, after which the choice of the green stimulus led to the reward.

The rules of this conditional discrimination task can be described as follows: "Choose red if pecking had just occurred; choose green if pecking had not occurred". The choice of the red or green comparison stimulus would thus be analogous to the question, "Did you just peck or did you just refrain from pecking?" One can thus use this question to test the semantic knowledge of the animal, the learning of the associative rule. The preceding presentation of the vertical or horizontal line should accordingly arouse an "expectation" that, in the case of a correct answer—pecking or not pecking—a decision would have to be made between red and green, and the correct choice would be rewarded.

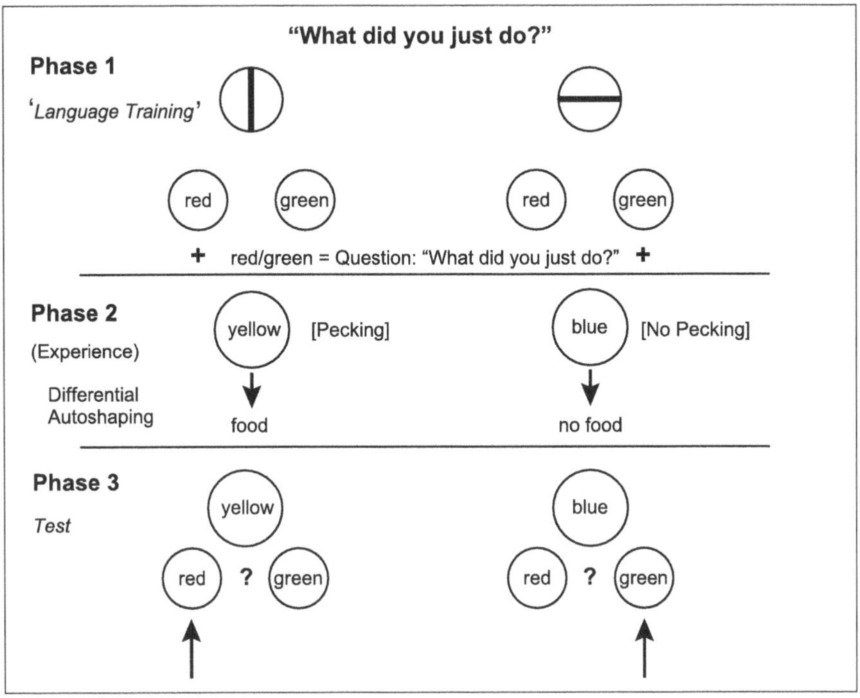

Fig. 6.1 The unexpected question test for episodic memory. (From [570])

The next step is to have the pigeon perform a behaviour that does not trigger a question-expectation. If I eat something today, I usually do not expect to be asked about it the next day. The pigeon experiment is simpler because the behaviour being asked about was again: "Did you just peck or did you just refrain from pecking?" To create a situation in which pigeons peck or refrain from pecking, the researchers use the principle of "autoshaping". Based on Pavlovian conditioning, pigeons start pecking at a light spot that appears just before food delivery. Conversely, they do not peck at a light stimulus that never appears in temporal connection with food.[1] This well-known fact was exploited by Zentall and co-workers [569]. They trained the pigeons in two phases. First, the pigeons learned a symbolic matching-to-sample task according to the principle "if vertical, then red; if horizontal, then green". Then, they were given the opportunity for self-conditioning, whereby a yellow light stimulus was used in combination with food and, after a short time, also by the pigeon's

[1] The principle was called *autoshaping* because it is a form of conditioning in which an individual who frequently receives food after a light is switched on then conditions itself to peck at the light, even though the pecking is not necessary to receive the food. The emergence of this conditional response (the pecking) occurs (according to Pavlov's principle) due to the repeated sequence of conditional (light) and unconditional stimulus (food).

automatic pecking response. Alternatively, a blue light stimulus was never accompanied by food and therefore did not trigger pecking.

In the test, the two phases were combined (Fig. 6.1). First, the pigeons were subjected to either yellow or blue light stimulus, and then, the two comparison stimuli were presented in red and green. Accordingly, the birds were asked to peck for yellow initial stimulus and then choose red. Conversely, they should not peck at blue and then choose green. Most pigeons did exactly this.

To counter the objection that the pecking from autoshaping was not spontaneous but induced by food and could therefore have a semantic component, a second experiment was added. Now, autoshaping was instead dispensed with a new stimulus, a circle, which was shown in the test with the expectation that pigeons might peck at it. Which they did indeed do again and again. Equally new was the darkening of the monitor, on which the pigeons then no longer pecked. These two situations followed each other at random. Interestingly, the pigeons then chose the red comparison stimulus if they had pecked on the circle and chose the green comparison stimulus if they had not pecked on the dark screen. These results suggest the ability of pigeons to elicit knowledge about unique and singular experiences when they could not expect a prompt. This might then demonstrate that language is not necessary for creating episodic memories. However, it must be conceded that the pigeon only had to provide nonlinguistic information about those memories for which there were only two possibilities. Even more critical is the time component—the events took place only seconds before, for which working memory would also suffice. Likewise, they did not have to provide any information about where the event took place; everything happened in a small Skinner box. Under these circumstances, it is not surprising that these experiments convinced only a few doubters of the existence of episodic memory in animals.

Imitation and Episodic-Like Memory

Another objection was raised about memory of very simple behaviours, in the pigeon's case whether to peck or not to peck. Other studies have also tested laboratory animals on the basis of their memory of simple events such as object exploration or feeding. Humans can remember a great many and very unusual things, and animals' everyday experiences can also be more complex and content-heavy than moving or not moving. In the case of pets, this is compounded by the fact that they experience quite different things in contact with people. If you consider what a dog perceives in the human environment in the course of its life, it would be interesting to know what it remembers. Unfortunately, you cannot ask him about this, so you have to come up with other methods of assessment. Claudia Fugazza and colleagues from the Eötvös Loránd University in Budapest have developed an imitation experiment for this purpose [571]. First, dogs were trained to imitate arbitrary actions of a human on the command "Do it!" in a "Do as I do!" manner. To ensure that the dog learned about the movement and not the orientation to the object, two actions were always performed on the same object. For example, the demonstrator climbed on a chair or put both hands on the chair. After some time, the dogs learned to perform only the previously shown action on command—in this case, to climb onto the chair and stand on it with all four paws or to touch the chair with both paws while sitting on the hind legs. To ensure that the dog did not already have the expectation of imitating a demonstrated action in the memory test, another action was trained, namely, to lie down on another command. In the test, a new action, one that had not previously been included in the imitation training, was now presented. Then, the dog was led behind a screen for a certain time (either a minute or a whole hour) so that it could no longer see the target object. He was then brought back near the object and asked to lie down. This was followed by the command "Do it!". Only if the dog can remember what it has been shown before can it imitate faithfully.

An earlier study by the same authors has shown that dogs can remember demonstrated actions in the context of Do-as-I-do over 24 hours [572]. The difference now was that another activity was inserted between the demonstration and the command to imitate, which, following the pigeon study with the criterion for remembering, without being able to expect the request to do so. The dogs were even surprised when they suddenly heard "Do it!" command while lying down. This surprise was evidenced by the prolonged eye contact with the demonstrator (in the same way that small children watch spellbound when surprised, for example, during a magic trick).

In the sample video supplied with the publication, it is noticeable that the dog lies down immediately after coming out from behind the screen. He obviously already knows what to do; after all, he has been tested several times with the same procedure. Overall, the dogs seemed to have internalised the imitation game—this kind of training with an owner and lots of treats is greatly fun for dogs. Again, they were asked to repeat an action previously shown by a human. Lying down in between doesn't change the game, it is just another facet. The dog is being tested in the same room in which it was trained, and the object to which the test action refers, for example, the chair, is also in close proximity there. Thus, part of the experience he is supposed to remember is right in front of his eyes. Finally, one must impose the same restriction applied to pigeons, namely, that the action the animal is to remember was the last one shown before the memory interval. No other demonstrated action could interfere with this memory. In any case, remembering the last action shown is easier than remembering the third or fourth last action.

Is There Nonlinguistic Evidence of Journeys into the Past?

In episodic memory studies of pigeons and dogs, remembering was highly contextualised and embedded in explicit training with few choices. According to Tulving, two of the three criteria for episodic memory [567], where and when, were not available at all. The arrow of time into the past ended each time at the last relevant action, the pecking behaviour or the demonstrated action of a person, always at the same place. Contrary to what the Hungarian researchers' video suggested, the dog does not suddenly step on a red ball on its way into town and has to remember this experience later and at a different place. He did not have to remember when and where he saw a person putting her hands on an armchair on the way to the test lab in the university building. Even though the actions of the memory test had not been shown before, they were asked of the same person, in the same room, in view of the same object and as part of a well-trained sequence that recurs in the same form.

More than 20 years ago, dolphins were already confronted with a similar yet much more complex task [573]. First, the researchers trained two dolphins to perform an action that was new or had not been performed recently using a specific hand gesture. These dolphins were used to repeatedly performing new jumps and movements under and above water. In fact, they learned this abstract command and made an effort to always show something new. The dolphins were also trained with another hand gesture to always show the last action performed. Finally, the two commands were combined. The gesture "Do something you have not done before?" and the corresponding new action were immediately followed by the gesture "Repeat the last action!" To be successful in this, the dolphins had to use their own behaviour as the basis for the second action rather than using the gesture from the previous trial as the basis for the current response because the previous gesture only showed that the dolphins could perform a relatively new response. In fact, the dolphins were able to perform this sequence correctly. In a sense, they fulfilled the criterion proposed by Thomas Suddendorf and Janie Busby [560], namely, the re-enactment of events through mime as nonlinguistic evidence of subjective time travel. However, even the dolphins did not escape the previously mentioned problem: as with the pigeons and dogs, the content of the memory was only the last event, which moreover did not need to be located.

Critics of animal studies always emphasise that people can remember episodes of their lives that happened years ago [574]. It is not just last actions or events in a strict, experimental corset with few possibilities but also things embedded in a vast sea of other events and experiences. With a specific question or a small hint, the memory of an experience can be brought out of the memory. An experiment with 15 chimpanzees and four orangutans comes significantly closer to meeting this requirement [540]. Their task was to remember search actions with tools. In an earlier experiment, two tools were hidden in two places in four individual trials, with these hiding places alternating unpredictably. Three years later, the animals were now asked to find these tools again. Only one of the 15 apes did not search in the two correct locations. To check the likelihood of searching in these places without any experience doing so, a control group was tested in the same way. These apes had not participated in the earlier experiment, so they were clueless. Not a single ape searched—as one might have guessed—in either place. Therefore, the animals in the first group remembered events that had happened only four times and 3 years earlier. In another experiment, the researchers tested whether the apes were also able to remember singular events, and the results were convincingly successful.

An interesting aspect of this study is the way memory was elicited. Unlike people, who can be asked specific questions, provided cues had to be specific enough to distinguish the event from other similar events. Due to the unique combination of rooms in which the apes were tested, the experimental setup and the experimenter fulfilled this function. However, unlike in the imitation experiment with dogs, the monkeys were also tested by the same experimenter in other tasks (for example, tool use, object selection paradigms), the same rooms were used for other tests with other experimenters in other experimental setups (for example, cooperation tasks, tool use tasks), and other experimenters presented the objects used in other experiments. Thus, each feature was not specific enough on its own; only in combination were the features sufficient. The animals therefore had to combine these elements and ignore a number of irrelevant associations to be able to activate memory for the relevant event. The linking of relevant temporal and spatial components as triggers for the retrieval process is also an important feature of autobiographical memory in humans.

Previous studies have examined the memory of apes for singular events after a relatively long time. In the 1970s, several attempts were made to retrieve pieces of food previously hidden by humans. For example, in front of the eyes of a young chimpanzee clinging to the back of Emil Menzel, a well-known ape researcher at the Yerkes National Primate Research Center at Emory University in Atlanta (USA), 18 pieces of food were hidden one after the other in a 1-acre area. Afterwards, the chimpanzee was returned to its group, only to be let back into the area a few minutes later with all the other animals. In 16 such experiments, the chimpanzee quickly and unerringly found most (approximately 70%) of the hidden food, while the unknowing group members found only 5% through aimless searching [575]. This retrieval performance was subsequently surpassed by that of a 3-year-old male bonobo named Kanzi. In a 20-acre patch of forest at the Georgia State University Language Research Center field station in Atlanta, USA, various pieces of food were buried in 17 widely scattered locations in his presence. Four months later, he was allowed to lead a keeper, who knew neither the hiding places nor their contents, through the forest to find the hiding places. With photos and graphic symbols offered to him at certain positions along the paths, he was able to indicate the places and the food hidden there, respectively. His indications were always correct, and he moved directly and unerringly from one hiding place to the next [576].

Besides showing Kanzi's excellent (location) memory for widely scattered hiding places, the study was also the first to use a keyboard with graphic symbols (*lexigram*), a forerunner of the tablet, for symbolic communication with people. It contained a total of 256 symbols for objects, toys, places, requests, and actions. Kanzi and other apes at the Language Research Centre learned symbols partly by observing already competent animals and used them to communicate easily with their handlers and experimenters. For example, by pressing the symbols for "fetch", "Coke", and "refrigerator", they could tell their caregiver that they wanted a can of Coke from the refrigerator.

A chimpanzee named Panzee also acquired this pre-linguistic competence for 120 symbols [577]. At the age of 11, she was allowed to watch Charles Menzel, a scientist at Georgia State University (not to be confused with Emil Menzel), hide food in the outdoor area from the laboratory building [578]. The last time Panzee had been in the outdoor area where the food was hidden was 6 years earlier. Since her immediate access was blocked, she tried to tell a naïve keeper about the contents and locations of the hiding place, thus asking him to bring her the hidden food. There were symbols on the tablet for different types of food and for places in the outdoor area, and she also indicated with various gestures that she had knowledge of the location of hidden food. She held one hand in front of her eyes, a gesture she had learned while playing hide and seek with the keepers. With the outstretched hand, she repeatedly pointed in the direction of the hiding place and led the keeper to the door to the outdoor area. A special aspect of this study was that her memories were not initiated by others but by herself. She tried to tell the naïve keepers where food was hidden by means of symbols as soon as the opportunity arose (and she did not do this if nothing had been hidden before). More importantly, for the question of episodic memory, she often reported the hiding the next day (once 90 hours later) because that was when she would receive the tablet. Since her memories referred to one-off, episodic events and not to general knowledge about food hiding places, nor was Panzee herself at the places or hid the food there, and her memories were reported in an abstract manner instead of making a selection from a few alternatives, the existence of episodic memory in her was certain for Menzel [578].

A western lowland gorilla named King was also able to display two features of episodic memory, the person (the who) and the content (the what) of a food donation, in the *Monkey Jungle* in Miami (USA) [579]. He also learned to communicate using symbols—wooden cards with graphics. However, the corresponding study has a decisive advantage over the Kanzi and Panzee studies. A language-like report with symbols referred to a unique, non-repeatable event that had passed by the time of the communication and left no traces. Namely, at intervals, King was handed one of five different types of fruit. When he was asked 7 minutes later by his keeper which fruit he had been given, he was successful 70% of the time; if he was asked after a whole day, he even increased to 82%. In a second experiment, he was also asked which of two keepers had given him which of three types of fruit. The accuracy for the correct fruit was 55% and 73% after short and long intervals, respectively, which was significantly above the random probability of 33%. When it came to specifying the person, the gorilla was even more accurate, with an accuracyof more than 80% in each case.

The results of observations and behavioural tests with chimpanzees, bonobos, orangutans, and gorillas are impressive and confirm the assumption that animal species in the pyhlogenetic vicinity of humans are also capable of impressive memory performance. The danger here, however, is that one limits the search for evidence of episodic memory in the animal kingdom to our closest relatives, thus losing sight of other groups of animals, especially those at phylogenetic distance. So far, in this chapter, I have mentioned only vertebrates. But would it be possible to find episodic memories in invertebrates as well? In fact, research on octopuses over the last decade has yielded surprising results. The "what, where, and when" of past events are remembered when cuttlefish forage for prey. It may have depended on what and where they had previously eaten and how much time had passed since that meal [580]. Furthermore, they adjusted their foraging behaviour according to whether they had seen or smelled the prey on a previous occasion [581], and they also had expectations to change their behaviour: If they expected shrimp to be available later, they would eat fewer crabs [582]. In summary, these molluscs are indeed able to integrate experience over time to guide their behaviour.

Perishable Food

The examples mentioned so far contribute considerably to removing the doubt that animals other than humans also have long-term memories. These findings show certain, if not all, characteristics of episodic memory. An important limitation of these studies is that they have not yet addressed the controversial question of consciousness in relation to time. How is time represented in animals?

We humans imagine time as a long string on which pearls are beaded, so to speak. The pearls are the specific events that we remember and that are endowed with more or less detail. These are arranged on the string, if not in an absolute way, then at least in a relative way. We can remember (some) events in an absolute way; that is, we can determine their exact time in the past [483, 501]. However, even if we no longer know precisely when a certain event took place (on Monday of the previous week at 12 o'clock), we can often classify it in relation to other events. To do this, we use semantic knowledge (for example, I only eat breakfast in the morning and can therefore narrow down the time period to breakfast) and memories of related events. For example, although I cannot remember the exact year I attended my first scientific conference, I do know that it was early in my doctoral studies because I was presenting part of my master's thesis. I also remember that it was in Bergamo (IT) and that I was allowed to ride in the car of my PhD supervisor Rupert Riedl.

An episode like the one just described is enriched with many details of what happened, allowing me to make a past-life association without travelling on an abstract time dimension. Therefore, it has been argued that episodic memory does not aim to travel in time but rather to distinguish complex events from each other. This distinction, it was argued, should be based on the association of elements with a particular occasion [583, 584]. The temporal element (the when) of the episodic triad would thus be too narrow a concept and should be expanded to include other, nontemporal features that help identify a particular past event. Judgements about when an event occurred could be derived from a variety of cues. Cues can be information from the apparent distance of the event from the present—for example, memory strength or contextual overlap with the present—and ones that suggest the relative temporal sequence of events as well as associated contextual information (for example, semantic knowledge about elements of the content). Or to think even more radically: there is no natural time code in human memory at all, and therefore, a linear chronology, as suggested by the idea of mental time-travel, is an illusion [585]. Chronological information would accordingly be dispensable and would also relieve us of the obligation to provide episodic memory with evidence of conscious experience.

Can it at least be proven that animals have a relative, perhaps even a pearl-clutch-like conception of time? Can animals at least roughly locate events in the past? Do they have memories that, according to Tulving's original definition of episodic memory [567], store information about episodes or events and their temporal and spatial relationships with each other? In fact, these questions were asked in the first experimental study of episodic memory in animals. Clayton and her team asked their scrub jays the famous what-where-when question [529]. As already described in Chap. 5 (subheading "Planned Actions"), these jays are excellent feeders that not only place their food in hundreds of places during summer and autumn but also find it in large numbers in winter. So it made sense to check what the birds remember when they have to find the hiding places again some time later. Do they remember what they hid, where they hid it, and when? Do they know in combination, or do they remember individual places, specific times, and specific types of food independently? If the memories integrate all three features, are they flexible enough to be used in novel situations that were not foreseeable at the time of storage?

The experiment followed an elaborate plan (Fig. 6.2). Although the experiments were carried out with animals kept in the laboratory, with artificial hiding places ice cube trays filled with sand and forced procedures, the animals' natural abilities could be extracted. The jays were divided into two groups: a so-called Degrade and a Replenish group. All of them were given food in a sated state to hide in the containers. However, the two types of food differed in their popularity: jays particularly like wax worms (the larvae of the large wax moth, a moth of the caterpillar family,

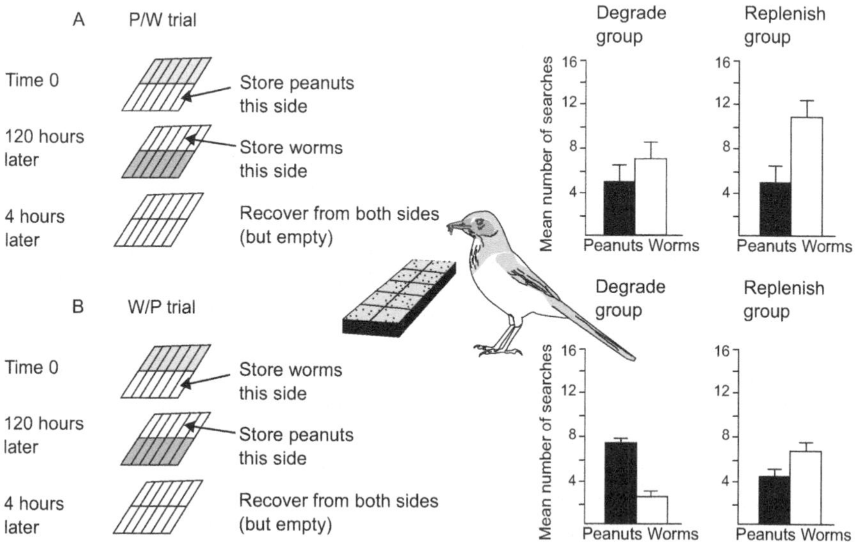

Fig. 6.2 The experimental design and the results of the jays in the Degrade and Replenish groups of the seminal study by Clayton and Dickinson [529]. The left side shows the time sequence of the experiments; the right side shows the results; (**a**) P/W trials (peanuts cached first, wax worms later); (**b**) W/P trials (wax worms cached first, peanuts later). (Adapted from [532])

are particularly rich in energy and protein and are therefore popular feeders) and not quite as fond of peanuts. In a pretraining session, the members of the Degrade group were able to learn that although the dead wax worms were still quite tasty after a few hours, they were inedible after 5 days.

In the first of two experiments, these animals were given a unique opportunity to cache wax worms and nuts one after the other, with one half of each ice tray covered. First, they were allowed to cach nuts on the left side of the container. Exactly 5 days later, they were given the opportunity to bury wax worms in the right half of the containers. So the caching was performed according to a P/W scheme: first, nuts (*peanuts*) and second, larvae (*waxworms*). Four hours after this event, the first test was performed. Once the animals had access to their hiding places again, where would they look? Would they do so randomly, in an innate pattern, or perhaps strategically? If they remembered where they had hidden their favourite food, they would dig it up first. In fact, 80% of the animals first chose the side with the wax worms and conducted an average of seven searches on the right side and only five on the left (Fig. 6.2).

The animals behaved quite differently when the wax worms were hidden first and the nuts 5 days later, according to the W/P scheme. Not a single animal chose the worm side first, and on average, fewer than three of the subsequent searches were on this side (Fig. 6.2). This large difference suggests that the animals knew that the wax worms were already inedible and that they should therefore make do with the nuts. To confirm this, one has to test animals that do not have this knowledge, and this was the purpose of the Replenish group. Its members inspected the worm side in the second case somewhat less often, but in both cases, all the animals were on this side first.

A third experiment was designed to test whether the animals were innately able to handle perishable food. Jays know that their cached food items are repeatedly stolen (pilfered) by conspecifics and therefore take countermeasures: they dissolve some of their hiding places early and create new ones [586]. Therefore, they should know that the more time that passes between storage and recovery, the more likely it is that the caching places will be pilfered. To teach the birds that the probability of the worm caching places being pilfered increases with time since caching, a third group of birds ("Pilfer") was pretrained: they had to learn that their worm caching places would be empty after 5 days; the wax worms were removed from the storage dish by the experimenter. This group should have behaved in the same way as the Degrade group when hiding nuts and wax worms and developed a stronger preference for the peanuts when they were allowed to hide the wax worms first; thus, 5 days would have passed before the worms could be found again. This was also essentially true. Although they showed no difference in their foraging behaviour after 4 hours (the wax worms were cached 120 hours after the peanuts) and showed an equally strong preference for the worm side as did the other groups, they changed their retrieval strategy after 5 days. If the jays in the Pilfer group had indeed learned that the likelihood of the caching sides being looted increased with the time that elapsed between caching and retrieval, by remembering relatively well how long ago they cached the wax worms, then their preference for the worm side should have been reduced in the 5-day experiment compared to the preference shown by the Replenish group. Indeed, 29% now made their first search on the side with the nuts. Remember: Not a single animal did this in the Replenish group.

In summary, the scrub jays knew something about what they had cached (i.e. wax worms or nuts), where they had cached it (i.e. left or right in the ice box), and when they had cached it (i.e. 4 hours ago or 124 hours ago). Moreover, they obtained this information from the experience of a single caching event. Thus, jays seem to be able to recall memories of specific, singular events in their past, namely, the caching of wax worms and peanuts at a specific time and place. Although the individual pieces of information were factual and the memories therefore contained semantic knowledge, they were singular and specific in their combination [529–532].

You could call these events episodes. But were they also experiences? Do the animals and their actions themselves appear in these episodes? Although the results fulfilled most of the criteria for episodic memory, they still cannot prove that the birds use autonomous consciousness in their memory of past events. This autonomic consciousness of self-made experiences, or own experiences, is a special aspect of phenomenal consciousness. The authors admit that this is probably not testable in animals because this form of consciousness has no obvious manifestation in nonlinguistic behaviour. Therefore, they call their findings evidence for "episodic-like" memory and, like Tulving, make "episodic" memory [567], the uniquely human phenomenon.

Some philosophers wonder whether there is in fact no nonlinguistic manifestation of subjective experiences in general nor of the specific consciousness accompanying episodic memories [73]. Does no elaborate theory of consciousness exist that would allow us to ascribe phenomenal consciousness to the jays on the basis of their cognitive performances? I address this question in detail in Chap. 10 ("Do Animals Have Consciousness?"). In any case, Clayton acknowledged the possibility that animals do not have conscious experiences despite considerable cognitive performance. It could be that animals do have intentional states, connect them, and are in states of knowing and remembering because of these connections, but that all these cognitive processes take place in complete or great "mental darkness".

How Much Time Has Passed?

Like humans, scrub jays are diurnal and orientate themselves visually. The individual depressions of the ice cube containers were therefore made distinguishable with coloured Lego bricks on the side so that the scrub jays could orient themselves by these when finding their hiding places again. Rats, on the other hand, are nocturnal animals whose visual system is not as well developed as that of feathered vertebrates. In contrast, the olfactory system of rats is excellent. With their fine nose, they explore areas to detect food. In doing so, they memorise many different smells, which often differ only in nuances. It therefore made sense to investigate the memory capacity of these animals on the basis of smell. In one experiment, rodents learned the locations of different foods within a radial, eight-armed maze [587]. Without visual cues, the animals had to remember the spatial arrangement and locations where the food was deposited. One arm of the maze had chocolate, which the rats preferred, and three other arms had other kinds of food. Later, they could revisit the maze to show whether they had remembered where the chocolate was. Two breaks of different lengths were given: a short 1-hour break and a long 1-day, 1-hour break. It turns out that they found the chocolate more quickly and more safely after the longer break than the shorter one. Apparently, the chocolate's location had been better stored in memory. Next, the chocolate was made inedible with lithium chloride during the longer break. The animals remembered this and began to visit the arm with the chocolate less often. However, if a grape replaced the bad chocolate in the same arm, the previous frequency of visits was immediately restored. The authors concluded from these results that the rats, like the scrub jays in the previous experiments, remembered; the what (the food), the when (found recently or a long time ago), and the where (the arm of the maze).

In another experiment by Jonathan Crystal of Indiana University (USA), it was shown that rats can combine different components of the chocolate-finding event in the maze (namely, where they found it, in what way—did they bring the chocolate there themselves?—and in what environment) into a unified, i.e. episodic, memory [588].

But how do they encode the when; the moment of finding chocolate in their memory? This again raises the question of how animals can locate events or activities in the past. In the triumvirate of Tulving's criteria for episodic memory, the when seems to hold a central position. Do animals have memories whose position—in a string-of-pearls model—can be determined in absolute or relative terms? The clues to exactly when something happened and how long ago were not distinguished in earlier studies of episodic memory in animals. The scrub jays and the rats only had to distinguish between a short and a long time since the wax worms were hidden and the chocolate was found, respectively; they did not have to know whether it was 4 or 124, 1 or 25 hours. Humans can [483, 501] at least for some particular episodes, store the exact time or reconstruct past experiences within an absolute temporal dimension.

How could animals determine when in a relative way? For example, by remembering how much time has passed since the event took place. This can be realised in different ways. For example, animals might remember how long ago an event occurred by estimating the elapsed time using so-called accumulators or circadian clocks, using their own behaviour, or using the strength of a decaying memory trace. Animal episodic memory may be realised quite differently in their brains than human episodic memory is in ours.

Laboratory experiments with rats were used to inspect the content of the when feature in animal memories. In two experiments, William A. Roberts and his team from the University of Western Ontario (Canada) asked whether the temporal cues rats used contained either the exact time of the event, how long ago it occurred, or perhaps both [589] (Fig. 6.3). The event the rats were asked to remember was finding an extremely popular food, cheese, in the eight-armed radial maze.

Training → Test interval	Group: When A	When B	HLA A	HLA B	When + HLA A	When + HLA B
9.00 → 9.30 Uhr	P	R	P	R		
9.00 → 13.00	P	R	R	P	R	P
12.30 → 13.00	R	P	P	R	P	R
12.30 → 16.30	R	P	R	P		

Fig. 6.3 Design of the first experiment by Roberts and co-authors on episodic-like memory in rats. Details in the text. (From [589])

On each day of training, the rats were allowed to enter four randomly selected arms. Three arms each contained two pellets of the usual rat food, but the fourth arm contained the cheese cube. In the test phase, the rats were returned to the maze. All eight arms were opened, and the rats were free to choose among them. The four arms that were closed during training now each contained two pellets, and the three arms that contained pellets during training were now empty. There were two conditions for retrieving the cheese cube; either it was replenished, for the test (R) or not, thus pilfered (P).

The rats were divided into three groups of ten animals each and were given different information. In the so-called When group, the indication of the exact time was relevant, and the indication of how long ago the training took place was made irrelevant and therefore useless (Fig. 6.3). There were two subgroups (A and B) in each of the three groups, which differed in the order of the two conditions (R and P) to check for any influence of the event order. If subgroup A had their training in the morning (at 9:00), the cheese was not put back down in the later test, whether that was immediately afterwards or 4 hours later. However, if the subgroup had training at noon (12:30), the cheese was put back in the test, again whether immediately afterwards or 4 hours later. The opposite design was for subgroup B; thus, a time-of-day effect (whether the "morning or afternoon" condition had an influence) could be ruled out. In any case, the When rats could only expect cheese in the cheese arm, and thus selectively visit it, if they had remembered at what time of day they found cheese there but not how long before. The animals in subgroup A should therefore remember the following: "If you had training at 9:00, you should not expect cheese later in the day; get the pellets from another arm. But if you had training at noon, you can get the cheese there again later in the day".

In the second group, the "How long ago" group, (HLA), the time that had elapsed since the training determined the presence of cheese in the later test. This subgroup A was thus guided by the following rule: "If you had training recently, you can no longer expect cheese. However, if you had training a long time ago, you can get the cheese now". For subgroup B, the other way around.

Finally, in the third group, both cues—when and how long ago—could be used (When + HLA). These animals were only tested at 13:00, with subgroup A being able to find cheese in the cheese arm if they had training at 9:00, and subgroup B if they had training at 12:30. They were therefore able to orient themselves to both time markers.

With respect to the evaluation of the tests, it is important to note that only the tests at 13:00 provided decisive data because only in those was the exact time (when) irrelevant. Indeed, if the rats were tested at 9:30 or 16:30 (the first and last rows in the figure), animals in both the When and the How long ago groups were able to use the time of the test as an indication of the replenishment of the cheese. Two measures were used: first, the probability of returning to the cheese arm within the first four arms visited and second, the mean rank of entry to the cheese arm in successive arm decisions. Early return to the cheese arm resulted in a high probability and a low rank.

The results were revealing. First, the rats in the tests rarely went to the arms that contained pellets during training. They either went to the four arms that had remained closed during training or to the cheese arm. Looking at the behaviour of the animals across all tests, particularly how often and when they went to the cheese arm, animals in all groups went to the cheese arm more often and earlier when it was refilled than when it was left empty. But this picture changes if only the tests at 13:00 are analysed. In this case, the When group failed—they made no distinction between refilled and empty cheese arms—while the distinction was still intact in the other two groups. Apparently, the rats were guided by the length of time that had elapsed since the training but not by the exact time of day of the training. The authors of the study considered that the animals remembered better when the two intervals between training and testing were more different and considered a long interval of more than 1 day in length. However, even in a second experiment with intervals of 1 and 28 hours, the overall picture had little change. Again, the animals in the When group were unsuccessful in the 13:00 test. Thus, these experiments suggest that rats cannot remember at what time of day an important event, namely, the finding of a highly desirable food, occurred, but only how much time had elapsed since then. These findings led the authors to conclude that rats do not have episodic memory, at least not one comparable to that of humans. Humans, after all, can state when an event occurred within a past temporal frame of hours, days, and years, and they can do so in absolute numbers, not just relatively.

Only Familiarity?

One could also interpret the results of the rat study in such a way that the animals do not care when exactly a certain event took place but only whether it ever happened at all. In other words, the rats' memory stores information about familiarity with events, objects, social partners, and other important things. It is not essential to locate the event precisely on the timeline. Human memory, on the other hand, for a previously experienced event, such as the perception of a particular object, may contain both a sense of familiarity with the object and a memory of that perception on a particular occasion. Memory and familiarity can be selectively affected and are thus likely to be based on different neural systems. Many tasks used to test memory in nonhuman animals test recognition, and success in doing so may rely solely on familiarity cues. Without the ability to selectively examine memory and familiarity, until recently it has been difficult to determine the neural basis of these processes in animals.

People have a phenomenal memory for images: When test subjects are presented with a list of 2,500 images, each shown only briefly, they distinguish seen from unseen images with over 90% accuracy in a test given 3 days later [590]. This remarkable ability is undoubtedly due to the highly developed visual system, which has proven to be particularly advantageous in the phylum of apes for the visual recording and storage of objects and scenes.

The familiarity of an image can depend on different characteristics. If a subject is presented with a series of faces of many different people, one after the other, some images will later seem more familiar because they were seen at the very beginning of the list, while others will show even greater familiarity because they came towards the end and, therefore, more recently. Other faces will even seem familiar because they have been associated with experiences or memories. The presentation of a historical figure, take Maria Theresa as an example, will evoke semantic or referential information about the reign of the Habsburgs or Theresian school reform. Furthermore, images might evoke episodic memories. The image of a close friend is likely to evoke multiple memories of shared experiences, specifically ones that occurred at different times and in different contexts. Recent experimental studies with rats have shown that some of these features of our own memories, typical of humans, also are within the long-term memory of these rodents. Rats also seem to be able to remember several episodic events associated with different contexts at the same time. How has this been shown?

We have already learned from studies with scrub jays that information about the what, when, and where of an event can be stored in combination. In addition, information about the existing circumstances, the "context of the event", seems to enter the memory. As soon as animals receive situational cues, memory of an event is evoked. For example, hiding places can be created during the day or at night. When hiding places are found again, animals retain their contents and location depending on whether the place is in the light or the dark. This was tested in an experiment by hiding two types of food, A and B, at the ends of an E-shaped maze but changing their position depending on whether the test was performed in a light or a dark environment. For example, A had been hidden by the experimenters in the left arm in the light situation and in the right arm in the dark situation and vice versa in B. The rats were then made to look for A in the test after they reached certain habituation conditions; they were constantly given only B and thus finally wanted food A. When the test was light, they preferred to search in the left arm; when it was dark, they preferred to search in the right arm. Therefore, they remembered under which circumstances where the food had been hidden [591].

Now, one might argue that this binary task is too simple compared to the natural foraging behaviour of animals. After all, rats can remember a large number of recently experienced odours [592], similar to how humans can store many visual impressions [590]. A large olfactory bulb with many finely branched connections to the forebrain and brainstem appears to enable this feat in rats [593]. Therefore, when asking whether rats can retain multiple episodic memories, one should use more than just two different odours. Analogous to the imagery test in humans, a study with rats tested whether they could remember many different odours along with their conceptual information [594]. The procedure used here, the "odour span task", is typically used to examine working memory in rodents, rats or mice [595]. It consists of an inverted matching-to-sample test in which the number of stimuli to be remembered increases over the course of the test, and only the choice of the stimulus that is newly added in each unit is rewarded—often with a small piece of chocolate. The animal must therefore remember which scents it has already smelled that day. This procedure is usually carried out in a circular experimental arena with wells ("food holes") that are either closed by lids saturated with different odourants or filled with sand mixed with odourants (Fig. 6.4).

In the memory test, the rats could distinguish 70 or more different odours presented in a single test unit [594]. The odours used were commercial fragrances or spices such as cinnamon, lavender, or vanilla. The rodents demonstrated enormous

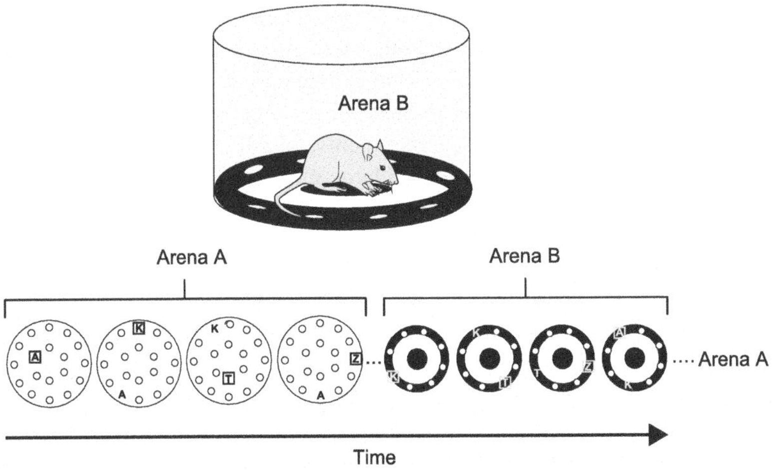

Fig. 6.4 Schematic representation of the sequence of new odours (shown as letters) in Experiment 1 by Panoz-Brown and co-authors [594] with two environmental changes (Arenas A and B). The "new" odour in each trial was rewarded (indicated by a dashed border), while the "old" odour was not. First, rats in Arena A were presented with 8 of 16 odours (only 4 are shown); then, in Arena B, the same eight and eight additional odours were presented as "new", even though these odours had already been presented in the other context. Finally, the rats returned to Arena A and were presented with those eight odours that they had not been presented there before. Each day, 16 odours were randomly selected from a pool of 40 odours, and the odours were presented in random order. (From [594])

working memory, as they selected a lid with a new odour or avoided lids with already-known odours, with an accuracy of more than 80%. Finally, to investigate multiple episodic memories, the researchers used two arenas (A and B) to simulate two different environments (Fig. 6.4).

The large Arena A was white and contained 18 wells arranged in two concentric circles. The smaller arena B had alternating black and white circles on its floor and contained eight wells equally spaced along the Plexiglas wall. In these arenas, the rats always had to choose between two lids placed on randomly selected wells. In total, 40 different smelling lids were used.

In the first of two experiments, the animals switched back and forth between the two arenas, from the large to the small and back, or vice versa. After they had learned to choose the new smell in the first arena, they switched to the second arena and were presented with different smells there again. The experimental test then took place in the first arena. Smells that were present in both arenas or only in the second were tested. Familiarity with these smells should make no difference. However, if they had also remembered in which environment they had perceived a smell, they were to choose the smell they could only smell in the second arena when they returned to the first arena. In the first arena, this would be new and therefore correct. With an accuracy of more than 75%, they chose significantly better than random. These results, which are indicative of multiple episodic memories, were further corroborated with various controls, which allowed us to rule out the possibility that the choice was based on a difference in relative familiarity. To increase the testing of the context dependence of memory, the rats were tested in a second experiment after switching between the two arenas five times and in the end even after 15 switches. When tested, which smell was new then and there, the animal could not predict. Thus, the rat not only had to have all the smells in its memory but also where it had perceived them. These experiments, which are costly for humans and animals, have shown that rats can store a large number of episodic memories and retrieve them specifically when the same event—the smell—is perceived in different contexts.

In humans, one area of the brain, the hippocampus, plays a crucial role in episodic memories. Patients with deficits in this region of the brain have great difficulty remembering past events or experiences [596]. In mammals, the hippocampus interacts closely with virtually the entire neocortex, from which high-level association areas receive highly processed information. Distinct oscillatory patterns specific to the hippocampus (called theta waves and sharp waves) observed on an electroencephalogram (EEG) during rest or sleep coordinate activity between the hippocampus and its association areas for the encoding and retrieval of information that contributes to episodic memories. Originally, namely, in fish, the hippocampus was assumed to represent space (the "where" function). Later, an increasing number of features and information about events were connected, enabling more extensive, vivid memories. Interestingly, unlike the mammalian hippocampus, the equivalent structure in birds receives only visual and olfactory input, not pre-processed information from the higher association areas. Additionally, oscillatory patterns known in mammals, which are involved in communication between the hippocampus and higher-order neocortical association areas, were not found in birds. This has led to discussions about the episodic memory of scrub jays [597].

As I mentioned in the sub-chapter on the power of imagination, in humans, there is a network in the brain that enables streams of memories and daydreams to take place [507]. This so-called default-mode network has also been demonstrated in rats [509]. If episodic memory is indeed dependent on a well-functioning hippocampus, indirect evidence could also be provided by means of targeted disruptions of this brain area [598].

In a 2018 experiment, Crystal's rat researchers expanded upon their previous elaborate testing procedures to include the effects of pharmacological disruption of the hippocampus [599]. The functional expansion referred to the consideration that human episodic memory is characterised not only by many "what, where, when, and in what context" memories but also by the ability to remember the sequence of events at the time. For example, my wife not only remembers the many events at our wedding in detail but also the exact sequence, even though our "most beautiful day" was three decades ago. She can play that day like a film in her mind's eye. Could rats also mentally play through memories of perceived scents in the experimental arena true to their temporal sequence and make appropriate decisions when suddenly prompted? Can they search their memory space so specifically that they find a particular bead on the temporal string of pearls according to the bead's position, even if all other information associated with that event is insufficient?

The American researchers tested five criteria for episodic memory in a single experiment. First, the animals should be able to reproduce the sequence of several unique events; second, they should be able to reproduce this sequence without the aid of non-episodic cues, i.e. semantic or memory alternatives; third, reproduction should be part of long-term memory; fourth, reproduction should be resistant to interference from other memory contents; and fifth, reproduction should be dependent on a well-functioning hippocampus. The trick of the study was that the rats were "asked" at an unpredictable time which scent was second to last or fourth to last in the previously completed sequence of scents dependent on the existing context. The sequences, which changed every day, consisted of five to a 12 different odours [599].

Because their lengths changed daily and unpredictably, the rats could not predict when the sequence would end, and when they would then be moved to another context, i.e. another arena. In the other arena, the rats had to choose the second to last or the fourth to last scent of the sequence they had just completed over another scent from the same sequence. Whether it was the second to last or the fourth to last scent depended on the environment; in one arena, it was the second to last scent, and in the other it was the fourth to last scent. The high accuracy of approximately 88% on average, which deviated significantly from chance, led to the conclusion that the rats memorised the scent sequences completely and, when asked which was the second or fourth last scent, ran through these sequences again, in reverse, in their heads (*episodic memory replay*). In any case, this would correspond to the human procedure of episodic memory.

The researchers also carried out various control experiments to confirm these results. For example, they interspersed tests in which the correct depression did not contain a reward to rule out the possibility that the rats were only orienting themselves to the smell of the chocolate (which was not the case). Furthermore, they sometimes extended the pauses between the last four scent samples, which affected working memory. However, the rats did not show a decrease in performance, which is another indication of episodic memory. Third, the researchers carried out interference tests, in which they were presented with 21 additional scents in a completely new, triangular arena between learning the sequence and the tests. This insertion should have interfered with remembering the sequence in the test if what was previously learned was incorporated into general familiarity-based memory. Episodic memory, on the other hand, which is based on unique combinations of content, places, times, and contexts, should not be vulnerable in this respect. Indeed, the rats were not affected.

Last but not least, the researchers examined the effects of pharmacological disturbances in the region of the brain, which plays a particularly strong role in episodic memory, on test performance. Recent neurobiological studies of the rat hippocampus confirm that the so-called *place* cells, which code for the location of an event by representing one area of an environment at a time, can indeed be activated sequentially [600]. These neurons also encode the current presence of emotions (hunger, curiosity, fear) and which phase of a complex behavioural sequence an animal is currently in. They apparently determine the position in a temporal sequence, whereby contents are variable. Since the tests addressed the storage and retrieval of scent sequences, the researchers manipulated the rats' hippocampus via a chemogenetic procedure. They implanted designer receptors (DREADDs) into the seahorse-shaped region of the brain, which could later be activated exclusively by certain designer drugs. This activation triggered temporary suppression of the hippocampus. As a result, the performance on tests to choose the fourth-to-last scent decreased considerably. In contrast, there were virtually no negative effects of hippocampal deactivation on other learning and memory tasks in which semantic knowledge was acquired. With these sophisticated experimental tests, the presence of episodic memory in rats could be confirmed alongside and independent of semantic memory. Of course, we must mention the usual caveat that this conclusion is based on the usual behavioural criteria but not on evidence of memories of subjective experiences. This note is important because, for Tulving and other scientists who followed this reasoning, the concept of mental time travel envisages a close connection between episodic memory of past events and thinking about personal future events. However, others have argued that the concept of mental-time travel involves subjective phenomena such as consciousness and therefore imposes unhelpful and ultimately unnecessary constraints on research of episodic memory and future thinking [584]. Therefore, alternative views of scene construction have emerged that could bring together work on episodic memory of past events and thinking about future or fictional events, and furthermore, make the results comparable between humans and animals. For now, however, one must say goodbye to the idea of mental time travel.

Chapter 7
Metacognition

In many countries, there is a popular TV show in which a contestant, after winning the opening question, takes a seat on a chair and has to answer questions of increasing difficulty one after the other. With each correct answer, the winning amount increases, but a wrong answer can lead to a considerable or even total loss of the amount won thus far. At the end, after the last and most difficult question, a million euros await as a reward. The answers are given according to the principle of multiple choice, which means that there are already four answers to each question, but only one of them is correct. As time passes, candidates become more insecure, not only because the questions become increasingly difficult but also because one wrong answer can make the large cash prize disappear. The tension is further increased by the fact that the person in the chair can also quit—that is, drop out—if they think they cannot answer the question correctly. In this case, they keep the money they have accumulated up until that point. Part of the show's fascination is the fact that people have to check their level of knowledge in front of an audience. They have to assess as accurately as possible how sure they can be of knowing the answer. Often, the players know the answer, but they are not quite sure. Sometimes, they have only a vague feeling that they have known the answer before or that the answer is "on the tip of their tongue". Therefore, to a limited extent, they are given the opportunity to seek help or to ask for more information. Either two wrong answers can be removed, or the audience or a known person can be asked.

Knowing what one knows and what one does not know is a very important mental ability that involves thinking not about an object or event in the environment but about something in one's own mental world. For a long time, it has been disputed among philosophers what it means to be able to think about thinking or to know something about knowing. René Descartes, for example, with his statement "I think, therefore I am", brought this form of self-referential thinking close to self-consciousness. Many have followed him in this, because knowing what the mind holds could be a prerequisite for knowing whose mind it is, which in turn is

L. Huber, *The Rational Animal*, https://doi.org/10.1007/978-3-031-60803-2_7

important for having a concept of the self, self-awareness, and, ultimately, fully reflective consciousness.

With or without self, this kind of abstract thinking, called "*metacognition*", could be an important first step on the ladder of consciousness. In any case, it rivals language and tool use in its potential to reveal commonalities and transitions between human and animal minds. We humans have no doubt that our journey into our own minds requires consciousness. However, what about animals? Can they even think about their own knowledge or memory content?

In a somewhat broader sense, this is about a special form of "*cognitive control*". Certain cognitive mechanisms ensure the control of thought and action in accordance with current goals and intentions. In the simpler case, this happens at the so-called object level. Representations of objects, other stimuli, or even events are created, and these steer behaviour towards a current goal or intention. This happens implicitly; that is, the bearers of these representations are not aware of these processes [601]. A higher form of cognitive control is based on the creation of representations at a higher level, the "*meta-level*". Here, representations are formed that do not represent properties of the external environment but rather properties of one's own internal processing. They contain information about a property of a cognitive process, for example, the variance in the firing of a group of neurons in the visual cortex. Such metacognitive representations make reference to a property of a cognitive process, for example, the reliability of the representation of a content just perceived [602]. The firing rate of neurons, for example, is apparently an indicator of the likelihood of the object representation being correct [603, 604]. Empirical evidence for this was provided by experiments with rhesus monkeys in which neurons in the parietal cortex reflected both the formation of a decision and the degree of certainty of that decision [602].

The big (theoretical) question is now whether (some) non-human animals possess metacognition. The secondary (practical) question is how to verify or even prove it. With humans, the primary means were and are the analysis of their verbalised self-assessment, the detailed reports of human introspective reflections. But this approach is not applicable to nonlinguistic creatures. Therefore, indirect methods must be devised. Interestingly, one has used the three most important characteristics of the previously mentioned TV show as test parameters: showing uncertainty, being able to drop out, and seeking additional information. Let us now take a closer look at the empirical evidence for this.

Communicating One's Own Uncertainty

The uncertainty was first tested by the American psychologist David Smith from the State University of New York (USA). He had trained a bottlenose dolphin named Natua in a Florida harbour to press a certain key—"High"—when a tone with a pitch of 2,100 Hz was played but to press the "Low" key when a lower tone was sounded [605]. This was no problem for the dolphin, as long as the tones were

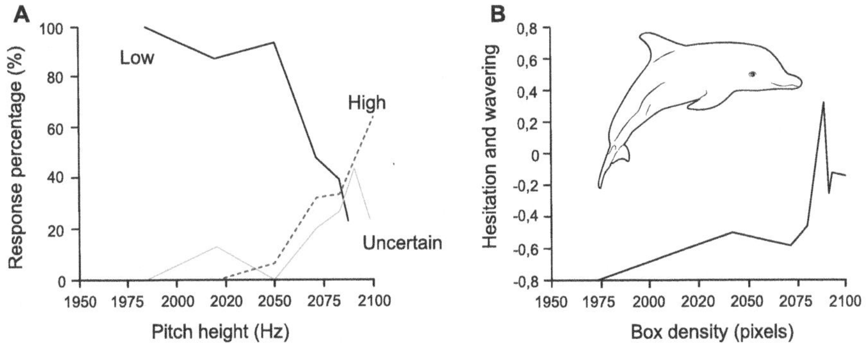

Fig. 7.1 (**a**) Performance of a bottlenose dolphin in discriminating tones. The thick line indicates the frequency with which the "Low" key was pressed, and the dotted line represents the frequency with which the "High" key was pressed. The pitch of the low tones was dynamically adjusted to manipulate the dolphin's perceptual threshold for discriminating low and high tones and to examine in detail its response pattern within this range of maximum difficulty. The *x*-axis indicates the pitch of the pre-recorded tones. The soft line represents the frequency of trials in which the uncertainty key was pressed. (**b**) The assessment of the dolphin's behaviour on tones of different frequencies that indicated uncertainty (hesitation, slowing, or wavering) in the eyes of human observers. (From [605])

sufficiently different. The dolphin swam quickly and unerringly to the right key and pressed it without hesitation. However, the higher and thus more similar to the 2,100 Hz tone the second tone became the more often Natua made mistakes. The error rate accumulated at tones around 2,086 Hz, which is only 0.11 semitones below the high tone, at the apparent perception limit of these animals. At this pitch, the dolphin not only pressed the two keys the same number of times (Fig. 7.1) but also showed an interesting new behaviour: it hesitated, slowed its approach, and swam back and forth between the keys (Fig. 7.1b). This was an obvious sign of uncertainty.

To avoid the accusation of anthropomorphic interpretation, Smith introduced yet another way in which the dolphin could express its potential uncertainty. He introduced a third key, which immediately switched off the sound and started the next but easily and without reward. At first, the experimenter demonstrated the effect of this key on difficult tasks, but soon, the dolphin understood its function and was using it regularly on tones close to the indistinguishability threshold (the thin line in Fig. 7.1a). This response, sometimes called the "*uncertainty response*" (UR), allowed the bottlenose dolphin to indirectly report or address the task's difficulty. This is because if animals can monitor their own level of knowledge, they should prospectively recognise difficult tasks as having risk of error and selectively reject them.

The results of experiments with a single bottlenose dolphin are insufficient for drawing general conclusions about non-human animals, so Smith and his colleagues designed a similar experiment with rhesus monkeys [606]. Here, too, the animals had to master a discrimination task of varying degrees of difficulty. The stimuli to be discriminated consisted of fields on a computer screen with varying numbers of white dots, which appeared in either dense or rather loose clusters. The animals were asked to use a joystick to decide whether the field—containing up to 2,950 dots—was very dense or sparse. They received food pellets for correct answers and a "time-out" for wrong answers, in which the screen remained blank for a few seconds. The monkeys seemed to take great pleasure in the computer game and became agitated when the blank screen interrupted them. As with the dolphin, they got increasingly worse as they got closer to 2,950 dots and reached chance level at approximately 2,700 points. However, they soon learned to address difficult tasks by selecting the third symbol with the cursor. Selecting a "star" was followed neither by a reward nor a time-out but immediately by the next attempt. Analogous to the bottlenose dolphin, the macaques used this option specifically for difficult tasks, namely, for pictures with 2,400–2,900 points.

Get Out in Time

Some professional colleagues were still not convinced that Smith had demonstrated the discovery of metacognition in animals with his dolphin and rhesus monkey studies. An alternative explanation for the choice to skip difficult tasks is that the animals simply wanted to prevent a time-out and move on quickly. This could be a simple conditioned response in which the monkeys learned the behaviour that would lead to the greatest reward. However, this reasoning has at least two weaknesses. First, it does not explain why the monkeys selectively chose to skip difficult tasks. Second, it does not explain why other animals—pigeons for example—do not choose to skip the same task. Finally, Smith was able to show through follow-up experiments that monkeys who learn to use the skip button in the original test can immediately transfer this response to a brand new task that creates equal difficulty and uncertainty. This means that the uncertainty response cannot simply be a conditioned one.

However, it could be that the uncertainty response is specific to discrimination tasks. Therefore, a group of researchers led by Robert Hampton at Emory University in Atlanta (USA) chose a different task to test rhesus monkeys with regard to uncertainty response [607]. They chose memory tasks based on the principle of delayed matching-to-sample. They showed monkeys a picture, and after a pause, the animals had to choose between two pictures, one of which was different from the previously shown sample image; the other was not (Fig. 7.2). To be able to choose the correct picture, namely, the one shown earlier as the sample, the animals had to remember this picture and grasp the principle of the task. The longer the pause was, the more difficult the task was because the longer the animals had to remember the picture. It was therefore to be expected that the memory of the sample would decrease with longer pauses and that the animals would make more mistakes. The logic of the experiment was logically based on the fact that animals that are able to recognise that the presence or absence of their memory can improve their performance by

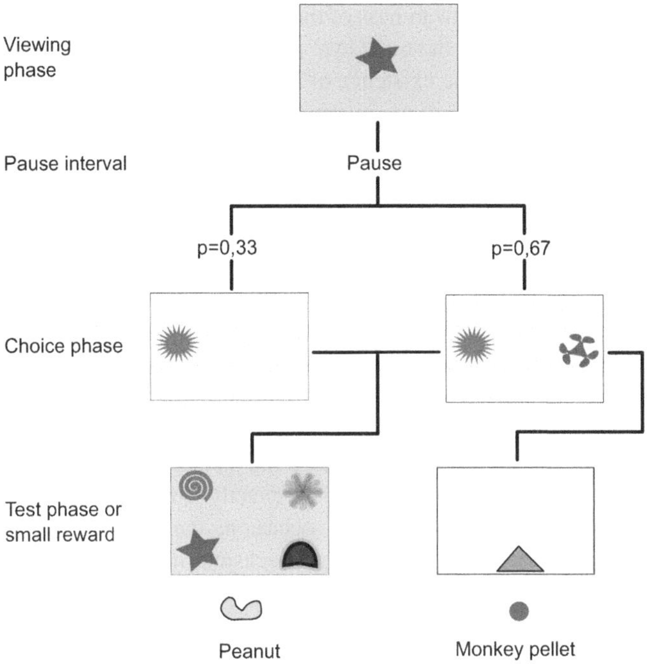

Fig. 7.2 Matching-to-sample procedure for assessing whether rhesus monkeys can report the presence or absence of memory. Each trial began with the presentation of a randomly chosen, colour clip-art image in the centre of a touch-sensitive computer monitor (in this case, a star). In the following pause, monkeys sometimes forgot this sample image. In the choice phase, monkeys were either forced to take the test (in one-third of trials, left) or were free to decide (in two-thirds of trials, right) whether to undergo the memory test (symbol sun) or not (symbol drone) to receive a little food pellet instead. If they chose the star in the test phase, they received a preferred peanut. (From [607])

being allowed to refuse memory tests, precisely when—as in the Millionaire show—they have forgotten the sample image or are unsure. Therefore, monkeys should perform better if they can opt out of the test instead of making expected mistakes.

Their performance was tested via a comparison between two conditions. In one-third of all trials, the monkeys had no choice; they had to choose one of the two comparison pictures to either be right and be rewarded with a treat or be wrong and take a time-out. However, in two-thirds of all trials, they could opt out by pressing another symbol. This response led them to move on to the next trial with a small reward. These monkeys also, at least partially, took advantage of this opportunity in a targeted way. Even more, they improved their performance on the trials where they could drop out but did not compared to the forced trials. Thus, the monkeys proved that dropping out is worthwhile if the task is too difficult. Moreover, they most often refused the tests when memory was attenuated by an experimental cue—extending the pause. Although only one out of two monkeys met these two criteria under all test conditions, the awareness of their own forgetting seems to be the most likely mechanism underlying the monkeys' abilities.

A recurrent objection to the existence of metacognition for exploring one's own state of knowledge is the use of information from the environment to show uncertainty or capitulate to a task [608]. The animal's response is then to this external stimulus, not to the internal representation of the memory for that stimulus. Hampton was aware of this problem; therefore, more recently, he undertook further experiments with rhesus monkeys. His aim was to show that animals are aware of their memory state—called "*memory awareness*" [609]. Being aware of one's memory is an explicit process that is different from implicit decisions that are not accessible to cognitive control and consciousness [610, 611]. Generalisation tests are a way to exclude the influence of external information specific to a given test. For example, animals might drop out because the test situation is novel or unexpected. One way out of this would be to change the strength of the memory. This does not change the novelty of the test situation but should change the metacognitive response itself if it is indeed based on memory awareness. This is exactly what recent experiments with rhesus monkeys showed [612]. On one occasion, memory of the sample was increased by showing it twice; on the other occasion, it was decreased by unprepared dropout of the sample. The tendency to drop out was altered accordingly, decreasing in the first case and increasing in the second. This cannot be explained by the always same unusualness or unpreparedness of the test situation; thus, accessing internal knowledge or a memory store is more likely.

However, the doubters were still not completely satisfied. Above all, arguments were put forward that allow for an explanation on a simpler level, namely, through mere associative learning processes instead of introspection [613, 614]. Monkeys could use both the strength of their memory and the speed with which they choose an answer as a cue for their decision. Thus, they would not need to check their memory; rather, they use only the speed with which they think of an answer or decision as a cue. Another non-metacognitive possibility is that the animals simply react to the fear they may feel in uncertain situations but without bringing their state of fear into consciousness and checking it there. Similarly, on the Millionaire show, instead of doing some introspection, a contestant might simply use the speed with which a first answer comes up, or her anxiety during a tough question, to make a decision. As we know, this is not the case, but it is a possibility and therefore at least conceivable for the monkey-like behaviour.

Unfortunately, studies of information seeking in monkeys have thus far failed to test alternative hypotheses in a rigorous manner, and reports often end with the vague conclusion that some species may have memory consciousness [615]. In any case, there is a lack of concrete and precise determinations of which cues subjects—whether humans or animals—use to solve metacognition tasks.

However, associative explanations and simulations are also still too indeterminate. For example, they cannot explain why there are such large differences between the species that have been tested thus far. For example, pigeons can fail at metacognitive tasks based on the matching-to-sample scheme, where they can improve their performance with selective trial skipping [616], as shown by rhesus monkeys [607].

Moreover, associative models that could explain performance across paradigms and species would be unacceptably complex. Even if some metacognitive performance in humans is based on simpler associative processes [617, 618], this does not necessarily mean that all non-human animals can make do with these mechanisms. One way out would therefore be to create experimental setups and control conditions that require a precise determination of the information that the subjects use to solve the metacognitive task. In essence, this would be discrimination tasks, as they have been traditional in psychophysics for a long time. This is because the reaction to internal signals, such as the strength of the memory, should be determinable in the same way as external signals, such as light or sound. Thus, subjects would have to be rewarded to respond one way when the memory is strong and another way when it is weak. The response to skip is an example of this [607]. Appropriately designed experiments could also measure whether the strength, content, or ease of memory recall controls the discrimination response. In any case, it is important to distinguish whether animals act on the basis of publicly observable ("external") stimuli or on the basis of private mental events (through introspection and metacognition).

Betting on Your Own Knowledge

Another approach to showing that animals rely on metacognitive information, rather than environmental cues or non-cognitive cues from their own bodies, concerns the flexibility of its use. Humans can use metacognitive information quickly and without prior training and can transfer it to new situations. Most importantly, they can do so in both retrospective and prospective tasks. Retrospective tasks are those in which information from past events contributes to decision-making. These are, for example, confidence judgements such as "I am sure I made the right decision" or memory judgements such as "I remember the time when I learned this" or judgements about the source of information such as "I know because I heard it on the radio". Prospective tasks are those in which judgements are made about confidence in future events, such as feelings of knowledge ("I am sure I can give the right answer").

Nate Kornell and colleagues from Williams College in Massachusetts (USA) developed a very elegant method for testing whether animals are capable of making such judgements about their confidence in their own knowledge. They had monkeys bet on how good the decision they had previously made in a memory or perception test was [619]. First, Kornell trained rhesus monkeys to judge which of two lines on a touch screen was the longer (or shorter) one. Then, they were taught to show their confidence in their answers by betting on them. The monkeys had a certain number of virtual coins displayed in the corner of the screen and were allowed to choose either a symbol on the screen that represented high risk or a symbol that represented low risk. They had to make this "bet" after they had made their decision in the length estimation but before they received any feedback about it. If they played with low risk, they received a guaranteed but small reward. On the other hand, if they played with high risk because they were sure of what they were doing, they received a large reward if they were right beforehand but nothing if they were wrong. With these bets, they had to communicate how certain they were of the decisions they had made before.

The second task consisted of quantity estimation. The animals had to learn to choose either the larger or the smaller set of objects in a configuration. In addition, as a third task, they were given a memory test in which they had to indicate whether an image was present in a previously shown image sequence. In these tests, too, the animals had to bet with their virtual coins. In all conditions, they were very accurate, betting with very little risk when their decision was actually very likely to be wrong but with high risk when their decision was likely to be right. This procedure allowed the animals to be "polled" about how confident they were in their decisions without having been trained to do so.

A few years later, these experiments were continued; this time, the rhesus monkeys from the first experiment were also confronted with prospective questioning [620]. The animals had to submit their bet after seeing the task but before they had decided (Fig. 7.3). They had to indicate their confidence in their ability to solve the problem. Here, too, they were usually right. It is also important to note that they had to make their metacognitive bets at a time when the stimuli of the task were not present. This approach allowed them to rule out the possibility that their bets were

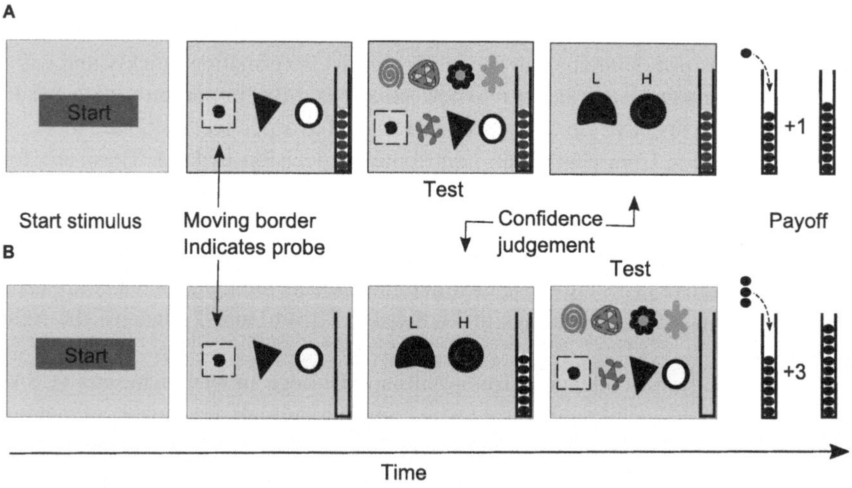

Fig. 7.3 Experimental design for retrospective (**a**) and prospective (**b**) confidence judgements. Rhesus monkeys first had to press the start button. Then, they saw the image to be remembered (the sample, framed). In the later matching-to-sample test, the animals were asked to select the sample image they had seen before. Either after this test (A) or before this test (B), they were asked to bet whether their choice was right (A) or would be right (B). If they chose the high-risk symbol (H) and the decision before or after was right, they received three coins; if the decision before or after was wrong, they received nothing. If they chose the low-risk symbol (N), they always got one coin, regardless of their test performance. If they received a certain number of coins, they received a food reward. (From [620])

based on these stimuli. Their confidence judgements therefore appear to be based on knowing something about their own knowledge state and subsequent decision-making process. These experiments have shown that the animals can form metacognitive judgements very flexibly, independent of the problem and external stimuli. The monkeys transferred their metacognitive monitoring abilities to a new memory task, replicating the transfer abilities described a few years earlier [619].

It Is Better to Look Than to Guess

Another approach to determining metacognition is to examine whether—and if so, what—information or cues that can improve performance are obtained by the animals without being instigated or even forced to do so [621]. The logic behind this approach is that this information search is started only if the task seems too difficult or if there is too little information available to solve it, which in turn means that one's own knowledge is insufficient. The first study on this topic was conducted by Josep Call and Malinda Carpenter at the Max Planck Institute for Evolutionary Anthropology in Leipzig. They set chimpanzees, orangutans, and 2.5-year-old children a search task in which coveted objects—food for the apes, stickers for the children—were hidden in one of two (or three) tubes [107]. The square tubes, 30 cm long and 5 by 5 cm wide, were open on both sides and oriented so that the subjects could see into them if they oriented themselves appropriately (Fig. 7.4). After first inspecting the tubes to ensure that their function was understood, the experimenters hid an object in one of the two tubes and let the apes or children choose. The researchers varied whether the subjects could see the hiding, whether they could see through the tubes, and whether there was a delay between the hiding and the choice. The key measure of behaviour in terms of checking one's knowledge and obtaining additional information was the comparison between overt and covert hiding. Indeed, most apes and children looked into one or both tubes when they could not observe the hiding because a screen had blocked their view (Fig. 7.4a). Interestingly, they often chose correctly even when looking only into the empty tube, apparently because they could conclude that the target object had been hidden in the other tube. Looking was easy to detect because the tubes were low down, so the subjects had to bend down to look inside. Subjects also looked more often when there was a pause between the presentation of the tubes and their choice, presumably because they could use the extra time to examine the contents of the tubes closely. Yet, they hardly ever looked if they could see beforehand which tube the object had been placed in. These findings correspond exactly to expectations with regard to an efficient search strategy, suggesting that the apes and infants oriented themselves according to their own level of knowledge on what they had already seen, before they made a decision. Unprompted, they also sought further information before taking action, like detectives who did not witness the crime.

A few years later, Call repeated the search game with great apes, introducing further controls to rule out alternative explanations [621]. An alternative hypothesis

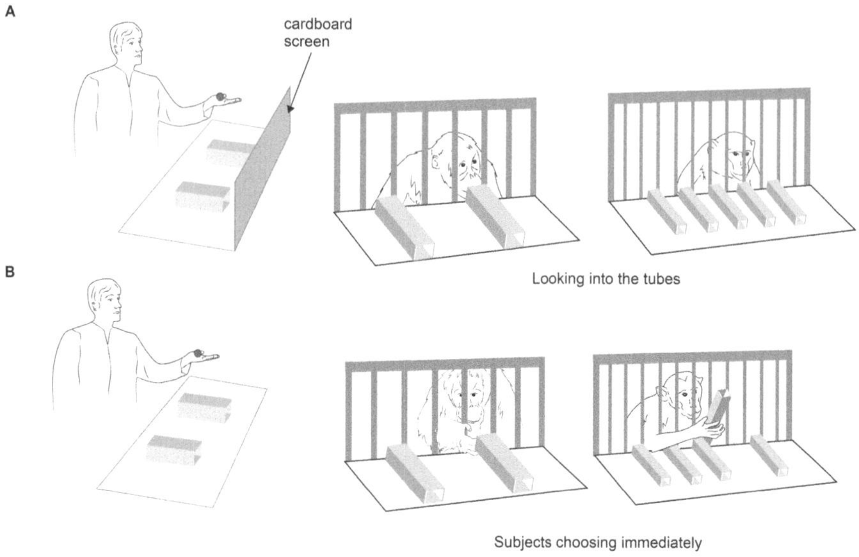

Fig. 7.4 The behaviour of primates (here, orangutan and rhesus monkey) when searching for hidden food. (**a**) If the primates could not observe the hiding of a piece of food, they first looked in the tubes before making a decision. (**b**) They did not do this if they could see the hiding process. (From [107, 622])

is that of response competition [622]. According to this hypothesis, the apes should decide without hesitation if the tendency to reach a reward is strong. Call now introduced two different strongly preferred food types. Since the animals looked even more often in the tubes at the preferred food, which should have promoted a stronger tendency to choose without looking, the response competition hypothesis could be ruled out. Furthermore, Call manipulated the costs of looking and found that these costs were also taken into account by the animals when making their decisions. Thus, the results of the search game suggest that animals consider three factors: the cost of obtaining additional information, the value of the target object, and their level of knowledge. The combination of these three factors fulfils three important criteria for metacognition: complexity (of the decision), flexibility, and control [623]. The apes also demonstrated flexibility because they did not need to see the food piece to make their decision for one tube or the other. After looking into the empty tube, they immediately chose the other tube without looking into it (Fig. 7.4b). They thus drew a logical conclusion, namely, one based on the *inference by exclusion* principle.

It is not only great apes and infants who behave in the search game as if they metacognitively decide to seek further information. Rhesus monkeys also kept looking in a comparable task to increase their success with this measure [622]. Capuchin monkeys, on the other hand, behaved differently [615]. Not all of them reliably looked when they could not observe the hiding, and they remained unaffected by the increase in the cost of looking, unlike great apes [621].

Metacognition as Interpersonal Cognition Control

The broad definition of metacognition as "cognition about cognition" is often used [624, p. 170], but what does it mean exactly? In general, as mentioned in the introduction to this chapter, one could speak of mental processes that contain information from another cognitive process and exert a controlling influence on it. However, this definition is too broad for some psychologists because it would probably include any kind of cognitive control. Those who value the uniqueness of human metacognition define metacognition more narrowly as control processes that use multiple representations with metacognitive content, that is, representations of a property of a cognitive process. Although all metacognition is a form of cognitive control, only some forms of cognitive control are metacognitive. Cognitive control can also be guided merely by object representations, for example, values or abstract goals [625, 626], but few control processes use metacognitive representations; for example, by internally representing that inconsistent actions have been set and therefore require the use of cognitive control processes for decision-making [627, 628].

One way to distinguish between two different types of metacognition (a simple one, which also occurs in non-human animals, and a higher one, which occurs only in humans) is to look at the interpersonal or suprapersonal level (Fig. 7.5). A model for this system was proposed a few years ago by philosopher Nicholas Shea of King's College London (UK) and coauthors, including comparative psychologist Cecilia Heyes of the University of Oxford (UK) and neuropsychologist Chris Frith of the Wellcome Trust Centre for Neuroimaging in London (UK) [629]. While the simple form, System 1, concerns only one individual, the higher form, System 2, controls processes of control between several individuals.

In accordance with the more general theory of the dual-control system in the human brain [630], System 2 is based on a distinct form of cognitive processing that

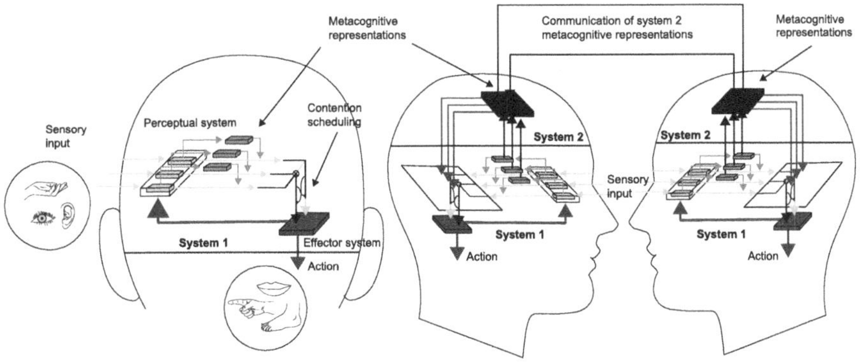

Fig. 7.5 The two systems of metacognition shown schematically. In the left head, only System 1 is shown; in the right head, also System 2 is shown. While the metacognitive representations of System 1 contain the representations of perceptions, those of System 2 contain the metacognitive representations of System 1 and communicate these to other individuals or are influenced by them, thus contributing to coordination. (From [629])

bundles various functional features, all of which are characterised by dependence on capacity-limited, generalised working memory [80, 630, 631]. System 1 works quickly by running many processes autonomously and in parallel and is little affected by general working memory usage. System 2 handles problems serially, requires more time for processing, and depends on the general utilisation of working memory. Its efficiency correlates with general intelligence (IQ). While the representations of System 1 are formed implicitly, those of System 2 are characteristically explicit, i.e., conscious.

Within an individual, cross-domain learning processes can quickly find the best way, based on experience, to prioritise inputs from and outputs to competing sensorimotor systems (Fig. 7.5). Such processes rely on metacognitive information—for example, the variance in the distribution of working processes in the brain, the duration to complete a processing task, or the activation of incompatible response tendencies—to achieve the best trade-offs. These control processes improve over time with experience, just as these processes can learn to use any other kind of relevant cue. The result will be a form of control implemented within System 1 processing that relies on metacognitive representations [629].

The proponents of this theory seem to implicitly assume that the various metacognitive performances of animals that we have discussed in this chapter are implicit, automatic learning processes using internal cues, i.e., they do not require introspection or conscious reflection on one's level of knowledge or the strength of one's memory. A comparison to the TV show, or to the student who only answers after she has thought about whether she truly knows for sure the question asked by the teacher, would therefore not be appropriate. Consciousness, which is inherent to human beings, occurs only in System 2. This system is capable of coordinating sensorimotor representations between two or more people and using metacognitive information located somewhere within the various interacting people [629].

The special function of metacognition in System 2 is the selection of metacognitive information for transmission while controlling the sensorimotor systems of two or more people involved in a common task. They thus serve suprapersonal cognitive control (Fig. 7.5, right). Metacognitive representations of System 2 are derived from information in System 1 but exist in a form that is available for oral reports. For example, the reliability of a sensory signal can be reported in terms of confidence, which improves the formation of a judgement about shared percepts by giving more weight to the more confident observer [632].

According to the theory of the two metacognition systems, the special benefit and evolutionary advantage of System 2 lies in its role in group decisions [629]. Apparently, joint decision-making about perceptions and other matters is much more effective when participants share (verbally) their metacognitive representations about them [632, 633]. When deciding who to trust more in a court case, jurors make use of the trustworthiness of the witness and other metacognitive representations, such as the calibration of trust and accuracy.

The suprapersonal or interpersonal coordination of metacognitive content is reminiscent of the so-called "*theory of mind*" [634] which I will discuss in the next chapter. This refers to a cognitive ability by which one can grasp the mental content of others and, equipped with this knowledge about others, manipulate them in one's own favour—quasi-*Machiavellian* [56]. However, proponents of the dual metacognition theory point to important differences. On the one hand, representations first concern one's own cognitive processes and not those of another. On the other hand, the mediation of metacognitive representations, not the capturing of certain mental states of the other through perspective taking [629], plays a central role. Put simply, theory of mind ability allows individual A to conclude from perspective taking that individual B has seen a rabbit. In contrast, System 2 metacognition allows Individual A to communicate that his visuomotor fixation on the rabbit is currently quite reliable. This ability made it possible for prehistoric humans to engage in more complex cooperative projects—think of communal hunting, for example—and coordinated tasks, for example, knowledge and labour sharing in the clan. Selection pressure for this ability is plausible for early human species, but—according to the authors—it is not for non-human creatures [629]. For them, System 1 would be perfectly sufficient.[1] If this theory and the evolutionary scenario derived from it are correct, it would still be a question whether non-human animals possibly possess other suprapersonal abilities. Although incapable of verbally exchanging metacognitive content, they would still be able to take the perspective of others and thus gain access to their knowledge, wishes, and even thoughts.

[1] However, this is challenged by recent findings of cooperation-based concepts in male dolphins. The researchers were able to show, on the basis of behavioural data collected over 30 years in combination with sound playback experiments on 14 males that they form a social concept of cooperative team membership at the second-order alliance level (at the first level, males cooperate in pairs or trios, at the second level they cooperate in teams of up to 14 males). This provided evidence of a classification of social relationships in a non-human species in which nonrelated individuals form an extensive, multilevel social network with a high degree of cooperation [635].

Chapter 8
Mind Reading

Gordon Gallup of the State University of New York in Albany (USA), the inventor of the mirror test for self-recognition in animals (see Chap. 10), has argued that self-consciousness also implies the attribution of consciousness in others [636]. The self, he argues, is social at its core, understanding itself from its position in the social matrix [637]. Empirical confirmation of this view comes from studies of patients. People who have difficulty imagining the thoughts and desires of others also have difficulty seeing into themselves, introspectively reflecting on their own actions, or anticipating the consequences of their own actions [638]. Of course, it is also conceivable to have self-awareness and yet not be able to think or feel one's way into others. However, such a consciousness would be empty, trapped in a solipsistic mind, unable to exchange with others, coordinate its actions with those of others, or even cooperate with them. Yet, this is precisely the hallmark of human reason. Human culture is inconceivable without a "*theory of mind*".

But what is the advantage of a theory of mind? That through it one can see that another person's knowledge or beliefs are different from our own knowledge and beliefs? I know what is behind the rock, but the other person does not, because from over there they cannot see that there is a scorpion sitting here. Having a theory of mind allows us not only to interpret but also to manipulate other people's behaviour through their beliefs. If he is my friend, I can warn him about the scorpion. If he is my enemy, I can tell him that there is nothing to worry about. The latter is called tactical deception. As Chris and Uta Frith of University College London (UK) rightly argue, human interactions overwhelmingly involve the exchange of true or false knowledge for good or ill [639].

Interpreting others is an everyday routine, and for some people, it is the primary content of conversation. Our language is rich, with hundreds of terms to describe the psychologisation of others. We speak of desires, intentions, thinking, imagining, and believing, and we describe our actions using a wide variety of verbs, such as suppose, expect, doubt, suspect, etc. (for epistemic states) and want, wish, hope, must, should, etc. (for desires). Even 3-year-old children begin to use such terms

L. Huber, *The Rational Animal*, https://doi.org/10.1007/978-3-031-60803-2_8

and the associated everyday psychological schemata [640]. Children can use these to discuss why others act as they do, what they are likely to do in the future, and so on.

This tendency to interpret the behaviour of others is so strong in humans that we often overdo it. We attribute thoughts, desires, and goals not only to other people but also to animals and even inanimate objects such as dolls and cars. People even "*anthropomorphise*" simple geometric figures that move on a computer screen, attributing willful actions to them. A famous experiment on this topic was published as early as the 1940s. Subjects watched a short cartoon with a square open on one side and various triangles and circles that moved in different ways [641]. A typical description of what happened, replicated many years later, described the events as a chase, in which a large triangle first chases a small triangle, and then both chase the circle, causing it to become frightened and finally turn away in a huff [642, p. 473]. A description of a similar event by an autistic person reads quite differently. It describes movements and collisions of geometrical figures in a purely formal manner [643, p. 840].

When I watch the movie, I can hardly escape humanisation; even in such obviously nonliving objects, I seem to recognise intentions and even emotions. I find myself attributing intrinsic movements to the figures when I write that they "moved in different ways".

Sally and Anne

Children develop the ability to attribute mental states to others as early as the second year of life [644], but this development is far from complete thereafter [645]. A number of tests have been developed with the aim of determining at what point people can relate their own beliefs about certain events to those of others. This is particularly easy to test when they differ. In fact, children as young as 2 years old can imagine the absence of knowledge about an event in another person, for example, if that person appears only after the event has occurred [646]. Even more intriguing, however, is the question of when children acquire the competence to attribute a belief, which is different from their own, to another person. This is about the child's understanding of another person's false belief. This ability develops slowly. It is not until around the age of 5 that children's interpretations of other's beliefs are able to be communicated and also explained to others [647].

To check this in children who have not yet developed full linguistic competence, one must develop clever tests. With his colleague Josef Perner of the University of Sussex, Heinz Wimmer of the University of Salzburg created a test for children that became a classic in the study of theory of mind [648]. Unlike other tests, this *false belief task*[1] importantly does not contain a test question with mental verbs

---

[1] In German language literature, the following expressions are used synonymously: false conviction, false conception, false assumption, and false belief.

(e.g., "What does Maxi believe?"), since a mere pointing gesture suffices as an "answer". For this purpose, children aged 3–9 years in Salzburg kindergartens and summer schools were shown a story in the form of a puppet show. First, the children watched as a little girl hides her favourite book in a certain place. Then, she leaves the scene, and a little boy finds the book and hides it in another place. When the girl returns, the children are asked to point to the hiding place where the girl thinks her book is. None of the 3- to 4-year-olds, around half of the 4- to 6-year-olds, and 86% of the 6- to 9-year-olds pointed to the correct place. These results indicate that it is not until the age of 4–6 that the ability to attribute a belief to another on the basis of one's own experience emerges—even though one is well aware of the reality that differs from it. These children can therefore distinguish between belief and reality; they understand that there can be beliefs that do not correspond to reality.

Even when a day-to-day expectation takes a sudden turn—instead of sweets, a sweet box suddenly reveals crayons—4-year-old children begin to attribute a false assumption to an unsuspecting newcomer [649]. Interestingly, autistic children behave differently in both cases. In a kind of "*epistemic egocentricity*", they only concede their own assumptions to the other person [650]. Children with autism, up to the age of around 10 years old, expect that the girl will look for her book where it was last hidden. For more complex problems involving a white lie (a more prosocial form of dishonesty to spare others from unpleasantness and to maintain relationships) or a double bluff, autistic children take much longer than neurotypical children to interpret them, and some are never even fully able to [639, 651]. In a particular form of autism, Asperger's syndrome, one part of the brain, the left medial prefrontal cortex, is not activated in tests requiring one to mentally empathise with other people (such as those described earlier), while in neurotypical brains it is [652, 653]. Apparently, impaired perspective taking (in addition to known idiosyncratic perceptions and behaviours—stereotypic movements, compulsions, repetitive behaviours, difficulties in voice modulation, language deficits, and pain tolerance) is a distinctive feature of autism.

There is great interest in identifying the "*social brain*", a network of brain regions specialising in the social domain, not only in relation to autism but also, more generally, in understanding human thought. Mind reading, or, in technical language, "*mentalisation*", is just one of a series of problems that our brain faces. When measuring brain activity during the performance of a variety of tasks involving perspective taking and mind reading, two regions are regularly located in the human brain: the medial prefrontal region (paracingulate cortex) and the temporoparietal junction in the superior temporal sulcus [654]. The medial frontal region is also involved when subjects think about their own mental states. The temporoparietal transition, on the other hand, seems to play a special role in using perceptual cues to identify the actions and intentions of actors. In the future, it should be possible to study the corresponding brain regions in non-human animals during the execution of similar mental processes with the use of imaging techniques. Then we will see whether the functional similarities also correspond to neuronal ones.

Psychologists and Simulators

That some non-human animals are capable of at least one basal form of mind reading, even without language or culture, was postulated early on based on functional considerations. Some influential primatologists hypothesised that primate intelligence was primarily an adaptation not so much to physical problems, such as foraging for food and catching prey, but to the particular complexities of social life. For example, to form optimal coalitions in an inherently dynamic society [655–657]. In the now famous essay "The Social Function of the Intellect" [657], Nicholas Humphrey of Cambridge University (UK) explained that primate social life is like a game of chess, usually won by those who are socially competent enough to anticipate their opponents' moves. Particularly in complex primate societies, this second-guessing requires both distinguishing the visible behaviour (called surface actions) of others and recognising their states of mind and thus outwitting them socially in "*primate chess*" [658]. As the mechanism responsible for this, Humphrey proposed what we already discussed in metacognition, namely, using direct access to individuals' own psychological processes with the help of the "*inner eye*" to then simulate the corresponding processes of their group members [659]. This idea was later developed by philosophers and psychologists into the "*simulation theory*" [660, 661].

In a very analogous vein, Oxford ethologists John Krebs and Marian Stamp Dawkins see the value of mind reading in the manipulation of others, especially in pursuit of their own ends [662]. In intraspecies communication, therefore, not only is honest information exchanged but also dishonest information in order to deceive and mislead others when it is in their own interest to do so. However, unlike Humphrey, these ethologists did not propose a simulation with the inner eye as a mechanism, but rather they proposed that the regularities of the victim's behaviour are internalised by the mind reader, who then can optimise his behavioural decisions with regard to the probable future reactions of his victim [662].

The anticipation of the behaviour of others is thus based on the collection of a large number of experiences of how conspecifics behave in certain situations, on the inference of general tendencies and on extrapolation into the future. Later, this hypothesis contrasted with that of the inner eye as an alternative main hypothesis of mind reading [663] and was even applied to metacognitive introspection of one's own thoughts and states [624].

Another impetus to the question of mind reading in primates came from reports of deceptive behaviour. Richard Byrne and Andrew Whiten of the University of St. Andrews (UK) compiled well-documented episodes of deceptive behaviour in baboons, which were consistent with Humphrey's postulates and closely related to investigations of the attribution of false presumptions in children ("Sally and Anne") [664, 665]. They later compiled an extensive database of such reports contributed by field primatologists and concluded, based on recurring patterns, that many great apes and other animals seem to be able to take into account what others can or cannot see [666]. Deceptive behaviours appear to be based not only on monitoring and manipulating the visual attention of others [667] but also on the ability to judge very accurately how to hide desirable objects or parts of one's body from the gaze of others [664, 666].

Does the Chimpanzee Have a Theory of Mind?

Plausibility arguments based on careful ethological observations and theories based on functional-logical analyses are one thing; hard empirical evidence is another. Therefore, how could it be shown that primates can actually enter the minds of others or anticipate their future actions? Forty-six years ago, David Premack and Guy Woodruff of the University of Pennsylvania (USA) already asked whether chimpanzees have a theory of mind and set corresponding tasks for this purpose [634]. In their view, an individual has a theory of mind if it imputes mental states to itself and to others. The somewhat strange designation of this ability as a theory of mind (or "*theory of the mental*") was chosen to emphasise that these are inferences because such states are not directly observable. Moreover, it could be used to make predictions, as with a scientific theory, namely, predictions about the behaviour of others. Now, to determine whether chimpanzees can impute mental states such as desires or beliefs, they first showed an adult chimpanzee a series of video sequences of a human actor struggling with a variety of problems. Inspired by Wolfgang Köhler's intelligence tests [100], they initially presented simple out-of-reach problems, such as trying to reach a banana. Later, they showed videos with increasingly complex problems, such as a man who could not free himself from a locked cage, who was shivering with cold because of a defective heater, or who could not listen to music because the record player was not plugged in. After each video, the chimpanzee received several photos, one of which contained the solution to the problem, such as a stick for inaccessible bananas, a key for the locked man, or a burning wick for the defective heater. The chimpanzee did in fact choose the correct picture in most cases, which supported the assumption that the animal had recognised the problem in the video, understood the actor's goal, and could choose the means that led to that goal.

Premack and Woodruff suspected that the chimp imputed at least two mental states to the human actor, namely, intention or purpose on the one hand and knowledge or belief on the other. The chimp seemed to understand that the man wanted the banana and knew how to get it. It could be argued, however, that Premack and Woodruff were able to explain only an understanding of goals by the chimpanzee based on *physical* problems. Instead of the chimpanzee's ability as a "psychologist", as they thought, it is a competence as a "physicist", already assumed by Köhler, that was put to the test here. It is questionable whether it is necessary to put oneself in the place of a human being to understand that a stick is the best means of fishing for a banana lying out of reach. This left open the question of whether the chimpanzee can impute mental states to others at all.

Nevertheless, this pioneering work triggered a wave of research and debate that tended to spill over first to children and then to other animals, such as corvids and dogs. Answers to the original question varied widely between research groups over the decades, and 40 years later, there is still no clear or even uncontroversial answer. One advance that has been made in recent years is the realisation that there are many different ways animals can assess others, interpret them, predict their behaviour, or attribute certain mental states to them, such as desires, preferences, or assumptions.

I Know What You Know

While in the wake of Premack's and Woodruff's work, developmental psychologists—as shown above—investigated whether children can impute (correct or incorrect) *beliefs* to others, primatologists initially focused on the attribution of states of *knowledge*. In the simplest case, the knowledge of another can be inferred from whether or not that person has seen something. If one perceives that the other is a witness to a hiding event, one should be able to assume that the other knows the hiding place thus has knowledge about the hiding place and can become an informant for an ignorant person. "I know what you know" is a metaphor for this particular mental attribution.

Again, Premack attempted to answer the question of whether non-human creatures were capable of doing this with chimpanzees interacting with humans [668]. Four young chimpanzees were allowed to watch as a piece of food was placed in one of two containers. However, they could not see which of the containers was filled because a screen prevented this. During the hiding process, the chimpanzees could also see two human trainers, only one of whom could observe where exactly the food was being hidden, as the other's view was blocked. Afterwards, the chimpanzees were allowed to tug one of the two trainers towards them with a string. While the witness to the event opened the hiding place and rewarded the chimpanzee with the food, the other trainer chose the empty container, and the chimpanzee remained unrewarded. While two chimpanzees consistently chose the informed trainer, and did so after only 24 trials, the other two wavered in their decisions. These ambiguous result weakened the conclusion that the chimpanzees were able to take the trainers' point of view.

Independent of this study, Daniel Povinelli of Yale University (USA) developed the idea of testing with true and false informants and published groundbreaking experiments that entered the literature under the title "*Guesser or Knower*" [412]. At first glance, he and his coauthors from Ohio State University (USA) believed that they confirmed the results of Premack's two successful chimpanzees [669]. Three younger animals between 7 and 10 years of age (Sheba, Darrell, and Kermit) and the older female (Sarah) were tested. Sarah was already well-known in the professional community due to her participation in studies investigating linguistic abilities for more than two decades [670], her simple mathematical concepts [671], and her ability to attribute in the famous work of Premack and Woodruff [634].

Again, the animals witnessed the hiding of food, but the experiment subsequently changed. One experimenter, called the "Knower", hid the food in one of four containers covered by a box, blocking the chimpanzees' view. The second experimenter, the "Guesser", was not present during the hiding process but joined afterwards. Now, both experimenters positioned themselves behind the four containers and pointed to one each, the Knower to the one where he had previously hidden the food and the Guesser to one of the other three. The chimpanzees were then allowed to choose one of the containers. If they chose the right one, they got the food. If they chose another container, the Knower opened his container and revealed the food inside, but the chimps went away empty-handedly.

The animals were tested ten times a day in a row for several weeks. In the sessions, the two experimenters kept changing their roles so that the chimpanzees could not simply follow the identity of the Knower to choose the correct container. The experimenters also randomly changed the positions of the hiding place and the container to which the Guesser pointed.

Furthermore, in a subsequent test, the procedure was changed to exclude further explanations. Now, both experimenters stayed in the room during the hiding, and a third experimenter hid the food, but the Guesser had a paper bag over his head so that he could not see anything.

The results of these tests were judged very positively by the three authors, and this method is often adopted as such up to the present day [e.g. 672]. However, a closer look, especially at the learning curves, provides less reason for euphoria. Sarah, who was very experienced in learning tasks, still remained at chance level after 300 attempts and failed the subsequent test. Sheba and Kermit needed more than 100 trials, and Darrel needed more than 250 trials to clearly favour the knowledgeable one. But all three animals, even after weeks of daily training, showed a very modest preference—varying between 60% and 70%—for the Knower in both tests. (It is just a pity that the authors do not inform us when the animals began to show this behaviour within the 30 trials, namely, right from the start or only later).

In the following years, Povinelli conducted further tests with changing coauthors. He found that 4-year-olds, but not 3-year-olds, passed the tests previously developed for chimpanzees [673]. Rhesus monkeys failed to use the communicative cues of a knowledgeable informant to find hidden foods [674]. However, later tests with free-living rhesus monkeys used a competitive rather than a communicative situation with humans and were more successful in understanding what a human competitor could and could not see [675]. The free-ranging monkeys on the island of Cayo Santiago are known for being keen to get hold of food that humans bring to the island for behavioural experiments. Although they are afraid to get too close to humans, they occasionally try to steal food. In doing so, they seem to assess the level of attention of those they steal from. This was confirmed in six tests with variations in what the humans could see. In each case, a monkey was confronted with two experimenters, each of whom had a grape on the table, but only one of them could see it.

Around 20 monkeys participated in the individual experiments. Each monkey had 1 min to steal the grape. In the first two tests, one of the two humans turned his back to the grape; in the third and fourth tests, one looked to the side—either with his head or only with his eyes; and finally, in the fifth and sixth tests, both humans held a tablet in front of their bodies but only one in front of his face or eyes, respectively, the other in front of his chest or mouth. In all six variants of the test, significantly fewer monkeys stole the grape from the human who had it within his field of vision. Clearly, in a single trial the monkeys were using their knowledge of the connection between body or eye position and attention to spontaneously make the right decision. The authors of the study explained this by saying that monkeys have a natural sensitivity to the gaze of others [675], not least because they have specific neurons that enable them to detect the gaze direction of others [676].

Does She Know or Is She Guessing?

The special sensitivity to the gaze of others seems to be a widespread property of the mammalian brain. Around the same time as Povinelli, my former colleague, Christine Schwab, undertook similar tests with dogs. Dogs too were tempted with very tasty food, in this case, a piece of sausage. Dog owners were invited to see if (and if so, when) their dogs would take the sausage in front of them, despite being forbidden to do so [677]. Unlike in the rhesus monkey study, we did not count whether the dogs succumbed to the temptation but rather how quickly this happened, depending on how strongly their owner's attention remained on the sausage after the prohibition had been issued. Either their gaze was directed directly at the sausage, to the side, into a book or away from the sausage. In the most extreme form, the person left the room, leaving the dog alone with the sausage. Not surprisingly, the less attention the owner paid, the quicker the dogs succumbed to the temptation. Nevertheless, it was interesting to see that the dogs used not only the view of their owner's face or open eyes to assess attention but also the direction of their gaze.

At the time, my collaborators and I did not focus our investigation on the question of mind reading but, as in much work thereafter, on the ability to track gaze as a form of shared interest [678]. Furthermore we have been interested in the relationship between dogs and humans, with the dogs' knowledge of our gestures [679], our identity [680], our emotions [681], and, more generally, our behaviour [682]. More recently, we have returned to the question of how dogs assess our attention and whether they do so with an understanding of the importance of vision as a basis for knowledge in others. We were inspired to do so by Michelle E. Maginnity and Randolph C. Grace of the University of Canterbury (New Zealand), who had taken up Povinelli's paradigm of knowing and guessing for testing with dogs [683].

In the first of four experiments, they repeated the original experimental condition with a "Knower" experimenter who hid the food and a "Guesser" experimenter who left the room in the meantime. Like the chimpanzees, the dogs had to choose one of four containers, two of which were marked by pointing gestures from the experimenters. To determine whether the dogs were more oriented towards hiding or less towards informant knowledge, they conducted the experiment again with both people staying in the room during the hiding event. In both conditions, the dogs preferred the container pointed to by the Knower, who was also the hider, i.e. even when the person who was not hiding remained in the room and watched. Apparently, their choice of informant was strongly influenced by the activity of hiding. At least, however, presence during hiding was also important because more than 80% of the dogs chose the hider in the very first trial when the Guesser was absent but only 50% when he was present [683].

Under further experimental conditions, the authors manipulated the informants' knowledge while a third person hid the food. Similar to the experiment with the rhesus monkeys [675], they manipulated the vision of the hiding by covering different parts of the face or by different directions of gaze; while the knowing person either held her hands to her cheeks or attentively watched the hiding process, the guessing person either held her hands in front of her eyes or stared at the ceiling. Needless to say, the two people switched roles several times, and the container that served as the hiding place and other experimental parameters were also randomly changed during the course of the experiment. To check whether pointing to the correct container itself provided a hidden clue or whether the dogs found the hiding place with their nose, appropriate controls—single trials in which the informants did not point to a cup or in which all the cups were empty—were also run for this purpose. In both experimental conditions, the dogs' preference for the knowing person varied between 60% and 70%, averaging just over 60% across all 24 individual trials. While this result was not exceptional, it was still statistically significant. Eleven out of 16 dogs (69%) chose the person with her hands on her cheeks, and only 5 chose the person with her eyes covered in the first trial. However, this spontaneous preference for the knowing person is probably attributable to how visible her eyes were, as this is an important signal of attention for dogs [677]. In comparison, there was no statistically significant preference for the Knower in the first trial with different gaze directions (54%).

Despite the meagre results in the test involving different directions of gaze, the dogs' behaviour, which overall showed a spontaneous ability to distinguish between correct and incorrect informants, aroused astonishment among experts. After all, their performance was significantly better than that of the chimpanzees in Premack and Povinelli's early tests. We recall that only 3 out of 4 chimpanzees distinguished between true and false informants, and that was only after well over a 100 trials [669]. Later experiments with these apes were even predominantly negative [684]. Other animal species, such as capuchin monkeys [685] and pigs [686], failed to attribute knowledge in analogous tests to those described earlier. This challenged researchers to re-examine the dogs' ability.

To dispel doubts about the way the results were obtained, it is extremely important to replicate the results. Unfortunately, this happens far too little (also) in comparative cognition research. In the case of some animals, further complications arise from the fact that unintentional cues are given from human experimenters that can be used to solve the task (see the famous case of *Clever Hans* [687]). Only very sophisticated test procedures and the use of as many control experiments as possible can help [679].

At the beginning of our experiments on perspective taking in dogs, my Master's student Amélie Catala replicated the first experiment by Maginnity and Grace [683], the comparison between present and absent raters, with the same number of dogs and trials [688]. We were amazed at how similar the results were to those of these two independently conducted studies. The mean frequency of choosing the Knower did not differ by more than 2% in both studies!

In the New Zealand study, the dogs performed best when the Guesser either left the room or covered his eyes. Just significant, but clearly weaker, was the preference for the Knower when both informants' eyes were open and visible to the dog during hiding. Apparently, dogs have difficulties taking the informants' perspective to recognise who can see something and who cannot. This approach is necessary when both informants look with their eyes open but not when one of them has his or her eyes covered or leaves the room. The ability to look at the world from the point of view of another is called "*geometrical gaze following*" [689]. Dogs follow the gaze of humans into space [678], but this orientation response may rely on a relatively simple mechanism to align their view with that of another individual looking at something [690]. However, dogs are particularly sensitive and willing to learn human communicative cues, such as pointing and gaze following [reviews in 682, 691, 692, 693]. Indeed, at least some dogs (approximately one-third in the corresponding study) are able to follow human gaze behind a visual barrier in a foraging context [694].

Geometric gaze following enables dogs to look for something they cannot see from their own position. Such overcoming of the egocentric perspective could be tested with stealing ability. In a test, dogs were required to infer that a human could see them from the geometry of space, even though they themselves could not see the human. Although dogs seem to understand how barriers affect the perception of others [390], they did not pass this stealing test in a visual context. However, dogs were able to conceal their theft in the auditory context by keeping their vocalisations down [695]. It is possible that dogs rely only on what they can perceive themselves when they are asked to judge what humans can see and hear. However, the success in discriminating the Knower from the Guesser in the study by Maginnity and Grace [683] suggests that they can use certain obvious clues about what people see—such as open eyes or presence at crucial events.

The question remains whether they can also deduce what humans see or not from indirect cues. This is positively supported by the following study: when dogs were tempted to steal sausage in the presence of the human who had forbidden it, they took into account the brightness around them—dogs only stole if they remained within the dark [696].

We then devised a second test to examine whether, in dogs, judgements of knowledge of two competing informants can go beyond the directly observable differences between them [688]. Although this approach would not yet provide full-fledged evidence of *mind reading* as opposed to *behaviour reading* ability [697], it would at least bring us an important step closer. In this new test, both informants were asked to behave identically while food was hidden by a third person: both informants looked in the same direction, namely, at a specific point on the wall, but differed in which side of the hider they were positioned and thus whether or not the correct hiding place was in view (Fig. 8.1). Looking at the dot on the wall was intended to prevent the informants from differing in that the Knower would follow the hiding event with her gaze while the Guesser would stare into the distance, thus indicating a difference in behaviour. Another important feature of this arrangement was that the food was not visible to the dogs, so they could not simply use the eye-object line [698, 699]. To identify which informant could know about the hiding place, they had to deduce what the humans could see from their line of sight, i.e. perform geometric gaze following. Amélie also implemented all the important controls to prevent dogs from using unintended cues, such as differentially confident pointing of the two informants, scent traces, or positional cues, to solve the task.

Averaged over all 24 trials, the dogs preferred the Knower in just under 62% of the trials, which was a statistically significant difference to the Guesser. It was remarkable, however, that, already in the very first trial, 82% of the dogs chose the container of the Knower and that the dogs did not increase performance in the course of the further 23 trials. These two values rule out associative learning of a possible cue, such as the positional relationship of the Knower and the hider. Clearly, the dogs recognised from the outset, by taking the perspective of the two informants, who were potential witnesses to the hiding, and thus trustworthy informants. Of course, this ability need not be based on a flash of inspiration; we assume that it was acquired in the course of the dogs' previous lives through the cumulative growth of their experiences [700]. Like they learn the emotions of humans by learning their facial expressions [681], dogs could also learn what the direction of people's gaze or the position of their body means.

With the interpretation of looks in a cooperative situation, the "mind reading" of dogs is at a lower level of representation than the distinct ability to attribute the concept of "seeing" because the latter should also include the ability to infer without cues [701]. So, do we find evidence in the animal kingdom that this ability to attribute knowledge without cues, i.e. by inference based on one's own experience, was acquired?

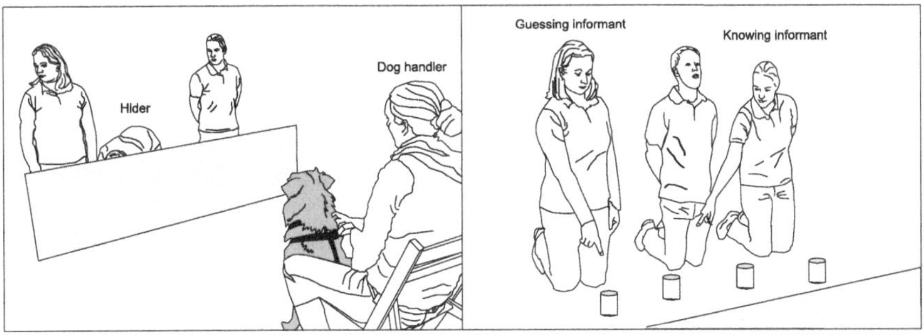

Fig. 8.1 Two informants look to the side while a third person (the hider) hides food invisible to the dog. Then, the two informants each point to a different cup. The dog must now decide which clue to follow to find the food. (From [688])

Machiavellian Mind Readers

After further tests with—albeit very young—chimpanzees in Daniel Povinelli's laboratory revealed rather sobering results, the animals could not distinguish between a sighted human and a blindfolded one [684], and Povinelli became a sceptic of theory of mind in non-human beings [412, 702], and other laboratories devised a new approach. It was already known that apes could follow an experimenter's gaze towards specific targets [703] and that they also follow the gaze behind visual barriers, even if this requires them to change their location significantly [689]. Chimpanzees also know that a barrier blocks vision unless it has a window and they can judge whether or not they can be seen and move to a visible position in the context of begging for food [reviewed in 704]. It is thus considered certain that they are able to distinguish different states of attention using features of body and facial orientation, with facial orientation appearing to be more important than the eyes of humans [705].

A new approach to mind reading made use of this knowledge, but the crucial step was the change from a rather artificial, everyday situation to a context much more familiar to chimpanzees. Instead of confronting them with humans in search of help, they were to demonstrate their ability to compete for food with conspecifics. Brian Hare and his colleagues at the Max Planck Institute for Evolutionary Anthropology, Josep Call and Mike Tomasello, devised a series of elegant tests designed to determine the Machiavellian spirit in chimpanzees that had been theorised decades earlier [reviewed in 704, 706, 707]. Competition for important resources in the strictly hierarchical community of chimpanzees, especially for subordinate chimpanzees, was thought to trigger their highest cognitive abilities, including their understanding of what others see or have seen and therefore what they know and consequently will do next.

In several tests, the researchers focused on the strategic behaviour of a lower-ranking and therefore subordinate chimpanzee towards a higher-ranking and therefore dominant-conspecific in competition for coveted food [708]. If there is only one piece of food for both animals, the dominant animal will monopolise it, and the subordinate animal will go empty-handed—or risk becoming the victim of physical aggression. However, what happens if the subordinate animal sees a piece of food that the dominant animal cannot see because there is a partition blocking its view—or better—if there is also a second piece of food that both animals can see? Is the subordinate animal able to use this knowledge advantage strategically by preferring the first food? Since the two competitors were in opposite compartments from where they could see into a central testing room, the subordinate chimpanzee had to take into account what the dominant competitor could and could not see, even though the dominant's view differed from its own. The subordinate chimpanzee was allowed into the arena with a small lead and showed a clear preference for taking the piece of food that was not visible to the dominant chimpanzee. Apparently, the subordinate had taken the perspective of its dominant counterpart by recognising which food piece the other was seeing, deducing where it would run to, and then strategically choosing the other food piece for itself.

Alternatively, on a cognitively simpler level, the subordinate animal may have learned over time that it takes longer for the dominant to reach the food item near the partition. Likewise, the subordinate could have simply learned that the dominant animal—for whatever reason—ranges to the openly visible food piece. Finally, it may have learned that it is better to wait until the dominant animal has chosen one food piece and then run to the other. With a series of control conditions in further experiments, the Leipzig researchers were able to rule out these possibilities. For example, they made the piece of food invisible to the dominant animal not with partition but with darkness or they let the subordinate chimpanzee choose in front of the dominant, so that it could not orient itself to the dominant's behaviour. To test whether the subordinate was only orientating itself to the partition, the researchers replaced the opaque barrier with a transparent one. With this change, the subordinate's preference for the food behind the barrier disappeared. Accordingly, the results of the entire series of experiments were considered by the Leipzig primatologists to be convincing and consistent in terms of understanding what another sees and knows. Together with the results of additional control tests and numerous observational details of both the subordinate and dominant animals' behaviour, the conclusion was obvious: chimpanzees can resolve such competitive situations in their favour by taking the visual perspective of a competing individual, which is different from their own perspective.

As expected, however, not everyone was convinced. Above all, it was repeatedly objected that the supposed mind readers would orient themselves on some details of the current competitive situation but not on their knowledge of the other. As we will discuss in more detail later, the main alternative hypothesis is that chimpanzees apply certain behavioural rules that determine how to respond to the behaviour of others—in this case, orienting behaviours. Orienting behaviours are those that mark or accompany attention towards an object: glances, body orientation, and expressions. These rules for corresponding counter-reactions are either innate or have been learned in the course of an individual's life, sometimes even in the course of an experiment. In any case, the application of these rules does not presuppose an understanding of the perception or even the knowledge of others.

To disprove this alternative hypothesis, experiments in which chimpanzees were supposed to show that they understood what their competitor had seen, and therefore known, in the immediate past were devised. This is because chimpanzees were required to orient themselves on their memory of the other's behaviour and dispense with cues in the current situation. The Leipzig primatologists set out to investigate whether chimpanzees understand what another actor sees, not only what he or she is oriented to, i.e. his or her "superficial behaviour", but also what he or she registers in his or her environment, which is different from that of the mind reader, so that it influences his or her current or future actions. This thus concerns knowledge on the part of the competitor about where food is, although the food can no longer be seen, and referred to by using an orientational response. With the absence of an orientational response from the competitor, innate or learned responses by the tested individual can be ruled out. Understanding false guesses, as has been studied in young children, is based on this reasoning.

The Leipzig primatologists changed the original design of their chimpanzee competition in two ways [709]. In one version, the dominant chimpanzee either saw the place where the food piece was placed because its door was open at the time (informed condition) or the dominant did not see it because its door was closed at the time (uninformed condition). In another version, the dominant always saw the original hiding place, but then the food was moved by the experimenter to another place to be hidden. In some experiments, the dominant witnessed this change (informed condition), while in others, his door remained closed, and he could not see the change (misinformed condition). This second situation was modelled on children's experiments of false belief. Thus, the first situation compared informedness with uninformedness, while the second compared informedness with the transmission of false information. If the subordinate chimpanzees understood only whether another was informed or not, they could solve the first task but not the second, since they would not be able to predict where the dominant with the false belief would go. The results revealed that the chimpanzees behaved in this very way. They were able to distinguish between the informed and uninformed conditions, demonstrating an understanding of their competitor's ignorance, but they failed to distinguish between the uninformed and misinformed conditions. (They did not immediately run to the new, second hiding place where the experimenter moved the food).

Feathered Mind Readers

So far, I have reported mind-reading experiments with humans, chimpanzees, and dogs. This could lead one to think that the ability to take on perspectives is, if not unique to humans, unique to mammals. Primatologists even tend to assume special primate intelligence because only this group of mammals is said to have acquired certain cognitive abilities that have generally been summarised as Machiavellian intelligence [56]. Although the social intelligence hypothesis was developed for primates, it is based on general evolutionary principles and should therefore apply to other groups facing similar social challenges. Support for this argument is provided by evidence that the relative size of the neocortex correlates with group size in other mammalian groups as well [710]. If this is so, one might ask whether this relationship also holds for other animal groups, at least within vertebrates, provided that they are not only large but also individualised groups and have a functionally similar neuronal structure. Indeed, some bird species fulfil these criteria. The most widely studied, as we have now often seen, are corvids [454, 711–713]. While corvids have a brain structure functionally corresponding to the six-layer neocortex of mammals, the nidopallium, also called the avian cortex [714], they also have several species that are characterised by complex communities. Probably the most studied in this respect are the Western scrub jays [715] and the common raven [713, 716].

Both bird species are known for their sophisticated strategies in hiding food. In addition to the task of retrieval, which requires good spatial intelligence, there is the social challenge of intraspecies competition. These animals not only create hiding places, also known as caches, themselves but also sometimes raid the caches of conspecifics. This sets off a spiral of strategies and counter-strategies, as predicting the next move of the other gives an advantage. This ability did not emerge by chance. Predicting the behaviour of individuals is a feature of social interactions of highly evolved communities. More than that, there has been growing evidence that the degree of sophistication, which animals can achieve in predicting behaviour of others by integrating the observable features of others' current and past behaviour, is linked to the degree of complexity in the social life of the species [717, 718]. It was thus natural to examine the extent to which this informational integration requires the representation of others' mental states, as has been demonstrated with chimpanzees. In particular, the game of hide-and-seek in foraging corvids could form the context that, as in chimpanzees, gives rise to the ability to distinguish between what is seen and what is not seen or between knowledge and ignorance.

In corvids, there are some species that build up food depots, either for the day because they cannot keep everything with them or for the upcoming winter months when they can only find little food. Eurasian jays, for example, build extensive stores of acorns and other nut fruits. The foraging habits of the North American pine jay have been particularly well studied. This bird specialises in foraging the seeds of various pine species, which it hides in the ground throughout the year for later consumption [719]. These hiding places contain, on average, only 3–4 seeds, yet a single pine jay can store up to 33,000 seeds in more than 7,000 individual depots in 1 season [720]. To retrieve at least a large proportion of these—over 80% of which is estimated—in winter, this bird needs remarkable long-term spatial memory [721]. Seed caches can be found with extraordinary accuracy up to 9 months later and even when they are buried under up to a metre of snow. This is amazing when you consider that pine jays create caches in less than 30 s [722], make thousands of them each autumn, return to them months after they were made, and they are located in a substrate that has changed considerably between the time of creation and retrieval. Interestingly, these jays regularly hide more than they actually need. This represents a kind of insurance against forgetting, seed theft by other animals such as squirrels or woodpeckers, and above all by conspecifics.

Protection from looters is very important, as food reserves are a matter of life and death during the harsh winter months. For this purpose, the animals not only create surplus caches but also actively protect them from thieves. The greatest danger of theft comes not so much from accidental finds by animals of other species but from conspecifics that have been able to observe the creation of food depots. This is especially true for gregarious species where the animals live in larger communities because here it is particularly difficult not to be observed when hiding food due to cohabitation. A relatively simple strategy is to hide where the density of conspecifics is lowest [723]. A smarter strategy is to wait until the potential thieves are distracted or cannot observe you because something is blocking their view [724].

Not only in the wild but also in captivity, birds can be seen creating caches and protecting them from the watchful eyes of others. They have a good memory for their own caches but also for the caches of others. Therefore, they can plunder other caches after conspecifics have disappeared, and they do not have to fear confrontation with them [724–726]. It is interesting to note that the same bird can be both a hider and a looter and thus understands both sides of the hiding game [727].

Another protection strategy against scavenging seems to be the repeated changing of caches. This strategy could be used because it either confuses the potential scavenger or presents him with special memory problems. This tactic could also have evolved by chance, but it could also be intentional if the birds were able to deceive conspecifics. The latter explanation, however, would require the birds to assess the other's state of mind or level of knowledge to be able to manipulate him. But are birds even capable of this? Can they, as chimpanzees have shown in the above experiments, recognise different levels of conspecific knowledge?

In Nicola Clayton's laboratory in Cambridge, several sensational experiments were carried out to this end. In the first test, together with her student Joanna Dally and colleague Nathan Emery, she experimentally tested the protective foraging strategies of Western scrub jays [728]. As we learned in the chapter on episodic memory, these birds also hide food in the laboratory, using ice cube trays filled with sand and Lego bricks as distinguishable hiding places. Some birds, the "hider" group, were given the opportunity to hide wax moth larvae in two different dishes. A second group of birds, the "observers", could watch from the neighbouring cage. There were two different test conditions that always had one hider and one observer, but one condition had a partially transparent partition and the other had an opaque one (Fig. 8.2a). Therefore, in the first condition, the observer could see one of the two shells, but in the second condition, he could see neither. In fact, in the first condition, the hider showed a preference—in around three-quarters of all cases—for the bowl hidden behind the barrier but no preference when the observer could not see either hiding location. Thus, these birds seemed to show a similar level of stealth to chimpanzees, namely, to take the conspecific's view into account for deciding their own actions.

However, the authors cautiously noted that the result could also be explained in another way. Instead of considering the other's view, the hiders could simply use their own view as the basis for their actions. Following the motto "out of sight, out of mind", they hide where they see no one else. An elegant way to verify this is to partially obscure the hiding locations (Fig. 8.2b). While the hider could always see the observer clearly, the hiding locations were darkened at a ratio of 1:4. Again, the hiders were tested in the same two types of conditions; and again, hiders only showed a preference for the bin in the dark in the observation condition [729].

Finally, the scrub jays also exhibited very similar behaviour when the containers were at different distances from the potential observer (Fig. 8.2c). Here, the jays preferred the container further away as a hiding place and again only in those trials where the conspecific was not excluded from the hiding process by a partition wall [728].

The strategies for hiding can be adapted not only to the presence of a competitor but also to its social status. As we have seen, in chimpanzees, the strategy of preferentially grabbing a piece of food that a competitor could not see was especially common for subordinate versus dominant animals [709]. In scrub jays, too,

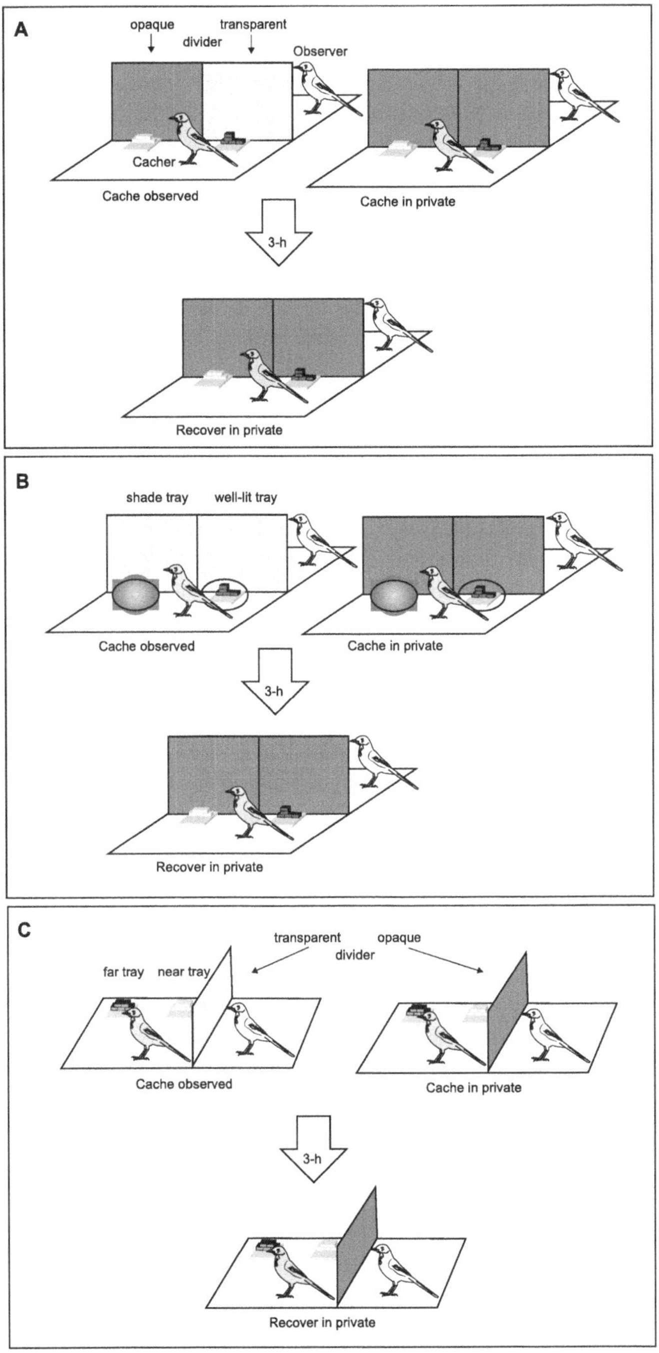

Fig. 8.2 Schematic of the experiments on the hiding strategies of scrub jays. The hiders were always tested in two conditions: in one (left), an observer could (partially) see the hiding space; in the other (right), this was prevented by an opaque partition. If the conspecific could see something, the animals were more likely to hide in those containers that were covered by a privacy screen (**a**), were in the shade (**b**), or were further away from the observer (**c**). Further explanations are provided in the text. (From [715])

dominant animals have the advantage of being able to successfully defend their food against conspecifics. It would therefore be interesting to see whether subordinate scrub jays use their strategies flexibly depending on the status of the competitor. Would they also unpack their best defence strategies if the opponent is dominant to avoid possible interactions that risk injury? Conversely, individuals should react more tolerantly if the only witness to the hiding is their mate because if they do not see the mate as a threat, they should refrain from investing energy in implementing defensive strategies. This is exactly what has been observed in the field; both with common ravens [724] and scrub jays [730]. Finally, this was also experimentally verified. When confronted with their mate and, in comparison, with a dominant and a subordinate bird, the jays significantly more often preferred the farther container over the nearer one as a hiding place with the dominant observer (and significantly less pronounced with the subordinate bird), but not with the mate and likewise not when they could hide behind an opaque partition wall [586]. This suggests that they discriminate between mates and other individuals and refrain from protective behaviour in the presence of their mate. Thus, the specific risk posed by a particular observer—and not just the presence or absence of an observer—seems to determine whether or not strategies are deployed to protect the food caches of jays.

Inspired by observations of natural food hiding behaviour, another laboratory experiment was devised. In the field, jays change their caches repeatedly by retrieving the food and hiding it again in a new place. It was assumed that this was a protective strategy because it confused potential thieves. This behaviour, reminiscent of the "*shell game*", makes sense, especially when you cannot hide your caches completely or permanently. Dally and her colleagues also devised an experiment for this purpose [728]. They had scrub jays hide under two conditions, which differed in the predictability of the observer's position and thus what the observer could and could not see. In the "restricted" condition, one hiding place was always in the observer's field of vision, and the other was always out of sight. In contrast, in the "free" state, the visibility of the hiding places depended on the observer's current position and thus carried a greater risk of being found. The jays were found to change their hiding places up to six times in the "free" condition but hardly at all in the restricted and predictable conditions. It is therefore reasonable to assume that confusing the opponent is a strategy in response to the unpredictability of how well another individual could have observed the hiding.

This assumption could also be confirmed by the fact that the jays were much more likely to change hiding spots if they were observed by a conspecific while hiding. By this tactic of constantly moving food, jays significantly reduce the likelihood of theft, as observers cannot rely on their observations to specifically raid a hiding spot.

In nature, it is rather rare that only one observer is around when a jay is hiding food. Much more often, these sociable animals have several conspecifics around and thus potential thieves in the vicinity. This raises the question of whether the jays keep track of who was watching when they were hiding food. This situation was created in the laboratory at the University of Cambridge by having two observer jays present at different times to see another jay create caches [586]. In the first test, a hider was allowed to hide in succession in different ice cube containers in the presence of one observer each—in container A for observer A and in container B for observer B. Three hours later, he was allowed to return to his caches. He was allowed to collect his food under one of four conditions: either under observation of A, under observation of B, without the presence of an observer, or else in the presence of a third observer (C) who had not watched the hiding. As predicted, the jays were very selective when changing their caches. They fetched food from container A in the presence of observer A and from container B in the presence of observer B and then moved the food. In the presence of a new observer C, only a little, if any, cache changing was performed.

Although the results of this experiment suggest that the jays remembered who was present when food was hidden, it is possible that the hider is merely responding to certain cues from the observer. For example, observers might be more attentive to a location where they have seen a bird hiding food. Thus, observer A would be more likely to stare at container A, which might prompt the hider to change that hiding place. To check whether the jay was only using purposeful looking as a cue, another experiment was conducted in which a control observer was made to look at a container (because he had previously seen a control bird hiding there) to see whether the experimental animal took this looking as a cue or not [586].

Crucial was the comparison between two conditions, the "Observed" condition and the "Observer Control" condition. Initially, the procedures in both conditions were identical (Fig. 8.3). As before, the hider was allowed to hide food in succession in different ice cube containers in the presence of one observer for each hiding event. During each event, both containers were accessible, but Plexiglas strips covered one of the containers to prevent hiding there (symbolised by a circle in Fig. 8.3). Hiding therefore had to be done in the other container (in container B for observer A and vice versa). Thus, both bins were filled after these two hiding periods.

In the "Observed" condition, a finding test followed 3 h later with observer A. Now, the hider should remember that observer A watched him hide in container B (symbolised by a rectangle in Fig. 8.3) and that he should therefore relocate the food. However, since observer A remembered this hiding place (B) and was now also watching with particular interest, the simpler explanation would be that the hider moved the food that was being looked at by the observer. To invalidate the explanation of the observer's gaze, a control condition was introduced in which, after the two hiding phases, a control observer was given the opportunity to observe a second hider hiding food in hiding place A (container B was locked). When the original hider returned after 3 h in the presence of the control observer, he or she saw the latter staring at container A. If he would now use this observer's gaze as a hint, then he should relocate food B. However, this did not happen; the hider rather relocated from both containers unselectively. The results of these experiments therefore support the hypothesis that jays remember who was watching them during their hiding and that they use this information to reduce the risk that observers pose to their hiding places.

Taken together, these results on Western scrub jay foraging suggest that these birds have the ability to read minds. Obviously, they are sensitive to what

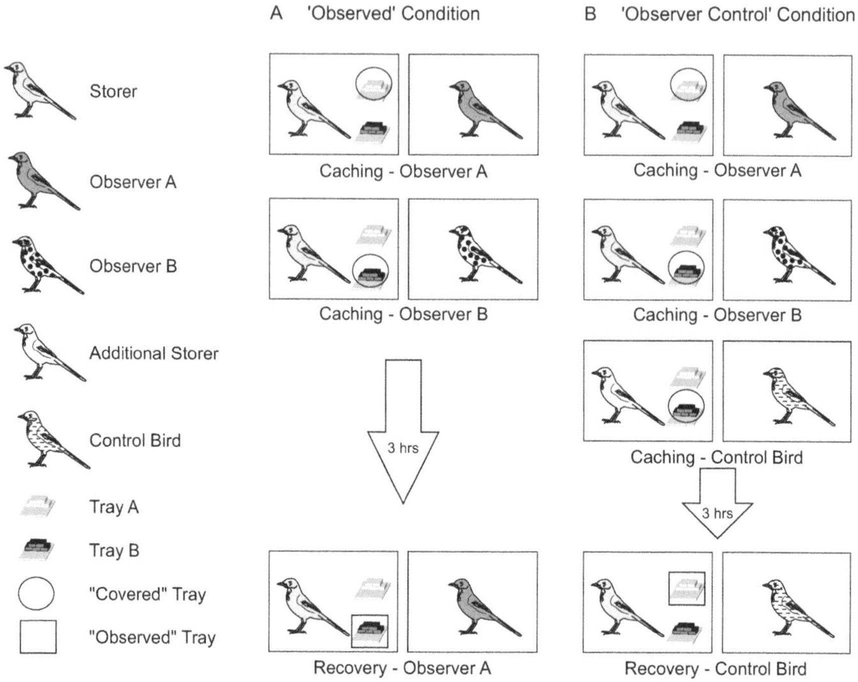

Fig. 8.3 Schematic sequence of the two experiments conducted by Dally and coauthors [586]. Column A shows the first experiment with the "Observed" condition, and column B shows the control experiment with a control observer. Explanations are provided in the text. (From [715])

information conspecifics, who sometimes act as thieves, and who have and do not have when creating food deposits for the winter, and they flexibly use various strategies to defend and protect them. They take into account conspecific presence and their identity, whether they are superior or inferior, known or unknown, and have a partnership relationship or not. To do this, they must recognise different individuals, remember them, and use that to take appropriate protective measures. The fact that they can do this even when these observers are not present at all means that they cannot use the current behaviour of competitors as a cue for their strategic actions. Changing hiding places in the absence of competitors is not a blind precautionary or confusion strategy but rather a very deliberate action, taking into account the circumstances at the time of hiding and afterwards. Keeping an eye on the competition does not mean merely reading their behaviour. These strategists are instead combining a wide variety of information gathered in different circumstances and at different times to make specific action decisions. This makes the use of innate behaviour and simple conditioning mechanisms highly unlikely.

Plausibility is one thing; proof is another. Nicola Clayton and Nathan Emery wanted to precisely understand and specifically examine the role of experience. They had noticed that not all animals moved their food caches, but it was especially those birds that had a previous record of being thieves [731]. Therefore, an obvious question was whether it is easier for birds in perspective-taking tasks to first experience the situation from the thief's point of view. With the controlled rearing of three groups of scrub jays, the researchers wanted to test the influence of experience in the two activities of hiding and stealing. One group, the "observers", was allowed to watch another bird hide; another group, the "thieves", was allowed to clean out the hiding places (and just listen to others hide); and a third group was allowed to do both (observe and steal). Then, all the animals were allowed to create food stocks in different containers twice: once in the presence of a conspecific another time in private. The question was now which animals would use the confusion strategy "change of hiding place". Interestingly, the pure observers hardly ever did so, but the members of the other two groups did so frequently and selectively. Birds without stealing experience did not use anti-theft strategies, even when they watched others hide their food.

The fact that only experienced birds used the confusion strategy has a number of important implications [715]. The first is that this behaviour cannot be innate; otherwise, all scrub jays should use it. Simple conditioning can also be ruled out, as the birds had never received positive reinforcement or punishment for changing hiding places. Thus, it is most likely that the jays used information gained during their own hiding and stealing to predict whether the hiding places were safe or not and thus employed the appropriate protection strategy. This anticipation of future risk also lends further fuel to the idea that they can plan for the future. Finally, Emery and Clayton [732] also argued that the importance of one's experience suggests that experienced jays are capable of both planning for the future and projecting experience. Specifically, the experience of stealing from a hiding place allows the bird to predict how another individual might think or behave—in this case, what the potential thief might do.

When birds with stealing experience are given the opportunity to protect their own hiding places, the thought pattern of stealing seems to be activated; they project onto another jay that is in the observer position and start switching hiding places. In humans, this ability to take another's action as an example has been traded as a form of introspection or deception. For the Cambridge cognitive researchers, this brings the scrub jays as close to a theory of mind scenario as has ever been observed in birds. Other researchers believe that it is hardly conceivable that jays must be the only ones in the bird world to do this. At least other corvids should have similar abilities.

Raven Intelligence

Thomas Bugnyar, my long-time colleague at the University of Vienna, was already interested in the strategic interactions of ravens in his doctoral thesis. Both in the Austrian Alm valley and in the US state of Vermont, he has observed ravens hiding pieces of food for themselves and from others, apparently very cleverly.

From the observations of the competing ravens, one could already conclude that the birds are capable of perspective taking. However, Thomas wanted to know for certain and—spurred on by the experiments with human children and chimpanzees—started to conduct tests with aviary ravens [723, 727, 733–735]. Since these animals, as scavengers, show a natural tendency to also care about the behaviour of other species in possession of food—predominantly large predators such as wolves and bears—in order to get some of it themselves, it is easy for the human experimenter to take part himself. Thomas could therefore play the hider and let the ravens watch.

In an experiment on the ability to distinguish between knowing and unknowing competitors, he was thus able to determine precisely who sees what and when [736]. With curtains, he could manipulate the view of the ravens involved, not only concerning the hiding places he had created but also concerning other onlookers. By deliberately opening and closing the curtains, Thomas could thus determine when a raven could watch and thus what information it would receive about hiding places and about other viewers—similar to Povinelli and his human informants [669] or Hare and his competing chimpanzees [708].

Several tests were carried out [736]. Three ravens took part in all of them, sitting in separate compartments of a large aviary. In the middle was the test compartment, where Thomas, as an experimenter, first created the hiding places and then tested the focal animal. Two observers sat opposite this supposed mind reader in separate compartments on the other side of the test compartment. They were later allowed to enter the test compartment with the focal animal and thus competed for the food placed there. The role of the focal animal and the roles of the two competitors were always taken by different animals.

In detail, the logic of the first experiment was as follows (see Fig. 8.4). The focal animal, which had first observed the experimenter create two hiding places (V1 and V2) in succession, was then confronted with a competitor. The competitors differed in the amount of information they received about the hiding procedures. A competitor could either watch both, none, or only one of the hiding procedures and thus have complete, no, or only partial information about the hiding places. While fully informed competitors posed a high risk of looting any hiding place, partially informed competitors only posed a risk regarding the hiding place they had seen. Thus, if the focal animal was allowed into the test arena just before a fully informed competitor, it should hurry to excavate both hiding places, in contrast to the situation in which the competitor was completely unsuspecting. The two hiding places should also be chosen carefully when confronted with a partially informed competitor; the focal animal should then move immediately to the hiding place that was known to the competitor.

It was important—as in Hare's chimpanzee tests—that the focal animal received a head start, i.e. was allowed into the test arena slightly earlier. Only when it had made its first decision by touching a lid was the competitor allowed into the arena. The slanted positioning of the covers over the hiding places also allowed the competitor, but not the focal animal, to see each of the "hidden" pieces of cheese (Fig. 8.4). Since the competitor could see both pieces of food, he or she had no gaze

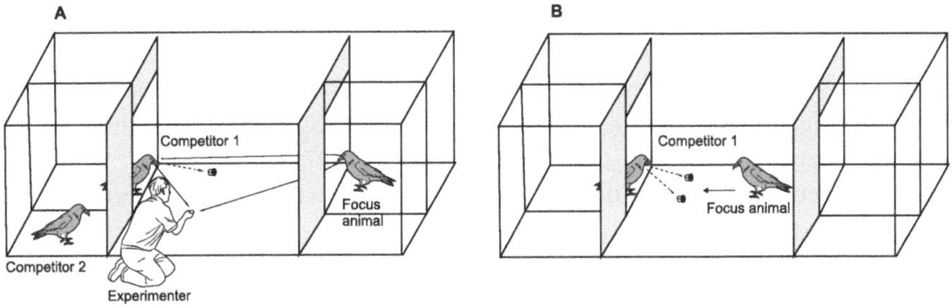

Fig. 8.4 Sketch of the experiment on the ability to distinguish between knowing and unknowing competitors in common ravens [736]. Left: the experimenter hides cheese in the presence of three ravens, two potential competitors on the left and the focal animal on the right. The opaque curtains on the doors of the rear compartments are raised so that Competitor 1, and the focal animal can see the hiding and each other; Competitor 2's view of the central test room remains blocked by the curtain. Right: during the test, the focal animal approaches the hiding places, while Competitor 1 remains openly visible but still locked in its compartment. The slanted positioning of the covers on the hiding places allows the competitor, but not the focal animal, to see both pieces of cheese. (From [736])

preference. This means that the focal animal could not rely on any behaviour or other signals, such as looks from the competitor, to make its choice. Together, these findings indicate that the focal animal could only act strategically if it remembered who had seen what before.

The results of this first experiment pointed strongly in the direction of actual perspective taking, but they were not conclusive. The focal animals were indeed faster when the competitor could watch. They also chose the hiding place that the partial observer could see, but only if it was the first of the two (H1). When the partial observer could only see the creation of the second hiding place (H2), the focal animals remained undecided. This suggests a specific memory phenomenon, namely, better recall of the first event in a sequence (the so-called primacy effect). However, this finding again supports the assumption that the focal animal referred to its memory when it was allowed into the arena. These results were then compounded by a second problem that Thomas became aware of. His experiment could not determine whether the focal animal's memory included the idea of what the competitors saw or just the simple fact of who was visible during the hiding events. The second, perspectiveless memory content would suffice to form the rule: "Watch out for those you saw when hiding was going on!" This rule would explain all of the results.

The incomplete evidence for strategic action based on memory or imagination of what others saw prompted Thomas to conduct a second experiment. Now, it was to be about competitors who were visible to the potential mind reader but could not see the hiding themselves. He managed this with a simple trick. While the visible competitor was standing on the ground in the first experiment, he now sat on a branch one and a half metres above the ground, and in half of the experiments, a curtain with a window was drawn over the observer's compartment. In these "window trials", the focal animal could see the observer, but the lower part of the curtain blocked the observer's view of where the pieces of cheese were hidden. Thus, the competitor was freely visible but still not an observer of the hiding processes. Would the focal animal understand this? Would it be able to imagine the competitor's view or line of sight to know that the freely visible competitor is nevertheless not an observer? A comparison with the other half of the tests, where there was no window curtain, was to decide that.

The results were clear: the two conditions with and without windows differed enormously. In the windowless trials, the focal animals showed a clear tendency to match hiding places with competitors; they chose hiding place H1 when tested in the presence of the competitor visible when H1 was used and H2 in the presence of the competitor visible when H2 was used. In contrast, there was no systematic correlation between the identity of the competitor (C1 or C2) and the first hiding place selected by the focal animal in the window trials. It is obvious that this result is a strong indication that ravens do not or at least not only base their strategies on the behaviour of others or their presence at the time of hiding but (also) on what the others could actually see. Therefore, the ravens must have used "*sight lines*" or "*eye-object lines*" [737]. For example, when the focal animal was confronted with Competitor 1 (C1) in the test, its choice had to be based on a memory of whether there had been an unobstructed straight line between C1's eyes and the first hiding place H1, or the second hiding place H2 before that. If this was true for C1-H1, the focal animal flew immediately to H1; if this was true for C1-H2 to H2, and so on. In any case, for this to happen, the ravens must have a sophisticated understanding of visual barriers and be able to judge the other's view in relation to their own.

In fact, it has previously been demonstrated that ravens possess the ability to track geometric gaze, i.e. to understand another's gaze behind a barrier [738]. That this ability is a higher cognitive matter has been proven by the fact that ravens develop it at the age of approximately 6–8 months [739], whereas simple gaze following has already been shown after 2–3 months of age [740]. I have already suggested this geometric gaze-following as a basis for perspective taking in the case of our experiment with domestic dogs, which revealed a preference for a knowing informant over an unknowing or guessing informant for finding hidden food [688]. In the primate literature, such an ability has been termed "*higher order intentionality level 1.5*" [665] or a social-cognitive schema for understanding the psychological states of others [741]. In child psychology, the same ability has often been considered a precursor or even an element of consciousness [742].

The theory of mind interpretation of the ability to accurately predict the behaviour and intentions of others, while foraging would also fit perfectly with the ravens' repeatedly demonstrated ability to strategically control their own intentions, such as by selectively withholding information [735] or, in other words, tactical deception [723]. Considering all the experimentally demonstrated abilities of ravens to behave tactically, especially when dealing with conspecifics while hiding or looting food caches, it seems reasonable to assume that they assess their counterparts based on memories of their counterparts' viewing opportunities for crucial objects during critical events. If this understanding of viewing opportunities is equivalent to an understanding of the other's perspective, we would have a basis for a genuine theory of mind.

For one of the harshest critics of theory of mind in non-human creatures, Cecilia Heyes, who has praised Thomas's two experiments as exceptional in the care and precision with which they were designed, interpreted, and reported, this would amount to "*minimal mind reading*" [743]. But she nevertheless hesitates to grant the animals this minimal version, let alone complete mind reading. Indeed, both versions of the experiment would be crystal clear evidence that the ravens did indeed use the eye-object line as an indicator of what other ravens saw and thus knew. Ravens could integrate associations between specific events—where is the other looking, what can I see myself, where is the food, when is the other looking where, etc.—at a higher cognitive level to reduce the complexity of social interactions by using abstract "*intervening variables*" that code for mental states in others. These variables then represent "*seeing*" or "*knowing*" [736]. Andrew Whiten considered such intervening variables a far more parsimonious explanation than remembering particular individual associations because the number of possible observable conditions in all possible situations in a raven's life would multiply [697]. A simple example given by earlier ethologists is thirst, which is a designation of a condition in the drinking behaviour of rats. Only three independent states (hours of deprivation, feeding dry food, injection with salt water) and three possible dependent variables (number of lever pressures in the Skinner box, volume of water drunk, quinine tolerance) would yield nine possible associations. Similarly, seeing would be a form of representation of conspecific states that is applicable to many situations in the context of food hiding and defence.

Cecilia is not convinced of this because it would still leave an interpretive gap open. She points to insights from cognitive science regarding alternatives in terms of cross-domain "*submentalising*" processes, i.e. those that lie below the understanding of mental states (in oneself or others) [744]. The effect of window manipulation would not be sufficient to conclude that the ravens must have used either minimal (eye-object line) or even complete mind-reading skills [743]. It is possible that in the tests of the experiments without windows, the ravens retrieved from memory the physical appearance of the competitor sitting in the compartment on the opposite side and that this image in turn indicated the place where food was hidden in the presence of this competitor. We humans also sometimes suddenly find a key again when we see a picture of the place where we lost it. These are encoding cues [743]. Since the choice of the hiding place was rewarded in the past, the focal animal henceforth always chose the hiding place that corresponded to this competitor. Cecilia admitted, however, that it is unlikely that the ravens learned this during the few trials of the experiment. But there would be no reason why they should not have learned this in the course of their lives, where there may have been many such competitive situations. In principle, however, Thomas also conceded this. Much more interesting, therefore, is how Cecilia could explain the results of the experiments with the window with mere rules of behaviour. She thought that the fact that the ravens in the window tests could not match the hiding places with the competitor could be an encoding specificity effect [567]. This is because in the window trials, but not in the windowless trials, the competitor appeared in a different context in the test (without a curtain) than in the hiding phase before (behind a window in a curtain). Therefore, it is possible that in the test, when the focal animal has to choose, it has fewer encoding cues available when retrieving from memory. Thus, the physical features of the competitor are less effective at activating a representation of the location of the appropriate hiding place, and the ravens choose randomly.

One may find this alternative explanation by no means more plausible or economical than simply assuming that the ravens imagine what the other sees (and knows). But it suffices to no longer have to necessarily assume perspective taking in birds. In the scientific contest, the opposing side was challenged to come up with new experiments again to also prove the last "*killjoy hypotheses*" [484] to reject them. Therefore, Thomas, together with a student, Stephan Reber, and an American philosopher, Cameron Buckner, who had already done much work on the question of theory of mind, devised another experiment [745]. The aim was to show that ravens do not consider the actual viewpoint of others but rather the potential viewpoint when they cannot see the conspecific at all. The questions were whether the birds protect their feeding caches from detection by others but only in response to the sounds of conspecifics and when a peephole is opened by a partition but not when it is closed (Fig. 8.5). What can be seen through the open peephole in the wall they have previously experienced themselves.

The experiment was conducted in two compartments of a large raven aviary separated by a wooden wall (Fig. 8.5). The wall had two functional windows that allowed the view from one compartment into the other. Both windows could be closed with covers. In addition, each window had a peephole that could be opened or closed. At the beginning of the study, all the ravens were familiar with the experimenters opening or closing the windows between rooms but were still unfamiliar with the peepholes. The first step was to determine the hiding rate of individual ravens that were given pieces of food in their compartment and could be observed by conspecifics (the windows to the adjacent room remained open) or not (the windows remained closed). Crucially, the peepholes were closed under both of these conditions. In the next step, each raven was individually trained to observe and later find human-made hiding places in the adjacent room through one of the two peepholes.

In the test, each raven was then confronted with two initial conditions (window open or closed) and a test condition in which the windows were closed but one of the two peepholes remained open. Specifically, under these test conditions, there was no raven at all in the next room. However, to acoustically simulate the presence of a competitor, sounds were played through a hidden loudspeaker that had once been recorded by a competitor in the next room.

The results clearly showed that ravens behave as if they could be seen during the test because they projected their peephole experience onto the potential competitor. Ravens finished hiding faster in the peephole condition and improved their hiding faster than in the unobserved condition. Apparently, they do not need direct cues such as glances to imagine observers. Thus, they understood the seeing state in a general, abstract way, for which they drew on indirect cues such as hearing and their

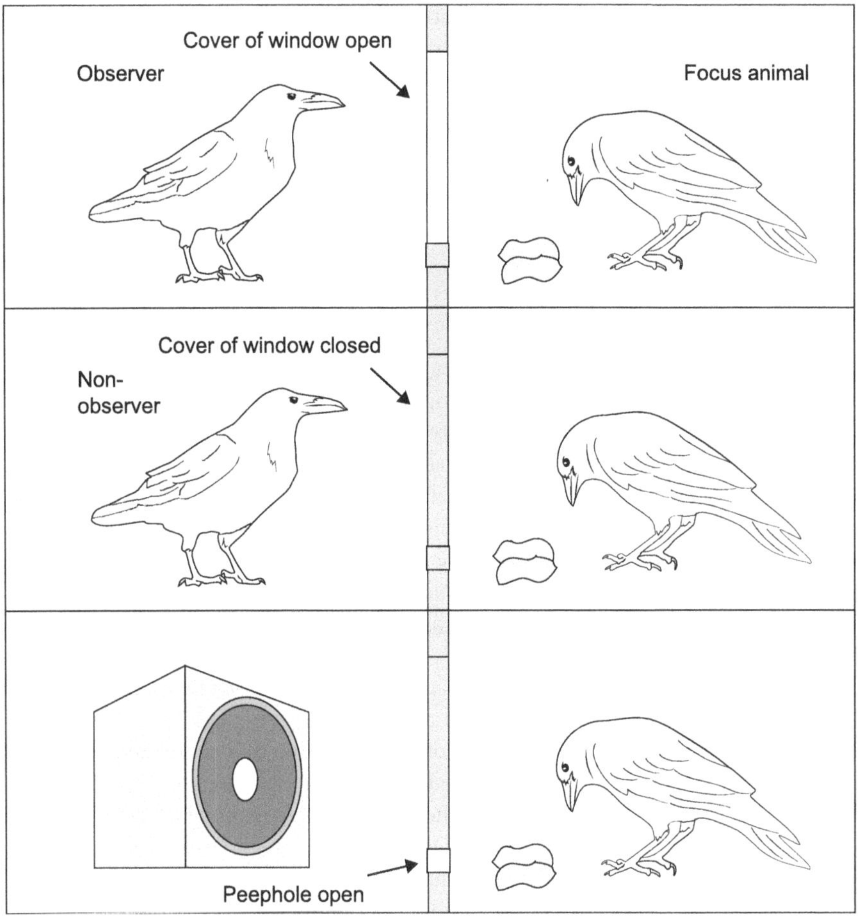

Fig. 8.5 Sketch of the experimental setup by Bugnyar and colleagues [745]. Top (observed hiding): the focal animal hides a piece of food from the observer's eyes because the cover of the window in the partition is open (white bar). Middle (unobserved hiding): the focal animal hides a piece of food unnoticed by a conspecific (non-observer) because the cover of the window in the partition is closed. Both observer and non-observer make sounds in their compartment that are audible to the hiding focal animal. Below (hiding with a peephole): the cover of the window is closed, but one of the two peepholes is open; therefore, the focal animal hides in the absence of behavioural cues such as gaze and gaze direction, but the presence of a conspecific is mimicked by the reproduction of sounds from unobserved hiding attempts. (From [745])

own experience. One might conclude that ravens cannot be reduced to mere behavioural readers, since they seem to have the capacity to be guided by their ideas. But the question remains whether they imagine a mental state of another, seeing as an ability or mental quality of the other, even as the current driving force of his action. When they hide, do they think "the other could see me" or rather "I could be seen"? Attempts have been made to approach this question by examining negation: can animals guess what the other does not know? Can animals imagine that others lack knowledge or are in error? As we have already discussed in the Chap. "Sally and Anne", the two developmental psychologists Heinz Wimmer and Josef Perner were involved in the development of the corresponding tests for children [648], and the first experiments with non-human creatures were again undertaken by the Leipzig primatologists.

Chimpanzee Chess

The experiments of Brian Hare and colleagues [709] have suggested that chimpanzees are inferior to human children in recognising false assumptions. Leipzig primatologists, led by Juliane Kaminski, were not satisfied with this assessment and devised new experiments in which chimpanzees and humans were tested in parallel and in the same manner [746]. The aim of the study was to test, comparatively, the ability to understand or recognise two related mental states of others: *knowledge–ignorance* and *false belief*. The tests were conducted in an almost identical manner with 10 chimpanzees (aged between four and 29), 12 children (aged 6), and 12 adults (aged between 21 and 41). In this comparative experiment, which later became known as "*chimpanzee chess*" [747], competition for coveted objects (favourite foods for chimpanzees such as grapes or pieces of banana, small toys for the children, and cardboard coins for the adult test subjects) took place on a table-like platform that was pushed back and forth between the two opponents.

There were three cups on the platform under which the objects could be hidden. In the first test, two cups were filled by the experimenter, and one remained empty. The hiding process took place in front of the eyes of the two competitors; however, one of them—we will call him the ignorant one—could only see how the first hiding place was filled with an object; we will therefore call it the known object—after which his view was blocked by a partition. The competitor saw everything and therefore knew under which cups the known and the—for the ignorant—unknown object was. After the hiding process, both competitors had the chance to retrieve an object by pointing to the cup of their choice. Who chose first depended on the test condition—once it was one, once the other—in any case, this choice remained hidden for the other because his view was blocked in the process. The crucial question now was how the competitors, equipped with full knowledge of the hiding places, would choose when they went first or second. In the case of the first choice, the ignorant individual should choose the hiding place known to him. However, if the ignorant individual only has his chance after the competitor, he should, to maximise his profit, choose one of the other two cups because he must assume that the competitor has already emptied the hiding place he knows.

The situation is somewhat different for the Knower. He should have noticed that the ignorant one only saw the first hiding place but not the second. If he only has his chance after the ignorant person, he should specifically choose the second hiding place under the assumption that the ignorant individual has already emptied the hiding place known to him. In the case choosing first, this consideration does not matter; here, the knowing competitor can safely empty one of the two hiding places known to him. In fact, this difference was found in all three test groups; it was more pronounced in adults and children but was also significant in chimpanzees.

These results are thus further evidence that not only humans but also chimpanzees are able to assess the knowledge of others in the restricted sense of "having seen something". Here, the animals correctly inferred what another had done a few moments earlier, based not on the other's behaviour but on what they had seen before. Since the current chooser can never see what the competitor has chosen, they must understand the competitor's mental state—what the competitor knows—and infer the competitor's choice of cup and avoid that cup thereafter. With this elegant procedure, it was possible to rule out the possibility that the behaviour of a chimpanzee (or a child) is determined only by the current behaviour of the competitor. It was also possible to invalidate an argument against earlier studies, which entered the literature under the title "*evil eye*". Here, a simple behavioural rule would apply, namely, to always avoid the "dangerous" object that is looked at by the other. The competitors of chimpanzee chess chose this dangerous object when they had their first choice.

After it could be shown once again that chimpanzees orient themselves to the knowledge or ignorance of opponents, similar to 6-year-old children and adults, Juliane and her colleagues used chimpanzee chess to test the orientation to false beliefs. This is when an individual believes something that is different from reality. Unlike knowledge or ignorance, understanding a false belief requires not only imagining *whether* the individual is aware of the actual, current state of their environment but also *how* they construct the world to be other than it is. In humans, this ability seems to be closely linked to language development [748].

In the second test by Juliane and colleagues [746], in some cases, the knowing competitor saw the experimenter mislead the unknowing one by first creating a hiding place in front of his eyes but then moving it when his or her vision was blocked. Sometimes, on the other hand, the experimenter opened the hiding place and took out the food but then put it back there. Does the knowing competitor understand and exploit these processes? Can he attribute to the ignorant one the ignorance of the deception and the consequent false conviction concerning the hiding place of the food? Will he logically assume that the ignorant person is wrong in his first choice when the hiding place has been changed in his ignorance and therefore the reward is still waiting when his turn comes? Conversely, will he assume that the ignorant person has taken the reward if the hiding place has not been changed because the experimenter has put the piece of food back into the original hiding place?

In this experiment, 3-year-old children and, as before, 6-year-old children were tested in addition to chimpanzees. In all tests, the unknowing individual received the first choice, and the knowing individual then had to decide whether to choose the container with the actual hiding place. Alternatively, he could choose a less attractive reward that was placed in a way that was visible to both opponents. All of these results confirmed those of earlier studies. The chimpanzees again showed an understanding of the opponent's knowledge state based on what the opponent could see.

However, they did not understand that what they saw could also lead to false knowledge or to false assumptions if they encompassed only part of what was actually happening. Specifically, the knowing chimpanzees chose the cup in which the piece of food was finally placed when their competitor, equipped with half-knowledge, had not seen the final placement. However, they did not distinguish between those two cases where the experimenter manipulated the hiding place in ignorance of the ignorant competitor, where the reward remained in its original location because the cup was only raised so that the ignorant competitor could know where it was still, or where the hiding place was changed so that the competitor could not know where the reward was. They could not take advantage of their competitor's false belief. The 3-year-old children, but not the 6-year-old children, also failed to attribute this higher mental state. This finding is essentially consistent with an earlier study involving orangutans as well as children and chimpanzees [749].

In developmental psychology, it has been recognised that the ability to take on perspective develops slowly and has therefore introduced further distinctions. John Flavell and colleagues from Stanford University (USA) have distinguished two stages [750]. In the first stage, children can distinguish their own perspective from that of others in that they understand that they can sometimes see objects that others cannot see and vice versa. In the second stage, they develop a deeper understanding of the other person's perspective because they understand that the same object may appear differently from another perspective [647]. Several Leipzig studies demonstrated that chimpanzees are capable of Level 1 perspective taking but not of Level 2 perspective taking.

A few years later, chimpanzee chess was used again to test whether chimpanzees are capable of knowledge attribution in an extended sense [747]. In particular, an attempt was made to exclude behaviours that might have influenced the animals' decisions. This was achieved by the fact that the two chimpanzees did not see each other at all during the test and thus had no chance at all to react to competitor behaviour. Their decisions therefore had to be derived from earlier perceptions and from their own experiences. Furthermore, they aimed to test a level of knowledge attribution between 1 and 2—not a false belief rather a false perspective.

Two experiments were conducted to assess whether chimpanzees can correctly predict a competitor's behaviour based on the competitor's perspective. Again, two competitors, A and B, sat opposite each other; this time, they were asked to choose one of two breadsticks in turn (Fig. 8.6). These were attached to an opaque panel, next to each other on the side, and facing B in such a way that A could only see the parts protruding from the top. Although both breadsticks were the same length, one stuck further out the top and therefore appeared longer to A. Would competitor B understand this deception or "false perspective" of A? If A could choose first but B did not see this choice (it happened behind a screen), would B choose the breadstick that appeared shorter to A? B initially saw the breadsticks, but when he chose a side, he no longer saw them. He had to remember the positions and then possibly make an assumption about which breadstick A might have chosen. As an important control for this social competition, there was another condition where there was no competitor.

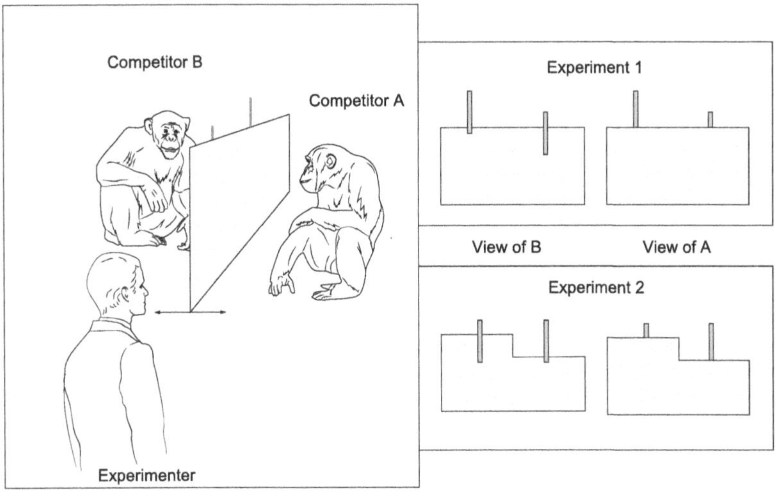

Fig. 8.6 Competition for breadsticks. The two breadsticks were the same length in both experiments, but they protruded further upwards in the first experiment. While they appeared to be of different lengths from the point of view of competitor A, they looked the same length from the point of view of competitor B. Nevertheless, unlike the children, the chimpanzees showed a preference for the pole that protruded further upwards when they were in the role of competitor B in the control without competitors. (From [747])

The results of the first experiment were ambiguous; although chimpanzees who chose second avoided the side on which the apparent "longer" breadstick was located more often in the social condition than in the no-competitor condition, like 6-year-old children, their choice in the social condition was nevertheless not significantly different from chance. In contrast, the children in the social condition chose the longer breadstick from the other person's point of view in only a quarter of the cases. To determine whether, in the no-competitor control, the chimpanzees chose the "longer" breadstick simply because it protruded higher, a second experiment was conducted. Although the illusion for A was maintained, the breadsticks for B now also protruded to the same height (Fig. 8.6). However, the results hardly changed; only the preference for the breadstick that was longer from the other person's point of view disappeared in the nonsocial condition. These results suggest that chimpanzees successfully predict what a competitor would choose when they are guided by their own preferences. However, when they have no preference for themselves and need to understand the other person's incorrect perspective, they cannot do so. In contrast, 6-year-old children can compare their own perspective, which corresponds to facts, with the competitor's false perspective and thus successfully develop an understanding of the latter's perspective: Level 2 knowledge attribution.

This study basically confirmed what several previous studies have found, namely, that chimpanzees have problems with contrasting perspectives [746, 751, 752]. They also fail to fool others with misleading information when hiding food from competitors [753]. Although chimpanzees can successfully attribute knowledge and ignorance to others, including informing group members of an impending danger [754], they did not seem to be able to detect misjudgements in others, even if this detection would give them strategic advantages. Apparently, this seems to be a truly (too) large cognitive challenge.

If imagination is not sufficient for assessing perspectives and appearances by combining different experiences, perhaps concrete experiences could help? For example, in the study with breadsticks, chimpanzees had no experience with the competitor's perspective, whereas in other studies on understanding the difference between appearance and reality (in "*appearance/reality tasks*"), they had previously taken the true perspective successfully [755, 756]. If a different perspective can be recalled from memory through concrete self-experience and is not generated first by the corresponding imagination, then chimpanzee perspective taking could also succeed at Level 2.

The Goggles Experiment

Although proposed decades ago, until recently, there have been few experiments aimed at directly testing whether animals can transfer their own experiences to others, thus revealing another's perspective and eventually another's mental states. As mentioned earlier, chimpanzees failed the bucket-over-the-head task in early studies discriminating between Knowers and Guessers, and the apes were no more successful after gaining experience with transparent and opaque views [757]. However, children as young as 18 months were able to transfer their own experience to their counterparts with transparent and opaque blindfolds [758]. Both types of blindfolds looked opaque from the outside, but it was still possible to see through one blindfold. After two groups of children tried either one or the other blindfold, their understanding of how an adult sees was tested in an eye-tracking task. An experimenter sat blindfolded facing the child and turned towards a target object to the left or right. The children who had tested opaque blindfolds followed the experimenter's gaze less than did those who had previously tried transparent blindfolds. Apparently, the children used their self-experience to deduce what the experimenter could see. (The experiment with children essentially followed Cecilia Heyes' suggestion [737], later refined [759], and entered into the literature as the "*goggles experiment*" [760].)

Katja Karg and Martin Schmelz, together with their Leipzig supervisors Josep Call and Mike Tomasello, repeated the children's experiment in a slightly modified form for their chimpanzees [760]. Instead of blindfolds, the Leipzig primatologists used yellow or blue face masks that were held in front of the eyes of the chimpanzees (during the experience phase) or the experimenter (during the test phase) (Fig. 8.7). In the first experiment, they tested whether chimpanzees, like children, were less likely to follow an experimenter's gaze when they saw him or her with an opaque face mask. Although chimpanzees have some flexibility in gaze-following behaviour (including geometric gaze-following [689]), the results were

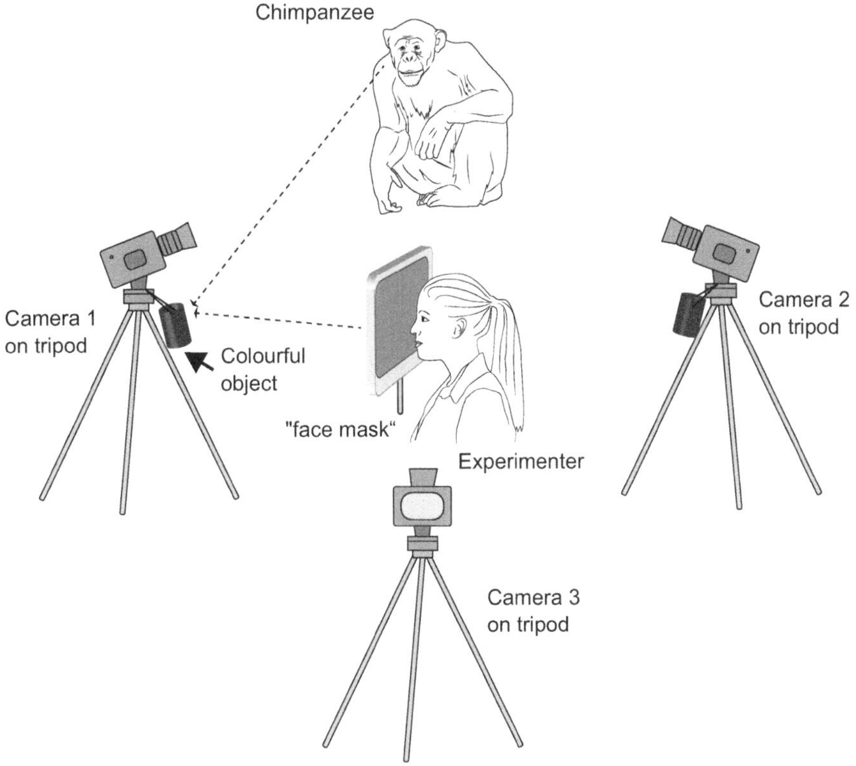

Fig. 8.7 Sketch of the experimental design of the first experiment by Karg and colleagues [760]. In the test phase, the chimpanzees sat opposite the experimenter, who wore a face mask and looked at the coloured object underneath the camera that records the chimpanzee's gaze behaviour. (From [760])

disappointing. Unlike 18-month-old children, chimpanzees exhibited no difference in gaze-following behaviour.

In the second experiment, the researchers chose a competitive task because it was already known that chimpanzees act more strategically in those than in cooperative tasks [761]. When Katja sat opposite a chimpanzee that had just been tested during training, she tried to steal peanuts from one of two boxes on the left or right side as soon as the chimp reached for them. She succeeded with the box that was closed with a transparent lid but not with the one with an opaque lid because, here, Katja could not see the chimpanzee's hand. The chimpanzees could learn which lid was transparent and which was opaque and what the consequences were in the competition.

In the subsequent test phase, the boxes had three different types of lids: opaque, transparent, and fitted with a black fly screen. While the opaque and transparent lids did not change their visual properties between their open and closed positions, the chimpanzees could only see through the lid with a fly screen in its open position but not when it was closed (the shallow angle of vision makes it opaque). In contrast, Katja could see through the fly screen from her perspective in both conditions. The animals were first familiarised with the visual properties of the lids when open, after which the lids were closed and the chimpanzee could choose which box to take the peanuts from. There were two pairings of boxes, either the opaque and transparent boxes or the opaque box and the box with a fly screen. A control without an experimenter was used to check whether chimpanzees had a general preference for the opaque box.

The results of the second experiment provided a first, albeit relatively weak, indication that the chimpanzees could pass the goggles test. The animals chose the opaque box more often in the competitions with Katja, although only just by chance: 55% in the pairing with the transparent box and just under 60% in the pairing with the box with a fly screen. In contrast, there was no preference for any box in the control without a human competitor. The primatologists interpreted these results to mean that chimpanzees can use their own experience with visual properties of lids to deduce in which box their hand remains hidden from the experimenter. The decisive factor for interpretation in terms of perspective taking is that at the moment of choosing between the opaque lid and the lid with a fly screen, both appear opaque from the chimpanzee's perspective, although they have previously had the experience of being able to look through the fly screen from a different angle. This means that the chimpanzees were apparently able to compete successfully because they projected their self-experience of an object—using an arbitrary feature such as its position—onto the human competitor to predict what the human could see.

Explanations based on learned rules of behaviour are unlikely in this novel and highly artificial situation. Rules could not have formed because the chimpanzees behaved the same way from the beginning, and there was no improvement over time. One could still argue that they did not form rules during the course of the experiment but rather earlier. However, the chimpanzees in this study lived on Ngamba Island, an island on Lake Victoria in Uganda that serves as a sanctuary for rehabilitated chimpanzees. Therefore, it is unlikely that the subjects were ever exposed to such experiences.

Another explanation would be orientation solely on the lids and their visual features. But this can also be ruled out because then the chimpanzees would have primarily chosen the box with the opaque lid even without competitors. Finally, lower-level behavioural reading is also ruled out as an explanation because the chimpanzees did not see their competitor during the competition for peanuts, and thus they could not orient themselves according to their opponent's body posture or line of sight.

Anticipatory Looking

In 2016, Christopher Krupenye from Duke University (USA) and Fumihiro Kano from the University of Kyoto (Japan), together with Satoshi Hirata, Josep Call, and Mike Tomasello, tested a total of 40 great apes (19 chimpanzees, 14 bonobos, and 7 orangutans) for implicit understanding of false beliefs using an eye tracker [762]. Implicit because the apes did not have to make explicit choices in terms of objects through body movements or hand signals. Instead, their visual attention was used to test whether they understood that others could have a false picture of reality. Previous studies have shown that apes look where they expect an event to occur, even before it actually happens. In this experiment, the animals were trained to watch short videos. The basic logic for these videos was taken from an earlier experiment on false belief with children [763]; only the content of the videos was changed. In two experiments, scenes were shown with two actors, a human and a figure resembling a gorilla ("Kong"). These scenes had already proven their worth earlier because they attracted strong attention from the viewers. Since it can be assumed that the scenes are very unusual and novel for apes, both in terms of the actors and their actions, reactions based on rehearsed behavioural rules are unlikely. First, the animals watched two habituation videos in which Kong either hid himself in one of two haystacks (Experiment 1) or a rock in one of two boxes (Experiment 2) while watching the human actor in the background. This was followed in each case by a test video in which the target object (either Kong himself or the stone) moved from the first hiding place to the second.

One time the human actor was present and thus witnessed the move, and another time he was no longer present. The (different) videos always ended the same way: the human actor disappeared behind a door, then the target object left the hiding place—Kong exited the haystack and disappeared, or he took the stone out of the box and disappeared with it—and finally, the human returned and searched for the target object. The video ended before he had decided where to look. This is the critical phase of the experiment because now the researchers measured where the ape looked ahead—to where the target object was hidden or to where the human actor is presumably looking for it. If the change in hiding place was observed by the human actor, he should search in the same place; however, in the other condition, he should not [762].

Of the 52 animals, 18 did not make an anticipatory gaze movement in one location or the other, and 37 looked at the correct hiding place that the human actor should have known about and searched. In the critical condition, where the human actor did not see the object change places, 19 animals looked to the place where the human actor would mistakenly assume to be the hiding place, and only 9 looked to the actual hiding place. Thus, more than twice as many apes seem to have been guided in their anticipatory looking by the human actor's false assumption. They were thus apparently able to imagine not only what he could see and therefore know but also what he thought he knew, even if this knowledge differed from reality. The perspective taking of these apes would thus have reached Level 2 [647] when they understood not only *that* the world may appear different to others but also *why*.

Although this result convinced even the notoriously critical reviewers of *Science*, doubts remained. It could be that the apes trusted a simple rule of behaviour, namely, that (human) actors are most likely to look where they last saw things. The possibility of applying this rule, which is either innate or has been inferred from many observations, has been put forward as an argument against perspective taking both in child studies and more generally in studies with non-human subjects [764, 765]. As already mentioned several times, one can almost always consider the possibility that subjects orient themselves to subtle behaviours, gestures, and looks of the actors to deduce rules about what they will subsequently do without having to recognise the content of their thoughts.

You can also guess what another person is thinking about and will subsequently do if you bring in your own experience of similar situations and project it onto the other person. This process is called perspective taking through projection of self-experience. As previously explained, the goggles test has been established as a simple variant for testing this ability [737]. Not only do children use their self-experience with different materials (goggles, barriers) to deduce what a blindfolded experimenter can see but also chimpanzees [751].

The combination of a goggle test with a gaze-following test has proven to be particularly fruitful for proving another person's understanding of a false belief. This was also the first [758] and repeated [766] study in human children. Although all children were shown the same video sequence at the age of 18 months, in which an actor could see an object being hidden in one place and was then blindfolded while the object was removed, only those children who had previously experienced the blindfold being opaque anticipated that the actor would later search for the object in its original—and now wrong—place. In contrast, children who had previously experienced the equivalent transparent blindfold anticipated the search to be at the correct, i.e. second, location [766]. These children obviously concluded from their prior knowledge that the actor could also see through the blindfold and therefore see the change of the object's position. An important detail in this experiment is that nothing in the actor's behaviour betrays this difference between the two conditions.

Recently, this goggle (or blindfold) test was repeated with 47 great apes (29 chimpanzees, 14 bonobos, and 4 orangutans), but they had relevant experience with a barrier instead of goggles [767]. While one group of 24 animals experienced the barrier as translucent, the barrier, which initially appeared identical, was opaque for a second group of 23 animals. Afterwards, both groups watched the same video in which, as in the children's study, a human actor was fooled or not, depending on the transparency of their barrier. In the video, the apes could see a human in a gorilla costume ("Kong") in the cage (in the foreground) hiding an object under a box in the left area of the picture. On the back of the frame, outside the cage, a human (actor) is seen watching the gorilla do this. The actor then moves behind a barrier attached to the grid wall separating the actor from the gorilla, which looks exactly like the one the apes were able to explore themselves before. While the actor remains behind the barrier, Kong takes the object out of hiding, places it first in a second hiding place on the right side of the picture, and finally removes it completely. The eye movements of the apes were measured at the end of the video, when the actor reached into the cage "ambiguously", i.e. straight down the middle through the bars. If the apes could use their experience with the barrier to infer the actor's perspective, those who perceived the barrier as opaque should attribute a "false belief" to the actor and anticipate his search at the old hiding place. More specifically, they should direct their gaze towards the first, left-hiding place.

This is what happened for those with experience of the opaque barrier [767]. However, this task was slightly more complicated for the animals that had come to know the barrier as translucent. They were supposed to assume that the actor could also observe all the processes in the cage. The actor should not only see Kong move the object from the first to the second box but also see him retrieve it from there and take it with him. Knowing that the object has disappeared altogether, individuals should not direct their gaze more strongly at any box when the actor puts his or her hand through the bars. Indeed, these apes' gazes were randomly distributed across the screen.

Helping

Although great apes passed the goggles test in an implicit way, namely, by anticipatory looking and displaying an understanding of false beliefs, experts are not yet convinced. There are still reservations on the part of philosophy but also on the part of cognitive research. What is lacking above all is an explicit test, one in which the animals show that they can go beyond implicitly predicting the behaviour of others—through anticipatory looking—and use their understanding of assumptions appropriately in real social interactions. Previous studies with great apes have been only moderately successful in this regard [747]. One reason for this could be that the apes had to recognise each other's beliefs to obtain food for themselves. This could distract them or require a greater degree of inhibitory control, possibly hindering the use of sophisticated social-cognitive skills. One way to prevent this would be to make the receipt of a reward in such tests not directly dependent on their choices but rather to reward choices independently. Leipzig researchers led by David Buttelman (MPI for Evolutionary Anthropology) have developed an implicit false belief test for this purpose. They first conducted this test with 1.5-year-old children [768]. Simplified testing methods enabled children under the age of 2 to pass the false belief test, which was previously possible only for 5-year-olds [765].

The infants were given the opportunity to help an experimenter open a box that contained a toy for them. As usual in such tests, there were two boxes and only one toy, which was first placed in the first box but subsequently retrieved and carried to the second box. In one condition, the experimenter could observe a second person move the toy in front of the children and thus know where the toy was at the end ("true belief" condition). In the second condition, she could not know the toy's final location because she left the room during the rearrangement ("false belief"). Now came the test for the children. In both conditions, the experimenter tried to take the toy from the empty first box after her return; only then did she turn to the second box. The crucial question was whether the children would help her open the first box.

In fact, the children passed the false belief test. In the false belief condition, the majority of the children did not help open the first box. The participants concentrated their help in opening the second box. In contrast, in the true belief condition, in which the experimenter had observed the rearrangement of the toy, the children helped to open the first box. The children apparently concluded that the experimenter had different motivations or goals for opening the first, empty box. A logical explanation for this could be that the children understood that the experimenter's attempts in the first case had to be directed at the toy but were guided by a false belief due to her ignorance of the shift; thus, she wanted to open the wrong box. To help here would be pointless. In the second condition, the experimenter had seen the relocation and should therefore know that the box was empty. Helping here could be worthwhile for another target unknown to the child. With differing behaviour to help between the conditions, the children showed that they had the ability to attribute different goals to the experimenter based on her true and false beliefs.

Great apes have also repeatedly demonstrated that they can anticipate the goals of others, for example, when confronted with intentional and unintentional actions [769, 770]. Thus, it was logical to also confront them with the test of helping that the children had passed. Chimpanzees, bonobos, and orangutans were to show that they too could infer different goals or motives to the actor. The apes could use their help in a targeted way by understanding the actor's various underlying assumptions [771]. In the first of two experiments, the majority of the apes were able to repeat the children's results. In the false belief condition, the apes might have simply followed the rule "if you do not know where the toy is, it is because you have been away for a while, look everywhere". So, the second experiment tested whether they could distinguish between false belief and ignorance. The animals did still behave differently in a corresponding ignorance control condition than in the "false belief" condition, showing that they actually assumed the experimenter's false belief in the latter. These results were seen by the authors as confirming the hypothesis that the apes could determine the content of the experimenter's assumption regarding the location of the object in the two conditions.

Clever Terriers

It is natural to think that solving the false belief task is, if not a purely human ability, at least one that we share only with our closest relatives, the great apes. To test this conjecture, in a new project on the evolution of the social mind, I and my colleague from the psychology department of the University of Vienna, Claus Lamm, started to investigate the question of the convergence of such abilities by comparing humans and dogs in the autumn of 2019 [772]. Previously, with two PhD students, Sabrina Karl and Magdalena Boch, we had investigated questions about dog–human relationships, empathy, and the understanding of goal-directed actions via functional magnetic resonance imaging in specially trained dogs. For this noninvasive imaging neuroscience technique, dogs are prepared for the environment and noise in an MRI scanner through sensitive, slow training, where wearing earplugs and remaining still for a few minutes is a major challenge [773]. Nevertheless, we had already been able to successfully test two dozen dogs with visual tasks—responding to the sight of people showing different emotions—in the scanner [774, 775]. These findings showed that dogs react differently to the sight of their owner than to the sight of another familiar person or a stranger and that the brain areas involved are exactly those that are also activated in humans during strong positive relationships (e.g. between mothers and children) [775]. It has also been demonstrated that dogs react sensitively to social-affective interactions between a human and a dog. For example, dogs show greater arousal and emotional evaluation in a situation that may be perceived as a potential threat to the bond with their caregiver, which has been interpreted as "jealousy" by some readers of the paper [776].

Our new project on the evolution of the social mind also focuses on the question of mind reading in dogs. In a behavioural test, we wanted to check whether dogs—after they had already provided clear evidence for perspective taking [688]—also passed the false belief test. My doctoral student Lucrezia Lonardo, my assistant Christopher Völter, and I were initially guided by the paradigm of Wimmer and Perner [648] in which active choice is supposed to distinguish between the true and false beliefs of a human agent.[2] To do this, we adapted a hide-and-seek task that was created to test the false belief understanding of children, chimpanzees, and orangutans [749]. We asked dog owners to let their dog choose between two buckets placed far apart in a room [777] (Fig. 8.8).

Initially, dogs perceived both the hiding of coveted food (sausages) by an experimenter (the hider) and a cue for the (supposed) hiding place from a second person (the informant). In one of these familiarisation phases, the dogs could only find the sausage in the opaque bucket if they followed the informant's cue—she walked to the bucket, lifted it up several times, and looked at the dog in the eye—because only the informant could see the hiding place, not the dog. The dog was able to build trust in the informant, who announced the hiding place through clear signs. In the test, however, the dogs themselves could see where the sausage was hidden. Analogous to the "Sally and Anne" paradigm [648, 650], the hiding place was changed by the hider, who moved the food from the first bucket to a second, distant bucket in front of the dog. The informant saw this change only in the true belief condition (Fig. 8.8) but not in the false belief condition, in which she was in the next room during the change.[3] We randomly divided the dogs into two groups and confronted them with one or the other condition in a single test (between-subject design). The central assumption was that the two groups could only differ in behaviour when they took into account the informant's level of knowledge.

[2] See the section "Sally and Anne".

[3] Note that in contrast to the previous false belief tests, we had the informant leave the room not only in the false belief condition but also in the true belief condition but with the temporal difference that the change of hiding place did not occur during her absence (nothing happened in her absence). Thus, a possible difference between the two groups of dogs cannot be explained by the fact of the informant's absence (the timing of the absence—before or after the translocation—was also checked with a control group and had no influence).

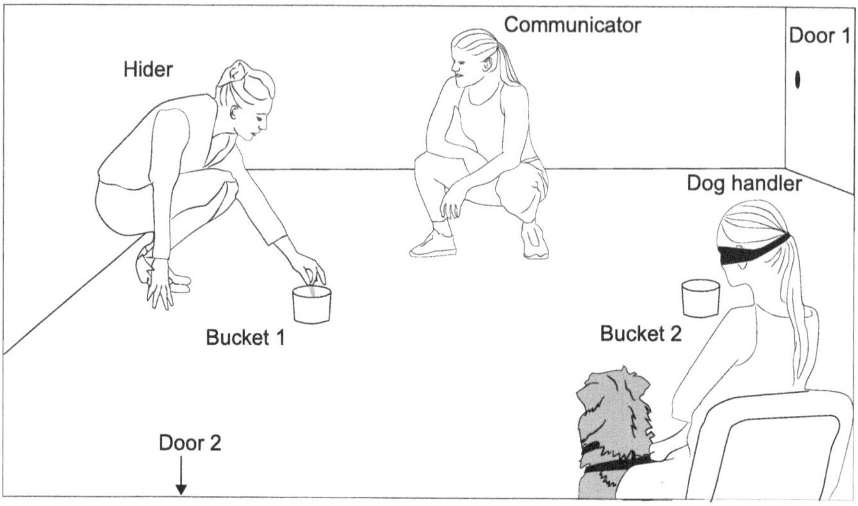

Fig. 8.8 Arrangement of the false belief test with dogs. The picture shows the true belief condition in which the informant observes the transfer of the piece of food into Bucket 2. The dog sees both the hider's transfer and the informant's observation. (From [777])

The difficulty in this hiding game, however, was that the dogs had seen where the food was at the end and, equipped with this knowledge, could successfully make their choice between the two buckets. If the informant in the false belief condition points to the first bucket, which is now empty but the dogs do not follow this hint, the simplest explanation would be that they follow their own knowledge. Therefore, we also had the knowing informant (in the true belief condition) point to the first bucket. Now, dogs could understand this information as a hint that there might be something else in the first bucket, possibly that it had been filled with even more or better food before and that it would therefore pay off to check this possibility. Only with this design feature could we—similar to Buttelmann and colleagues [771] in the helper study described earlier—establish a contrast between the two groups based only on the knowledge (or ignorance) of the informants, everything else being equal, from the dog's knowledge to the informant's hint. If a significant difference in the behaviour of the two groups was to become apparent, it could only be attributed to the difference in informant knowledge.

In fact, in this large-scale study involving more than 200 dogs, we found a significant difference between the two groups [777]. Although 62% of all dogs followed their knowledge and went to the second bucket, enough dogs (38% in total) were guided by the informant's hint and chose the first bucket. However, contrary to what might be expected from attributing a false belief, significantly more dogs followed the ignorant (48%) than the knowledgeable (29%) informant. However, this difference was significant only as a group effect across the entire cohort of dogs. When looking at the dogs according to their breed classification, it is noticeable that this difference was actually pronounced only in four FCI groups[4] (in Groups 2, 7, 8, and 10) and was not present in three (1, 5, and 6), and one (3) even had an inverse relationship. In fact, the terriers (FCI Group 3) decided quite differently. Only 4 out of 20 terriers in the false belief group chose the first bucket, following the informant's cue, whereas 10 out of 20 terriers in the true belief group did (the remaining 26 terriers ignored the cue and chose the second bucket).

Two conclusions can be drawn from these results. First, the majority of the dogs seemed to react to the time of the informant's absence during the rearrangement. This can hardly be explained in any way other than that the dogs took the informant's knowledge into account when making their decision; thus, they fulfilled a central condition for the theory of mind. Second, terriers apparently reacted like humans and (other) apes since four-fifths of them ignored the informants with "false beliefs".

In the previously described helper study [771], Buttelmann and coauthors concluded that apes and children were able to determine the experimenter's knowledge of the object's location and used this assignment and the experimenter's behaviour to infer what the experimenter desired. Whether the dogs also inferred more than the informant's knowledge, namely, her motives and goals for her "wrong" hint, remains open for the time being. Only further experiments can clarify this and the question of why the terriers deviate so strongly.

Although tests are becoming more sophisticated, determining the underlying motives of action in nonverbal creatures, whether infants, dogs, or chimpanzees, remains a challenge for comparative cognitive science. Although cognitive scientists are always devising new tests, by the time they are completed, crafty philosophers have already devised new objections. Kristin Andrews of York University in Toronto, Canada, is convinced that while apes knew how to help the person with a false belief (and infants), they may not need a concept of belief to do so [778, 779]. This leaves us with a tricky philosophical question: What do we mean when we say that primates recognise an agent's belief (or presumption or faith)? Philosophers tend to think that a belief must be understood as a kind of "*propositional attitude*" [4, 780]. According to this view, a belief is a functional representation of truth; that is, it can be true or false, and it is functional because—together with a desire—it prompts action. However, to determine how a belief, together with a desire, causes someone to act, one also needs information about how beliefs and desires work together to bring about a certain behaviour. This information can be gained through experience or by thinking about how one would act oneself in the relevant situation.

[4]The FCI (Fédération Cynologique Internationale) is the largest international canine umbrella organisation with 99 member countries. It systematises the breeds recognised with the CACIB (Certificat d'Aptitude au Championnat International de la FCI) by dividing them into groups and sections. Accordingly, the FCI distinguishes Group 1, Sheepdogs and Cattledogs (except Swiss Cattledogs); Group 2, Pinscher and Schnauzer—Molossoid and Swiss Mountain and Cattledogs; Group 3, Terriers; Group 4, Dachshunds; Group 5, Spitz and primitive types; Group 6, Scent hounds and related breeds; Group 7, Pointing dogs; Group 8, Retrievers—Flushing dogs—Water dogs; Group 9, Companion and Toy dogs; Group 10, Sighthounds. Note: no dogs from groups four and nine were tested in the study described.

Andrews now suggests that a belief, in the sense of a propositional stance that can be true or false and that, together with desire causes behaviour, is something that requires "*cognitive sophistication*". Given these supposedly high cognitive demands, some critics of the false belief task claim that animals can only solve this kind of problem by "submentalising", i.e. having a behavioural theory rather than a belief attribution. This objection has already been raised to the raven studies of Thomas Bugnyar [744] and to the primate studies on anticipatory looking [781, 782] repeatedly [783]. According to this, the great apes look at the original location of the target object before the rearrangement, not because they assign a false belief about it to the actor but because the wearing of the green shirt evokes a memory of the original location of the target object when the actor returns. It would thus be simple stimuli that determine the primates' looking, not attributions of mental states to other living beings. This argument of submental, cross-domain processes could be applied in this form to many, if not all, false belief tasks.

Andrews [778, 779] disagrees with this view and suggests a middle ground between belief attribution and submentalisation: apes can make a nonpropositional attribution of a mental state in their social interactions. Thus, this middle ground is a "*mentalist*" hypothesis—it grants animals access to the intentions of the actor, but it does not include a propositional stance with logic and sentence structure or the distinction between true and false. It does not matter how the experimenter represents the object or its location; it only matters that the subject recognises the experimenter's goal, something that both developmental psychologists and primatologists consider proven. The latter can draw on specific tests of submentalisation, which pick up on the green shirt memory argument described above [782]. When primates see the false belief test in the form of an animation with (green) geometric shapes instead of people, they do not anticipate nor look predominantly at the correct place [781]. This finding is also consistent with other studies in which primates distinguished well between goal-directed actions of humans and uniform movements of inanimate bodies [784]. The submentalisation hypothesis is thus as good as refuted.

The ability to adopt a propositional stance with logic and sentence structure is what distinguishes humans. In any case, this requires two cognitive ingredients that many have considered the crowning glory of the human mind: language and consciousness. We will now turn to these, with particular focus on the question of whether there are similar, if not the same, abilities in non-human creatures, before we conclude with a summary of the animal mind.

Chapter 9
Do Animals Have Language?

This question may seem strange when one considers the many forms of communication with which animals use: body signals, scents, sounds, their colouring, and many other methods to transmit information about themselves or the environment to conspecifics and non-conspecifics. Although it can be argued that the corresponding information is rigidly fixed as species-specific signals since it cannot easily be expanded into new meanings or statements, this does not invalidate the fundamental function of language as a means to transmit information and to (socially) interact. The following case of the honeybee's tail dance shows how cautious one should be with all too quick attributions regarding rigidity of form, one-sidedness of function, and paucity of knowledge transmission. For a long time, the "*dance language*" of the honeybee, discovered by the Austrian behavioural physiologist Karl von Frisch [785, 786], was used as a textbook example of limited animal communication. Today, however, we know that the dance has high information content[1] and a representational function, has an appeal function and a learning component, has dialects, and, last but not least, can also be passed on across species boundaries.[2]

[1] In the case of the waggle dance (there are also other dances), up to four types of information about a resource are communicated by the foragers to the nestmates remaining in the hive: the basic fact that resources have been found in sufficient quantity (as well as the invitation to gather there as well), the type of resource found (by smell and taste), the yield and quality of the resource (by intensity of the dance), and the location of the find as seen from the hive (by distance and direction). See [787].

[2] Video footage of a mixed colony of Asian honeybees and European honeybees confirms that the two subspecies have significantly different dance dialects, even when foraging in identical environments. However, when reared in the same colony, they can communicate with each other: Acc gatherers could decode the dances of Aml foragers to successfully locate an indicated food source. The authors point to the existence of a learning component in the honeybee dance language and even see the possibility of social learning. See [788].

L. Huber, *The Rational Animal*, https://doi.org/10.1007/978-3-031-60803-2_9

Language as a Tool for Communication

If we understand language as a tool for communication, we come full circle to the cognitive performance of tool use discussed at the beginning. Here, too, we could distinguish in the animal kingdom between rather rigidly fixed use and flexible, creative use, with the comparison to human language picking up speed at the creative end. If, with Plato [789], one considers speech as naming and naming as a tool, one can also recognise similarities with reference to the partly sophisticated methods of categorisation, as outlined in the opening chapter. However, Plato's naming serves primarily the purpose of instructing others and recognising the nature of things. This powerful tool is also found in the so-called "*organon model*" of the Viennese psychologist Karl Bühler [790], who refers in particular to the representational function of language. However, in addition to representation (with symbols), language, in its pragmatic function, also serves two other functions: appeal (with signals) and expression (with symptoms). While Bühler assumed that human language basically contains all three functions, one of which can become dominant at a time depending on the speech context, in non-human communication systems, one should assume functions that are initially separate and primarily limited to only appeal and expression.

Undoubtedly, human verbal language is a very effective means of communication. Human verbal language has great advantages, especially when referring to "*displaced*" objects and facts that cannot be shared in a simple way both here and now. Philosophers therefore emphasise that language, as a central means of interpersonal communication, is the bearer of meaning and transmission as well as the key to understanding the world and oneself. According to the definition by Edward Sapir (Yale University, USA), "language is a purely human and noninstinctive method of communicating ideas, emotions, and desires by means of a system of voluntarily produced symbols" [39, p. 8].

Language as a Medium of Thought

Viewed less philosophically but more cognitively and psychologically, language serves to represent or even transmit mental entities, such as concepts, the latter now thought by many psychologists and biologists to be independent of language [15, 791–793]. One therefore speaks of means of representation. However, if one assumes, as Humboldt did, that language is indispensable for all complex human activities and thought processes, one conversely comes to the insight that language is an "*afterthought*" means of communication between people and also that every conception of things and facts in the world is already linguistically structured. Things and circumstances are brought into contexts of meaning through the linguistic conception of the world. Language is thus—as Humboldt already suggested [42]—a medium of thought and of the conception of the world par excellence.

If one agrees with this cognitively charged determination of language in humans, one has to ask how it could have developed. Especially for biologists like me, it is important to know the evolutionary origins. In his comprehensive book on the evolution of language [794], my Viennese colleague Tecumseh Fitch from the Department of Behavioural and Cognitive Biology at the University of Vienna justified the uniqueness of human language. Despite its many similarities with non-human animal communicative means, human language enables us to represent and share unlimited thoughts with others. As such, it not only is crucial to all human societies but also has played a central role in the spread of our species from being small and peripheral members of the sub-Saharan African ecological community to the most dominant species on the planet today. Language is so central to humanity that it permeates all aspects of human cognition, behaviour, and culture. Fitch's conclusion is that there seems to be no communication system of equivalent power in the animal kingdom.

A Bag of Tricks

Language is an immensely complex system composed of many components that are studied (largely independently) in different disciplines. Although language is characterised by seamless interactions between these components, it is not a monolithic whole and can be better viewed from a biological perspective as a "bag of tricks" assembled through a process of evolutionary tinkering. Viewed from such a multi-component perspective, as advocated by Fitch [794], arguments about which component is the central feature of language are unproductive and misleading. What is useful, however, is to pursue the question of how the elements emerged and were assembled during evolution.

Comparison with the languages of other animals remains difficult. We share many components of human language with other animals, such as the requisite skills that include hearing, memory, categorisation, sequence learning, (movement) imitation, and vocalisation. But there are a few fundamental skills that distinguish humans (even from our closest relatives, primates), such as our vocal learning[3] or complex syntax. Focusing exclusively on primates, however, is misguided and risks overlooking many species that can teach us important lessons about human language. For example, the human capacity for complex vocal learning, which enables children to learn the words of their community's language, is shared with many birds, whales, and seals but not with chimpanzees or other primates. Even more, comparative linguistic research has shown that many of these aforementioned components of language have their own evolutionary history and are based on very different neural and genetic mechanisms.

[3] However, this difference is also sometimes questioned because, for example, it has been found that orangutans in captivity are capable of vocal imitation and can even incorporate completely new, human whistling sounds into their repertoire [218, 795].

To abbreviate a long history of research, one can very simplistically name three key components of human language, namely, signals, syntax, and semantics, in other words, the ability to articulate many different speech sounds, to form complex sentences, and finally to fill these with meaning. It is assumed that a "*lexical proto-language*" first consisted only of individually spoken words and that syntax was later added as an innovation.

An important question in the evolution of language is how its generative power arose. This power, which enables the communication of limitless thoughts and ideas, is a result of the combinatorial nature of human language: meaningless sounds can be combined into meaningful words and these into higher-level, meaningful structures.[4] The flexibility of these combinations transforms our motor and perceptual systems into an unlimited system of thought and communication. Wilhelm von Humboldt aptly characterised this as unlimited use with limited means [796]. Exactly how and why this combinatorial explosion, with its multiple semantic hierarchies, happened remains unclear. The development of human language had to be accompanied by the (cognitive) ability to decode or interpret what was spoken, especially in the case of ambiguous words or sentences. This can be successful only through mutual (hermeneutic) elucidation with a view to the upper systems (sentence) and lower systems (word, syllable).

Comparative language research has attempted to experimentally determine which of the abovementioned abilities also occur in non-human animal species and to what extent. Overall, the majority of scientists today tend to believe that we are dealing less with qualitative differences than with quantitative ones. However, this is not to throw the baby out with the bathwater and herald the end of human language's uniqueness.[5] Language may draw on a wide range of biological mechanisms shared with other species, but it is available only in its full form to humans. After all, language gives us the ability to express everything we can think about and to communicate those thoughts through a series of mutually intelligible signals. Many other animal species also communicate and share emotions or even knowledge, but the limitless expressiveness of human language sets it apart from all other communication systems.

[4] Linguists accordingly distinguish phonology and syntax, using different time scales. At the shorter phonological level, meaningless sounds (phonemes) are combined to form meaningful words, while at the longer syntactic level, words are combined to form sentences. With this linguistic terminology and distinction, animal communication systems can also be understood as primarily phonological or primarily syntactic.

[5] One should not be blinded by the strength of the term "*uniqueness*". Some people immediately think of it as a kind of God-image, but in reality, it means no more than a species-specific characteristic. Biologists with a basal understanding of evolution are not surprised to find unique or unusual features in the animal kingdom, think for example of the elephant's trunk, the bat's echolocation, or the "radar" in the electric fish. So, the fact that humans alone on our planet have a special language ability is nothing more remarkable, even more so where certain components are shared with other organisms.

Language, Narrow and Wide

The pluralistic, comparative multicomponent approach to language distinguishes between language ability in a broad and a narrow sense [797, 798]. In the broad sense, language encompasses all the mechanisms involved in language acquisition and use, both biological and psychological, and of these, many are shared with other animal species or with other human cognitive abilities, such as music or vision. Language in the narrower sense refers to a much more specific subset of mechanisms—essentially linguistic—that are unique both to humans and to language. Important in this distinction is the insight gained from modern evolutionary theory that only some aspects of language can be understood as adaptations for communication, while others arose more from cognitive, historical, or developmental constraints.[6] Noam Chomsky of Harvard University (USA) also emphasised the evolutionarily important distinction, but in his work, he essentially referred to the "*computational mechanisms*" central to syntax. Namely, these allow an unlimited set of structures to be formed by a finite set of rules based on a finite vocabulary [803]. Chomsky thus placed his focus on the complex cognitive system that underlies language and which is a property of an individual's brain. He called this system "*internalized language*" or "*I-language*", being the system of knowledge of language attained and internally represented in the mind/brain [803].

If we limit ourselves to cognitive elements of language and disregard the neuronal–anatomical conditions of sound production and perception in mammals and birds, there is at first glance a considerable discrepancy compared to humans. Although birds sing and chimpanzees grunt, humans write sonnets. Quite a few philosophers maintain this unflattering characterisation for animals.

However, is this perhaps less due to the animals themselves and more to our inability to understand them? Are we smart enough to know how smart animals are? Should we not use abstract sign systems that animals can learn and then use these symbols to assess the quality of their communication skills?

Language Training in the Laboratory

After the first attempts to teach monkeys and dolphins human language failed in the early 1960s [804], various laboratories used different techniques of two-way communication, including American Sign Language (ASL), with chimpanzees.

[6]The distinction between adaptations and *constraints*, which designate the persistent forces of the evolution of organisms, took place within the framework of the extension of the neo-Darwinian selection theory to a systems theory of evolution, as it was co-developed in Vienna by my teacher Rupert Riedl. See [799–802].

[805–808], an orangutan [809] and a gorilla [810] in the late 1960s and 1970s.[7] Then, the LANA project became particularly well-known to the public [816]. A female chimpanzee, Lana, was to distinguish between lexigrams, arrange words grammatically, and learn novel utterances at the Yerkes National Primate Research Centre at Emory University under the direction of Duane Rumbaugh.[8] Later, the bonobo Kanzi became known for his much more extensive communication with his adoptive mother, the researcher Sue Savage-Rumbaugh, as he learned to use many symbols with a lexigram tablet, to answer longer questions and, albeit in a very simple way, to ask questions himself [576, 817, 818]. In order to communicate where he was going or what he wanted, he first expressed his destination (place or object) with a lexigram and then an action (go, hunt, carry, give) with a gesture. Although his word combinations were not syntactically structured, they consisted of lists of preferred actions in Kanzi's preferred order.

Artificial language training was also carried out with dolphins, mainly to test their referential comprehension [819]. Song learning in songbirds and the amazing similarities between the process of song acquisition in these birds and human language acquisition inspired Pepperberg to train a grey parrot, Alex, to use referential language [112]. Unlike studies with monkeys and dolphins, which used sign language, plastic chips were used as naming devices as well as modified Skinner boxes. With these, Pepperberg trained her parrot to use the sounds of English to name objects, colours, and shapes. Alex was even able to ask simple questions, including what colour he himself was. This approach also provided insights into how language and complex cognition might have developed in our ancestors. However, some also realised the significance of creatures, separated by 300 million years of evolution, reaching some level of symbolic representation in parallel and even using basal syntactic rules to link symbols. Of course, some well-known cognitive researchers, led by Herbert Terrace of Columbia University (USA), were not convinced that the formation of sentences through explicit training was comparable to the language ability of a typically speaking human adult [820–822], and others even spoke of fraud [823]. In general, sufficient evidence remains that at least some animal species are capable of spontaneously generating symbols and meaning-generating concatenation of signs.[9]

Following these early investigations of symbolic aspects of communication, other psychologists began to use alternative "*two-way communication systems*" to

[7]The famous parrot researcher Irene Pepperberg from Harvard University in the USA, who gained worldwide fame and popularity far beyond the scientific community with her studies on the grey parrot named "Alex" [112], summarised the history of comparative linguistic research on nonhuman animals in a very personal essay [811]. Good overviews can also be found in [794, 812–815].

[8]The "lexigram" was a keyboard developed by Duane Rumbaugh and the Austrian Ernst von Glasersfeld with graphic symbols and a special grammar that Lana learned to use. This artificial language was called Yerkish, in honour of Robert M. Yerkes, the founder of the laboratory where the LANA project was conceived and carried out.

[9]Training animals to learn arbitrary signs and produce them consistently for simple forms of communication with humans still has its adherents, with other animals, such as dogs, now being trained to use lexigrams. See [811].

investigate various forms of cognitive processing that rely on symbolic representation. This evidence of language-relevant cognitive abilities, such as numerical concepts [824–826], relational concepts [827], rule-governed behaviour [828, 829], and symbolic equivalence [830], has accumulated in dolphins, sea lions, and grey parrots. On the question of syntax, some laboratories investigated the formation or at least the use of sequential and rule-governed behaviour in the form of ordered sequences of actions. Evidence of a basal understanding of syntax later culminated in evidence of the learning of artificial grammars in starlings [831] and cotton-top tamarins [832]. This focused particularly on the question of understanding recursive patterns, which is considered a key element of human language. However, linguists such as Chomsky maintained their criticism, which was already voiced when the language training of Kanzi and other monkeys occurred, that although animals can be trained to imitate some superficial aspects of language, this ability should not be confused with the rapid and effortless acquisition of language in human children nor with the generic production of language in a complex syntactic manner.

Analysis of Animal Communication in the Wild

While psychologists were advancing artificial languages and the testing of language-relevant cognitive abilities in non-human animals, biologists were attempting to decipher the species-specific "*code*" of communication of certain animal species in parallel [814, 815, 833, 834]. Their focus was on analysing the natural forms of communication by recording and evaluating signals of the sender and contingent reactions of the receiver. On the one hand, mechanisms emerged to increase the efficiency and reliability of information reception; on the other, signals emerged to increase the efficiency of communication and the reliability of manipulating the receiver to benefit the sender. Another research focus has been the analysis of all those factors that influence the quality of the received signals: how the signals are generated and transmitted; how they behave during transmission through air, water, or other substrates; how they are processed by the sensory and cognitive systems of the receiver; and which factors influence their consequences to fitness [835, 836]. The evolutionary history and the time period a species has spent in different signalling environments are also important. To answer all these complex questions, methods from many different fields must be applied and only by addressing Tinbergen's four questions [837, 838]—phylogenetic and individual origins, function and the underlying mechanism[10]—may we come to understand the design and evolution of signals and signalling systems.

---

[10] Niko Tinbergen, who, together with Konrad Lorenz and Karl von Frisch, was awarded the Nobel Prize in Medicine for founding comparative behavioural research, described in 1963 in an article entitled "On the Aims and Methods of Ethology" the four main questions of biology: function (what does the trait serve), evolution (how did it develop in phylogeny?), ontogeny (how did it

It would go too far to provide a comprehensive report here on the sometimes spectacular successes in uncovering animal communication systems, so the reference to corresponding overviews in the specialist literature shall suffice [811, 813, 834, 839–842]. In these reviews, one will not only be fascinated by signalling systems in the acoustic domain, for example, in the case of song birds [843–846] or also underwater in the communication of dolphins and whales [847, 848] but also by completely different sensory modalities, such as in the chemical sense [849] and, ultimately, by multimodal systems [850, 851]. The data collected in countless field experiments have provided detailed information about the form and function of animal signals, especially the contexts that trigger them and the information that conspecifics derive from them. Although these signals are radically different from the sounds of human speech,[11] their form and meaning can be illuminated using general tools of formal linguistics (phonetic form, syntax, semantics). In what follows, I would like to turn first to alarm calls, a class of communication signals that is not only particularly well studied but also of particular interest to us for reasons of function—due to the close connection with relevant events in the environment—and cognition, the underlying psychological–neuronal mechanisms operating in transmitters and receivers [852].

Alarm Calls

In monkeys, alarm calls usually refer to general dangers, in some cases to specific predators. In the late 1970s, the discovery of vervet monkeys' warning calls for three predators revolutionised comparative linguistic research. The research couple Dorothy Cheney and Robert Seyfarth from Rockefeller University (USA), who later worked at the University of Pennsylvania in Philadelphia (USA), discovered that these monkeys not only emitted sounds involuntarily out of fear but also sent a specific message about a particular threat. The researchers hid loudspeakers in bushes in Amboseli National Park (Kenya), played pre-recorded calls, and observed the reactions of the monkeys in their group. With these so-called playback experiments, they were able to show that each alarm call elicited a different reaction. The animals on the ground reacted to the leopard alarm by climbing trees, to the eagle alarm by looking up, and to the snake alarm by looking down. Together with a researcher of bird song, Peter Marler of Rockefeller University (USA), the researchers argued that neither caller characteristics nor context are systematic determinants of

develop in individual history?), and mechanism (how does it work?). In this essay, dedicated to his friend Lorenz on his 60th birthday, he emphasised the necessity of answering all four questions for a comprehensive understanding of a biological phenomenon, with the answers to each question providing complementary insights [837].

[11] I use the singular here in a very generalising and simplifying way, knowing that there are many (approximately 6000) natural (and constructed) languages that contain quite different sound patterns, not to mention different grammars.

response; rather, calls alone (even in the absence of actual predators) provide conspecifics with enough information to respond properly [853, 854]. One can therefore assume that the monkeys behaved as if each alarm call denoted certain objects or events in the environment—in this case, snakes, birds of prey, and leopards. With these sensational findings, Cheney and Seyfarth pushed open the door to understanding non-human communication—and thus also the mental world of animals [72].

Warning calls of different animal species can also communicate size, distance, location, or even threat level [852]. The small black-fronted titi monkeys provide a convincing example that a warning call can even communicate two characteristics of imminent danger at once. In their vocal sequences, titi monkeys communicate not only which predator is dangerously close but also where the threat is lurking [855, 856]. How do they do this? By producing sequences consisting of two main alarm call types that contain information about the type (mammal or predator) and location of a predator (in the tree or on the ground) in different parts of the utterance. This observation was the first example of a sequence-based alarm call system in a non-human animal [857, 858].

What does this finding mean for the evolution of language, especially for the evolution of syntax? One has to be careful here. First, it should be noted that phonetic structure (morphology) mostly consists of simple calls and only very rarely has compositional semantics. Likewise, the syntax is simple and rather limited in all cases that have been well elucidated so far [859]. A key question then is how the division of labour between semantics, pragmatics, and the environmental context (especially when context switches) works. For example, it has been shown that the sequence of black-fronted titi monkeys in which two calls are rearranged does not involve a complex syntax/semantic interface but rather a fine-grained interaction between simple call meanings and the environmental context.

Primate Linguistics

Since the vervet monkey studies, primatologists have sought to increase taxonomic breadth and record calls of other ape species using the same field methods and to analyse them using increasingly sophisticated techniques. In particular, the team led by Klaus Zuberbühler of the University of Neuchatel (Switzerland) and the University of St. Andrews (UK) has done great work in clarifying sound communication in certain Old World monkeys—the putty-nosed monkey, Campbell's monkey, and mantled guereza—as well as in a representative of the New World monkeys, the aforementioned black-fronted titi monkeys [859–861]. Thus, "formal monkey linguistics" has emerged [862], which has made great progress in deciphering natural, non-human communication systems, especially in recent years. Some biolinguists have specifically applied the analytical methods of contemporary linguistics, especially in the field of semantics and pragmatics. Data collected from field experiments have provided detailed information about the form and function of monkey calls, especially the situations that trigger them and the information that conspecifics derive from them [863]. In addition, the first insights could be gained into what the "*lexical meaning*" of individual calls is what syntax sound sequences have [864], how the meanings of the individual calls are combined [863], how calls or call sequences compete with each other when several are appropriate in a given situation, and, finally, how the form and meaning of signals have evolved.

Using long-term observations and predator simulation experiments on Campbell's monkeys, Zuberbühler's team have identified two sensational features of vocal communication, acoustic modulation and sequence variation [864]. The alarm calls of males consist of an acoustically variable stem that can be followed by an acoustically invariant suffix [865]. This suffix (*oo*) expands the meaning of the calls by transforming an eagle alarm (*hok*) into a general call for danger from trees (*hok-oo*) or a leopard alarm call (*krak*) into a general alarm call (*krak-oo*). In addition, Campbell's monkeys can combine different call types into different sequences in a context-specific way. These sequences are associated with group cohesion and movement, falling trees, neighbouring groups, non-predatory animals, non-specific predatory threats, and specific predator classes, all in a relatively flexible manner. The sight of eagles triggers four different sequences, whereas the sight of leopards triggers three, depending on how the caller learns of their presence. When linking sequences, Campbell's monkeys not only follow certain principles, such as certain transition probabilities, but also certain rules for adding specific calls to an existing sequence to form another or recombining two sequences to form a third. The scientists suggest that these primates overcame their limited vocal control through combinatorial creativity [859]. Because the different sequences are closely linked to specific external events, the Campbell's monkey calling system is perhaps the best example yet of a "*protosyntax*" in the animal kingdom.

Primates may be limited in their vocal flexibility, but this is not true of songbirds (or even whales). In contrast to the higher, *syntactic* level, where humans combine words to sentences—and to a much more modest extent Campbell's monkeys combine calls, the lower *phonological* level combines meaningless sounds into meaningful units (words). The ability to create new meaning by rearranging sounds is a fundamental part of language. Although we know enough about combinations of meaningless acoustic elements in bird songs, there is usually no evidence that the rearrangement of such combinations produces functionally different meanings. A few years ago, however, this was experimentally demonstrated in a bird from the Australian dry zone, the chestnut-crowned babbler [866]. Two tonal call elements, A and B, which differ in their spectrograms (temporal course of the frequency spectrum) are used in different arrangements to create two functionally distinct vocalizations. For example, flight calls (in the sequence AB) to stimulate an increase in attention and readiness to flee or prompt calls to stimulate begging during nest procurement (in the sequence BAB) and thus increasing the efficiency of food transfer to the offspring. Interestingly, the beginning sequence BAB stimulates changes in the behaviour of the recipients that are neither observed in response to the B element alone nor in response to the flight sequence AB. Accordingly, the researchers argued that the addition of the B element to the sequence AB changes the meaning of the call, just as an additional *phoneme*[12] changes the meaning of a word. Thus, the addition (or omission) of a contextually meaningless acoustic element at a single position creates a phoneme-like contrast sufficient to change the meaning between the two calls.

Nevertheless, attempt to draw parallels with human language should be taken with caution, as there are several key differences [867]. First, the elements A and B do not occur in any calls except those for flight and begging, whereas human phonemes are widely used in language. Second, these elements are acoustically isolated (separated by silence), whereas most human phonemes are fused into syllables. Third, the difference between AB and BAB is a case of presence or absence, while human phoneme contrasts are typically discriminative (such as "tip" versus "dip"). Finally, we must not overlook the fact that in human language, the same phoneme sequence is pronounced with countless different intonations and embellishments, conveying subtly different emotional and pragmatic intentions.

Meanwhile, there is increased interest in phonological aspects of animal communication, in sound sequences and in combinations, after a period of concentration on single calls or syntax questions. It is possible that there is similar or even greater combinatorial power in phonemic capacity in animals than in syntax-like constructions. After all, some of the most complex and variable animal communication systems involve sound patterns called "*songs*" in birds (and humpback whales). Hence, authors such as Marler [868] and Fitch [869] draw parallels with music, which lacks propositional meaning, and consider music to be perhaps an even more profitable source for understanding the evolution of combinatorial communication systems than "meaningful" alarm calls or calls for finding new food sources.

Social Communication

Alarm calls will continue to occupy a prominent place in the decoding of non-human communication because they are particularly good at revealing which aspects of the environment are significant to a caller and through which cognitive processes receivers extract meaning. Moreover, they are interesting from an evolutionary perspective because they present a certain paradox: what selection pressure could favour the evolution of conspicuous calling in the presence of a predator? However, alarm calls seem to be of secondary importance for studies on the origin of human language [852]. There is now a consensus that sound communication in the animal kingdom can not only contain quite different information than danger but also, in other contexts, is even more flexible, extensive, and finely structured.

[12] In linguistics, a phoneme is defined as the smallest unit in the sound system of a language that serves to distinguish meaning. The phonemes /p/ and /b/, for example, distinguish the lexemes (words) "pat" and "bat" from each other as a minimal pair; they do not carry any meaning themselves.

Human language is embedded in a network of social interactions and has primarily social functions. It is therefore natural to assume the origins of human language occur in complex primate societies.[13] Cheney and Seyfarth have therefore turned, after the vervet monkeys, to the baboons, known for their large and complex social systems. These Old World monkeys seem to know much about the social relations and dominance relationships of their group, and the social complexity seems to resemble human societies, at least superficially. In unusually large groups,[14] baboons are forced to form a mixture of short-term bonds for mating and longer-term friendships based on careful considerations of status and individual needs.[15] The central hypothesis is that key elements of human language arose from the need to decode and encode complex social interactions. In brief, social communication is the biological basis on which evolution built a more complex language.

If this hypothesis is correct, one would have to find features in baboon communication that are typical of human language and relevant to theories of language development. In fact, Cheney and Seyfarth identified six such features [874–876]. First, knowledge is representational because the receiver acquires specific information about a particular type of interaction between particular individuals in particular calls. Second, knowledge is based on discrete properties such as identity, gender, kinship, and rank. Third, by combining these characteristics, baboons can convey knowledge about social, hierarchically structured relationships. This requires a categorisation performance in which individuals are simultaneously ordered in a linear order and grouped according to matrilineal kinship. Fourth, knowledge is rule-governed and openly managed. Recipients can trust, for example, that calls are given only by subordinates to dominants. The classification is constantly updated because individuals can be added or dropped, but the underlying structure remains unchanged. Fifth, knowledge is not tied to a sensory modality. Although information about social interactions can be conveyed through sounds, individuals can also

[13] However, sophisticated communication systems in complex social relationships are also found in dolphins. Bottlenose dolphins invent arbitrary vocal designations for recognising individuals, monitoring individual behaviour, and tracking individual relationships [870, 871]. In addition, they can use novel, arbitrary signals to refer to objects, report on objects even when they are not present, and thus develop relational concepts [872].

[14] Baboons live in groups of up to 150 animals, consisting of a handful of males and eight or nine families of females (matrilines). The females remain in their natal groups throughout their lives, where they are placed in a stable, linear dominance hierarchy. They maintain close social ties with their kin and determine priorities for access to resources. The offspring acquire similar ranks to their mothers. The social structure of baboons can therefore be described as a hierarchy of matrilines, in which all members of one matriline are superior or subordinate to all members of another matriline. The ranks within and between matrilines are stable.

[15] Robert Seyfarth and Dorothy Cheney, together with many colleagues, have been able to decipher the essential cognitive mechanisms of how these primates use sound communication to modulate social dynamics, based on their decades of groundbreaking research on free-ranging baboons in the Okavango Delta of Botswana, which began in 1992. They wrote up their findings and conclusions in a best-selling book on comparative cognition research [873]. The book's surprising title refers to the famous quote by Charles Darwin: "He who understands baboon would do more towards metaphysics than Locke".

acquire social knowledge visually by observing interactions. Sixth, knowledge has a propositional nature, as rank reversal tests have demonstrated that receivers respond to a call sequence that includes an actor, an action and a receiver along the lines of "Sylvia threatens Hannah, and this causes Hannah to scream".

The sixth feature in particular shows the highly developed nature of "*baboon language*". Although actors and actions are not coded by subjects and verbs as in human language (Sylvia grunts aggressively, Hanna screams fearfully) and the order of the calls also seems relatively unimportant, the listeners interpret the call sequence correctly: Sylvia threatens Hanna. Cheney and Seyfarth consequently ventured the hypothesis that the propositional information that baboons receive when they hear such sound sequences involves an understanding of the causal relationships between a sender's threatening roar and a receiver's cry.

I have used the term "*baboon language*", albeit in inverted commas, because baboon sound sequences are similar to a sentence in our language in some ways. Although the authors concede that the calls of the two baboons cannot be described as even approximately linguistic in themselves, the receiver interprets the combination of the two calls in the way we interpret sentences. By combining their knowledge of call types (threatening roars, fearful cries, etc.) and the identity and position of the callers in the social network, they assume a causal relationship between the sounds of one animal and the sounds of the other. Thus, they acquire propositional information. It must be said, however, that although baboons have concepts that represent causal relationships and courses of action, they cannot articulate this information. They understand dominance relationships and matrilineal kinship but have no words for them.

With this observation, we return to the beginning of the book, where I discussed the sometimes complex and even propositional nature of animal categories and concepts. Apparently, as the aforementioned findings of the baboon experiments suggest, verbal language, as we use it, is not required for the internal representation of concepts, relationships, or courses of action. Thoughts can be represented, and thinking is possible even without language. Language does not have a monopoly on this, and it may not have been created for this purpose at all, as numerous linguists suggest [797, 877, 878].

Adapting Communication to the Knowledge of the Audience

Interestingly, birds and mammals do not always sound the alarm when they see a predator. In recent years, there has been increasing evidence that the production of alarm calls is influenced by the presence and composition of the "*audience*", i.e. potential recipients. Since calling is risky because one can be detected by predators, there are fitness advantages to alerting only when one is near other conspecifics who can hear the calls and when one is in the presence of kin and preferred mates rather than nonkin or even rivals. Field experiments have shown that non-human primates take into account subtle contingencies in their vocalisations, including context, the identity of the receiver, the identity of other individuals nearby, and the nature of their own recent interactions with the receiver and kin [879–882].

However, it would be even more advantageous if the discoverer of a predator could also make his or her calling dependent on whether the others already know about the danger. To be able to consider the level of knowledge that conspecifics have would require a certain theory of mind. Such abilities have been demonstrated in various situations and sophisticated laboratory experiments for a number of animal species, as described in Chap. 8. However, evidence was lacking in the context of natural communication in the field, until 2012. Catherine Crockford of the Max Planck Institute for Evolutionary Anthropology in Leipzig, Germany, and her colleagues published the results of a 20-month field experiment, which suggested that chimpanzees could detect whether others were aware of danger and adjusted their alarm calls accordingly [754].

Crockford and colleagues utilised their knowledge of chimpanzee societies (especially those in the Budongo Forest of Uganda) to create a sophisticated experimental design and technique. They confronted individuals with a dead, stuffed highly venomous snake and then recorded their vocal behaviour when other chimpanzees approached. These potential recipients either already had knowledge of the snake (because they had already seen it or because a previous discoverer had already given an alarm signal in response to the snake) or had not. Depending on what the calling chimpanzee, i.e. the sender, knew about the knowledge state of the receivers (regarding the snake), he or she gave an alarm signal or did not. From the caller's perspective, the receivers were ignorant, partially knowledgeable or knowledgeable. They were ignorant if the sender could neither observe the receiver seeing the snake (visual knowledge) or hear an alarm call if the receiver was within 50 m (thus thinking that the receiver also heard the call). If they could hear the alarm call but the visual confirmation was absent, the receiver was partially knowledgeable. Only if the receiver perceptibly had both types of information about the snake was she or he fully knowledgeable from the sender's perspective. A statistical analysis of the data on when senders produced alarm calls and when they did not revealed that the primary factor was whether the receiver had full or partial knowledge of the snake's location instead of their own risk or habituation to the sight of the snake. The senders most often called when the receiver had not seen the snake and least when the receiver had seen it.

Seyfarth and Cheney [883] emphasise the importance of these findings because they show that the attribution of knowledge to others has evolved in the context of a rich web of social interactions in which many different behaviours and memories of past events are correlated. Certainly, it is easier to study such complex yet subtle thought processes under more controllable laboratory conditions, but the weaknesses of field observations are more than compensated for by the much greater ecological validity of their results. In any case, Crockford and colleagues concluded that chimpanzees are sufficiently aware of others' perspectives to convey information to purposefully ignorant group members. It is therefore reasonable to assume that the integration of signalling and social cognition systems occurred in early hominoid lineages before additional language-specific features such as syntax emerged [884].

Non-human Versus Human Language

Before I attempt to answer the question of why such complex human spoken language came into being, let me again point out the contrast to what we know today about the most cognitively demanding animal languages. On the one hand, many bird species and some mammal species can form complex speech sounds, some of which even imitate the speech sounds of other species, demonstrably including humans. But on the other hand, the understanding of the meanings of such sounds clearly lags behind that of humans in these and other animals. The complexity of non-human sound patterns seems to far exceed the highest information content achieved. Some bird and whale songs can be as complex as human sentence constructions, but they have—as far as is known—only a very modest informational value. However, the reverse is also true: chimpanzees and, to some extent, dogs are amazingly good at learning to interpret information—from human vocalisations. Even then, however, they rarely form complex sound compositions (morphology and combinatorics). What is generally lacking in comparison to humans is the almost limitless creation of new meanings, which is then conveyed with an immensely high number of sound combinations. In this respect, a conversation with the chimpanzee Viki, the bonobo Kanzi, or the grey parrot Alex is relatively short and limited.

But why have the communication systems of even the most cognitively outstanding animal species remained relatively simple compared to those of humans? An answer to this question will be possible only if we consider function with its information content and also decipher the neuronal and genetic basis of language. In doing so, it is important to distinguish between questions that concern language as a communicative system and questions that concern the cognitive processing underlying this system, such as, above all, those that underlie recursion [798]. What seems certain is that each of the many different mechanisms involved in language requires different forms of neural computation. In humans, these processes have been found to be carried out in different ways and in different regions of the brain. However, scientists, especially from the young branch of neurolinguistics, are challenged to uncover the specific differences in neural computations in human and animal brains that are required to learn, process, and produce language. In doing so, they will not encounter the classical "*language organs*" in the left hemisphere, or in Broca's and Wernicke's areas, but will uncover a much larger network involved in language processing. It is becoming apparent that important processing and computational capacities have arisen for reasons other than in the course of communication. But after proving useful for communication, these were modified due to constraints involving both sensory—for example, the rapid processing of sound patterns—and cognitive centres, for example, conceptual and cognitive structures, pragmatics, and memory constraints [797]. Genetics and developmental biologists are challenged to investigate how these neural circuits arise at the developmental and genetic levels; here, insights into changes in specific genes such as FOXP2 will also play a role (a transcription factor found in all mammals [885] and birds [886]).

The Evolution of Human Language

The majority of both anthropologists and primatologists assume that the roots of human language go far back into the evolutionary history of primates. Within non-human primate species, the evolutionary history of calls can sometimes be reconstructed over millions of years. Although the evolutionary links between monkey calls and human language are still unclear, plausible evolutionary scenarios can be developed. For example, Seyfarth and Cheney hypothesised that hominids assigned meaning to the calls of other individuals and extracted rule-governed, propositional information from them [874, 875]. A crucial step in language evolution, they argue, occurred when individuals came under strong selection pressure to communicate their thoughts rather than simply extracting information from the calls of others.

Already, in early humans, who did not yet have a conventional language, a first form of shared intentionality appeared. Communal activities, such as hunting in groups, required joint attention and a basic understanding of common goals, which were communicated via natural, iconic pointing gestures. Due to the interdependence of cooperative partners, not only was information exchanged more and more effectively, but the intentions of partners were also reflected alternately and repeatedly. Socially recursive thinking set in—albeit still without conventional word language—and enabled increasingly sophisticated collaborations [887].

Later, it was assumed that, only with *H. sapiens* and *H. neanderthalensis,*[16] cultural influence considerably increased, collective intentionality and the need to communicate increased, and forms of communication became increasingly symbolic. In addition, in ever larger collectives, knowledge and skills had been passed on by means of instruction and were consolidated by certain forms of representation, such as drawings, representational and gestural symbols, and incipient language. Thus, knowledge could be passed down through generations relatively unaltered, and new knowledge could build on existing knowledge. This "*ratchet effect*" enabled cumulative growth and accelerated the multiplication and differentiation of culture [887].

Much later, only a few tens of thousands of years ago, language became increasingly sophisticated, references became increasingly abstract, and messages contained objects and events were increasingly distant (in space and time). At ever higher levels of mentalisation, the speaker referred to imaginary worlds, myths, deities, and ancestors. This progression made possible the highest levels of intentionality, a maximum of abstraction and decentring, imagination, and creativity [192].

Although it is difficult, if not impossible, to trace the evolution of human language precisely, one thing seems relatively clear: sophisticated word language did not emerge overnight, not by a lightning strike—following Konrad Lorenz's

[16] Today, anthropologists are generally of the opinion that Neanderthals could speak. The contrary laryngeal hypothesis has been definitively disproved by fossil finds. In addition, the culture of Neanderthal man seems to have been comparable to that of contemporary *Homo sapiens*.

concept of *fulguration* [84]—but rather gradually and in stages from a rudimentary to a more complex form. In the first stage, arbitrary signs and symbols were used. At the next level, signs or words were used to refer to a referent—the object meant—without the intention of triggering a certain behaviour and without the referent having to be present [73]. Finally, a certain signal was used without referring to anything, solely with the intention of achieving a certain effect in the recipient. In this sense, apart from the intention, which is still difficult to establish, vervet monkeys could be said to have simple language because they use learned, flexibly usable, arbitrary signs that do not resemble the referent and that trigger adaptive behavioural changes in the recipients.

Certainly, human word language is much more highly developed in its complexity and extended functions. Humans can argue with language; they can express statements with a wide variety of meanings, from conjectures, wishes, questions, and requests to firm beliefs, which in turn are embedded in a network of others' opinions. Many philosophers believe that only living beings who have beliefs and other propositional attitudes can be ascribed language. However, some of them go further and insist that someone can speak and even think when he not only has beliefs of his own but can also interpret the beliefs of others [888]. According to this, vervet monkeys would have no language because they have no beliefs in the strict sense, namely, propositional attitudes in the form of "I believe that…", and thus cannot interpret the beliefs of others.

The constitutive nature of the connection between speaking and thinking was postulated early on and regarded as a criterion of the special position of human beings. The phrase "man is only man through language" was formulated by Wilhelm von Humboldt [42, p. 44]. Since Descartes, the lack of linguistic ability has been considered the main objection to attributing thoughts and concepts to animals. It is, of course, undisputed that the language(s) of (present-day) humans enable a special form of thinking, namely, discursive, conceptual judgement, which is not merely based on assumptions but on convictions. Judgemental thinking, which can articulate itself linguistically and be the answer to a linguistically formulated question, therefore has a special status and is characterised by at least three properties. The most important property is that, by means of a judgement, reference can be made to something that not only lies outside the subject but can also be far removed from the here and now. This reference is made intersubjectively. It is thus objectifying, furthermore affirmative or negative, and it is true or false [13]. On the basis of this capacity for judgement, people can inquire about causes, draw conclusions, thematise facts, formulate problems, and articulate contradictions [43].

However, this "*lingualistic*" definition of thinking as speech and, subsequently, as symbolically composed judgement excludes the possibility of animal thinking and inappropriately narrows the concept of thinking. If the concept of thinking is designed in such a way that it simply excludes application to nonlinguistic living beings, then nothing at all can be regarded as evidence for the existence of such thinking [20]. What is then explained is what has already been established beforehand. One can try to escape this objection by saying that while it is not impossible for animals to refer affirmatively or negatively to something external without language, one adds at the same moment that the circumstantial evidence against it is overwhelming [13].

The arguments are based on the assumption that only the syntactic connection of at least two symbols in an affirmative or negative judgement opens up a semantic or intentional, true or false relation to being or not being "*things*" outside the judgement. What is decisive is not merely the connection of several symbols that characterises thinking in judgements but also their specific contents, namely, exact semantic and logical–syntactic structures. On the one hand, one needs a subject term that refers to something and, on the other hand, a predicate term that says something about the reference object by means of the subject of the sentence. In the sentence "The dog howls", the subject (dog) and the predicate (howl) are formally connected and thus form a judgement that can be true or false. Ernst Tugendhat and Ursula Wolf of the Free University of Berlin, Germany, started an investigation from the basic phenomenon of the predicative structure of human language, which has provided a decisive breakthrough to humans within cultural evolution [45].

Even if we think in a particular conceptual way, ultimately in judgements, it does not mean that all human thinking is reducible to this. This intellectualist perspective, taken by philosophers, overlooks how people proceed in everyday life, in solving simple and recurring problems, ones with incomplete information and in uncertainty. In turn, how adults think in everyday life and how children of preschool age think is also overlooked or marginalised. Not to mention the complex ability of human thinking did not fall from the sky but emerged evolutionarily in stages, thus building on simpler preforms [889]. Specifically, for this reason, as my esteemed colleague, the biology didactician Ulrich Kattmann from the University of Oldenburg, Germany, never tires of emphasising [890], the study of simple preforms can help us to better understand our complex language.

Chapter 10
Do Animals Have Consciousness?

So far, I have dealt with numerous animal thought processes that are good candidates for attribution of consciousness. These are thought processes that are usually not possible in humans without consciousness; they include logical reasoning, creative tool use and production, causal understanding and insightful problem-solving, volitional action, time travel (episodic memory, planning), metacognition, and perspective taking. (This list is not exhaustive; one could add imitation, teaching, mirror self-recognition, quantity estimation, abstraction, categorisation, and concept formation to the list.) But before I try to answer the question of whether animals act consciously in the aforementioned thought processes, I should first clarify what I mean by the term "consciousness". This term has many meanings and, above all, very strong connotations for human science and everyday language.

What Kind of Consciousness?

For a long time, the question of consciousness was a purely philosophical matter until human psychologists "got involved" more than 100 years ago. Sigmund Freud [891] did so with the distinction between conscious, preconscious, and unconscious and William James [892] with the physical, mental, and spiritual self. Around half a century ago, clinicians and neuroscientists joined in with an attempt to study consciousness as a brain process. More recently, scientists have begun to look at borderline cases, those concerning the question of consciousness in animals and machines. Therefore, a clarification of terms and a precise description of the

L. Huber, *The Rational Animal*, https://doi.org/10.1007/978-3-031-60803-2_10

phenomenon are important.[1] As the philosopher Tobias Schlicht [893] aptly notes, the interdisciplinary study of the mind not only holds out of the prospect of a more comprehensive understanding of numerous other phenomena but also of the danger of conceptual confusion. Is everyone truly talking about the same kind of consciousness? Most likely not, especially since consciousness has no monolithic meaning but is a "*mixed concept*" [894]. In the following, I would like to outline a few but what seem to be important distinctions that characterise modern theories of consciousness.

A central distinction is one between *global* and *local* states of consciousness, the former differing in cognitive, behavioural, and physiological terms and the latter in terms of content. Local states of consciousness include various kinds of perceptual experiences, pictorial experiences, bodily sensations, affective experiences, and emergent thoughts. These local states of *transitive*, intentionally characterised consciousness will concern us more in the debate about consciousness in animals, while global or *intransitive* (*being* conscious rather than *having* something conscious) states and levels of consciousness are more relevant in clinical research. However, the term "*level of consciousness*" not only is used to describe states associated with post-comatose disorders, strokes, anaesthesia, vegetative states, and sleep but also plays an influential role in theoretical and methodological contexts. For example, consciousness has not been described as an "*all-or-nothing property*" by some authors [895] but as a scale ranging from complete unconsciousness (death and coma) to vivid awakening [896]. Nowadays, it has become accepted that the level-based framework is untenable both for conceptualising global states of consciousness and for applying it to human infants and non-human animals and should be replaced by a multidimensional representation of global states [897].[2] Even if the dimensions of this state space are not yet known in detail, the diverse states must at least be described as regions in a space that cannot be spanned merely between contents and degrees of wakefulness.

Another important distinction that occurs in many theories is one between "*phenomenal consciousness*" and "*access consciousness*" [894]. The former concerns

[1] Philosophers in particular criticise the fact that natural scientists often use a very abbreviated or simplified concept of consciousness in order to reduce their research to an investigable level. In particular, it is urged that consciousness in its full meaning must also include the subjective, private aspect of conscious experience, instead of leaving an explanatory gap by omitting it. For it seems questionable whether "mental performances can be identified with brain processes or whether descriptions from the subjective first-person perspective can be reduced to those from the objective third-person perspective" [893].

[2] Limitations in content and cognitive/behavioural abilities resulting from the notion of a global state of consciousness must be understood in relation to the organism's abilities. A mouse that cannot communicate its perceptual experiences does not thereby have an abnormal global state of consciousness, unlike a human that cannot communicate its perceptual experiences because it experiences an absence seizure (brief loss and return of consciousness). Although nonhuman animals and infants may also have impaired global states of consciousness for the same reasons as human adults (e.g. sedation, epileptic seizures), there is no reason to believe that the global states of consciousness of infants and nonhuman animals when awake and alert are different from those of adult humans [897].

the experiential perspective, namely, experiencing "*what it is like*" to see, smell, taste, or feel pain. What makes a state phenomenally conscious is how it feels to be in that state [898]. Consciousness here refers to subjective experience as opposed to merely being awake and alert and responding to external stimuli. In fact, subjective experience is the stuff of novels, poems, and songs of our emotions and memories and could even be said to constitute the essence of being human [899]. Although it is difficult to imagine what it would be like not to be sentient, it is difficult or even impossible to know these sensations in other beings.

The impossibility of tapping into another's perspective of experience has been illustrated by philosopher Thomas Nagel of New York University (USA) using the example of the bat [900]. No matter how much we know about the brain of this animal species, for example, which neural processes take place during the echolocation-like perception of objects, we still do not know what it is like or how it feels for the bat to have such perceptions. But we can also only guess the experiential content of another person's mental or emotional state—such as the feeling of fear—and never fully grasp it. The quality of sensations, their subjective experiential content, also called "*qualia*", always seems to remain hidden from us.[3]

When a person says "I am afraid" while drawing on explicit memory content, such as specific situations in which they previously entered this state, this person is aware of their fear in the sense of access consciousness. Access consciousness thus refers to the availability of representations for use in thought, intentional action, and linguistic communication.

Unfortunately, these two concepts are often mixed together, which hinders our understanding of animal consciousness. Basically, however, the science of animal sentience is—according to the distinction made by Ned Block (New York University, USA)—more concerned with phenomenal consciousness, i.e. the ability to have primary representations of, for example, visual stimuli or painful stimuli, whereas comparative cognitive research focuses more on access consciousness and the ability to form secondary representations [901].

However, there are theories of consciousness, especially the so-called HOTs (*higher-order theories*), which follow classical philosophical concepts—for example, from John Locke and Immanuel Kant—and refer to the mechanism of inner consciousness as an inner sense similar to perception [902, 903]. The theories call for higher-order cognitive access for phenomenal consciousness. In other words, this kind of "*inner consciousness*" arises when one is aware of one's own mental representations. As I will discuss in a moment, this need does not automatically mean introspection, the act of deliberate and effortful self-observation, and evaluation of ongoing conscious experiences; rather, it can also be a relatively passive (cognitively "leaner") process of meta-representation that occurs unconsciously.

[3] Because of this impossibility of intersubjective access to qualia, some philosophers—such as Daniel Dennett or Paul and Patricia Churchland—advocate qualia eliminativism. The problem of qualia, they argue, is a kind of straw man to put reductionist or materialist perspectives in their place but in reality is merely the expression of outdated intuitions from metaphysics, for example, in the tradition of René Descartes.

Inner consciousness is intuitively equated with the capacity for metacognition, but this is not explicitly required in all HOTs [899]. In any case, it seems unlikely for most high-order theorists that animals (or even infants) have the conceptual sophistication required for higher-order thoughts ("I am in the mental state M") or HOTs (representations of representations) [904].

A classic proposal to enable phenomenally conscious experience through cognitive representation, as already discussed in Chap. 5, section "The Power of Imagination", comes from Endel Tulving [499]. He distinguishes between *noetic* (semantic) consciousness and *autonoetic* (self-)consciousness. The former produces higher-order states of consciousness, for example, consciousness of perceptual, conceptual, or memory representations. Noetic consciousness makes it possible to become aware of things that cannot currently be sensed. They are taken into use as facts to establish relationships or derive rules, taken together: semantic knowledge. As I will discuss in more detail, noetic consciousness is attributed to some animal species, particularly apes [899]. In contrast, autonoetic consciousness is attributed by most theorists only to humans. It is consciousness of oneself as the subject and owner of experience and refers to personal experiences. It is thus based on episodic memory, which is autobiographical—an explicitly conscious way of remembering through mental time travel. In this process, memories are virtually re-experienced. People can position themselves in time.

Tulving's two forms of consciousness seem to be preceded evolutionarily by more primitive, purely experiential forms of consciousness—of bodily states for example—that do not involve higher-order representation. They contain primary process experiences and are summarised as *anoetic* consciousness. These experiences are based on the contents of unreflective, implicit procedural, sensory, and affective memory. Basal affective consciousness was included in both the conceptualisation of Jaak Panksepp (Washington State University, USA) [905, 906] and the perceptual consciousness of Brian Merker (Uppsala University, Sweden) [907] on this anoetic level.

A third differentiation—and probably the most controversial for the question of animal consciousness—is that of types of consciousness according to the classical distinction of the individual's attention either outwards to the environment (consciousness) or inwards to the self (self-awareness). An organism that is awake and alert and thus "conscious", i.e. not asleep or in a coma, can process information from the environment and react appropriately to stimuli. It will experience perceptions, sensations, thoughts, etc., directly. It will be immersed in these experiences without being (itself) aware that such perceptual and mental events are taking place. Most animal species are thought to be unreflective actors in their environment. Humans also arguably spend much of their time in this unreflective state of consciousness, interacting with objects and people, talking, walking, or thinking coherently without monitoring these activities [908].

On the other hand, self-consciousness in the sense of self-awareness and self-knowledge requires inwardly directed attention, where the individual focuses not on the external environment but on the internal milieu; he or she becomes a reflective observer who processes information of the self. This self-information includes episodic (and thus autobiographical) memory content, self-assessment (metacognition), and, at least in humans, the ability to describe oneself. Consequences or byproducts *of* self-reflection are self-knowledge, theory of mind, self-esteem, and self-regulation. A linguistically competent creature can verbalise this self-information and tell of its experiences, current feelings, and future intentions.

One can now further differentiate this distinction and, within the framework of a developmental perspective, examine how gradually higher and higher levels of consciousness emerge in infants and children. The LOC (*Levels of Consciousness*) model by Philip David Zelazo (University of Toronto, Canada) [909] implies that with each higher level of consciousness, mental experiences become qualitatively richer and easier to recall, and (conscious) control over behaviour increases. This is also consistent with neurobiological models that describe a hierarchy of modular processes that construct an increasingly complex model of the world [910]. The brain would create such a model to interact more efficiently with the environment and to help simplify, regulate, and speed up subsequent encounters with and responses to the external and internal milieu. While the stage known as "*sensorimotor perception*" performs an analysis of incoming sensory information (including body signals) and enables the construction of a complex motor system, the stage of "*consistent awareness*" is where the organism creates a larger model of the world that is consistent with reality. Executive functions are used to integrate information provided by sensory systems and thus produce goal-directed behaviour. Here, the organism still focuses its attention on the environment and processes incoming external stimuli. Only at the last level ("*self-consciousness*") does the organism build a self-model that requires planning, evaluation, and self-monitoring. It evaluates new information from the perspective of a personal history, remembers the past, and projects into the future. In this respect, this self-consciousness is "*meta-representational*" [911].

From Core to Extended Consciousness

The neurologist Antonio Damasio of the University of Southern California (USA) has outlined in influential books an evolutionary development of consciousness, the most important distinction of which is one between "*core* consciousness" and "*extended* consciousness" [14, 912]. Both forms of consciousness can be described from a biological perspective and traced back to an underlying, purely biological substrate, the so-called proto-self, as their basis.

The proto-self is, as the name suggests, a precursor of the core self, but it is unconscious, a kind of "model of the body in the brain". Its main task is the automatic control of life processes by compensating for disturbances from the environment and thus ensuring that the body states remain within the narrow range of relative stability that is necessary for survival. This initial and restored state is called "*homeostasis*" in physiology.[4] According to Damasio, the regulatory mechanisms necessary in this process generate a kind of representation of the body in the brain, which continuously signal the state of the organism as a whole, thereby they do not generate consciousness but are indispensable for the systems that produce core consciousness.[5]

However, if the body is perfectly regulated and controlled by the proto-self to maintain a state of equilibrium, what is the need for consciousness? For Damasio, the reason lies in the guidance provided by the ideas or representations that the brain constantly updates in order to select those actions that lead to the sources of energy without which the renewal and maintenance of the organismal structures, at least of mobile species, would not be possible. These representations make it possible to make a selection from a repertoire of available action patterns and to optimise the execution of the chosen action. Imaginations also help to invent patterns of action that can be applied to novel situations. For Damasio, the pioneering innovation of consciousness was the possibility of linking the inner sphere of life regulation with the processing of ideas [14].

Core consciousness is not to be confused with (but presupposes) wakefulness or basal attention, and it is independent of memory, judgement, and language. Nor is it involved in processes such as planning, problem-solving, and creativity, when working memory is important. Rather, core consciousness is linked to emotions: both background emotions such as comfort or tension and primary emotions that accentuate the subject's actions, such as anxiety, fear, joy, anger, and disgust. The core consciousness is so strongly intertwined with emotions that their failure—in emotionlessness and apathy—also leads to the failure of the core consciousness. By the organism being able to respond emotionally to something external and represent this within, subjectivity in the form of core consciousness emerges from the proto-self. Whereas for Damasio [14] emotions remain unconscious as outwardly directed reactions observable by others; feeling arises as a conscious, private, mental experience of an emotion. Feelings are thus the perceptions of emotions [893].

The core consciousness consists of an implicit sense of self in the here and now, an "*inner sense*", as was also assumed by thinkers ranging from Locke, Brentano, Kant, and Freud to James. Thus, this core self is a "*fleetingly constructed*" subject limited to the present that carries the "*individual perspective*" and is constantly updated [893]. Even before Damasio, but in a similar vein, Gerald Edelman of the

[4] The term homeostasis generally refers to a state of equilibrium of a constantly changing, dynamic system and is also used in this way in other sciences and especially in cybernetics. In physiology, the term is defined as the constant maintenance of an internal milieu (target state), which comes about through a wide variety of automatically running regulatory processes of the autonomous nervous system—of temperature, pH value, and oxygen concentration. The term was coined, among others, by the Austrian systems theorist and theoretical biologist Ludwig von Bertalanffy as a special form of a steady state produced by secondary regulation [913].

[5] A much-noticed feature of Damasio's biologically based theory of consciousness is the fact that he includes the entire organism in his exploration of consciousness, especially its biological foundations, and thus does not remain purely neurocognitive. In other words, *embodiment* plays an essential role in his work [893].

Neurosciences Institute in San Diego (USA) distinguished in 1989 a *primary consciousness* as a present awareness of objects and events from a *higher-order consciousness* that is based on and dependent on the primary [914, 915].[6] According to these two authors, the first level of consciousness—the primary or core consciousness—is common to many animals, so it is not limited to humans. However, only extended consciousness (of a higher order) frees the organism from the "*remembered present*" [914] and extends the dimension of time into the past and future. Although Damasio believes that some other animal species also possess an extended consciousness of a simpler kind, it has reached its higher forms only in humans. Unlike core consciousness, which does not change as long as the organism is alive and does not depend on memory, thought, or language, extended consciousness develops only over the course of life, depends on conventional memory and working memory, and is further enriched by language at its highest stage of development. For Damasio, extended consciousness is everything that the core consciousness is, only bigger and better. Moreover, extended consciousness is capable of growing alongside the development and life experience of each individual, whereas core consciousness is genetically inherent and almost unchangeable.

The knowledge that people can access from their extended consciousness is constantly growing and covering a wide field in space and time. Above all, it is the many personal memories that we can access that determine our identity and personality. Therefore, the most important innovation of extended consciousness is the expansion of memory. While the time scale of core consciousness is only a fraction of a second, extended consciousness expands the time scale to hours and years. Long-term memory is responsible for this extended access, which allows individuals to hold at the ready, over a considerable period of time, many "objects" of the moment, whether perceptual, remembered, or imagined ones.

For Damasio, the two forms of consciousness also give rise to two corresponding forms of the self. While the core self, as already mentioned, is an ephemeral phenomenon of the core consciousness that is recreated for each object with which the brain interacts, the extended consciousness creates the actual self that gives the individual a sense of identity. The extended consciousness permanently holds the individual's particular facts, behaviours, and modes of existence and can refer back to the most important events in the individual's life in the autobiographical memory, and Damasio calls this the "*autobiographical self*". For this self, but not for the core self, language seems to be essential. A growing number of researchers believe that more sophisticated kinds of self-knowledge at least require language, more specifically, inner language [908].

Where does extended consciousness occur with the autobiographical self? For Damasio, only in organisms endowed with a considerable capacity for memory and thought does language not seem to be essential. Children acquire language in the

[6] Damasio emphasises that Gerald Edelman's primary consciousness is simpler than the core consciousness he proposes because it does not lead to the emergence of the self. In addition, Edelman's higher-order consciousness differs from Damasio's extended consciousness in that it relies on language and is limited to humans.

course of their first years of life, perhaps even a little earlier. Interestingly, Damasio thinks that apes such as bonobos also have an autobiographical self, and even some dogs, he dares to suggest. However, he says that extended consciousness should not be equated with intelligence. While the former enables the organism to have a broad horizon of knowledge and to present it in a clear and manageable way, intelligence has to do with the ability to successfully manipulate this knowledge in a way that new responses can be planned and executed. Thus, extended consciousness is a precondition of intelligence, whereas extensive working memory is in turn a precondition of extended consciousness. Together with focused attention, working memory contributes to extended consciousness, much like the concept of the "*global workspace*" described by Bernard Baars of the Neurosciences Institute in San Diego (USA) [916].

Only now does language come into play for Damasio. Equipped with language, extended consciousness allows humans to climb to the summit of their mental abilities. Damasio lists ten resulting mental processes and achievements [14]: (1) making useful tools, (2) taking into account the mental processes of others, (3) suffering pain instead of just feeling and reacting to it, (4) perceiving death in oneself and others, (5) valuing life, (6) developing a sense of good and evil, (7) taking into account the interests of others and the collective, (8) sensing beauty instead of just pleasure, and (9) perceiving a contradiction between feelings and (10), later, between abstract ideas. As I have tried to show in the main part of this book—and will summarise at the end of the book—at least a few of these achievements can also be demonstrated in non-human animals.

In summary, three (rather than two) levels of consciousness can be distinguished and contrasted with a general unconsciousness [908]: consciousness, self-consciousness, and meta-self-consciousness. Consciousness begins with the focusing of attention on the environment and, as a consequence, the processing of incoming external stimuli. At this level, Edelman's primary consciousness [914], Damasio's core consciousness [14], and Zelazo's minimal consciousness [909] are located. Self-consciousness, on the other hand, requires focusing attention on oneself and processing both private and public self-information. Here, one can locate Edelman's higher-order consciousness, Damasio's extended consciousness, and Zelazo's recursive consciousness. At the highest level—meta-self-consciousness—the organism becomes aware of being aware of itself, thus attaining an extended self [917].

Criteria for the Presence of Awareness

As aptly summarised by the philosopher Markus Wild (University of Basel, Switzerland), the important facets for meaning in human consciousness include states and processes that are at least as diverse as wakefulness, attention, intention, self-awareness, and sentience. Therefore, Wild [918] proposed the following subdivisions:

1. Vigilance awareness (being awake)
2. Cognitive awareness (directing attention to something)
3. Awareness of action (doing something with intention)
4. Self-awareness (being aware of one's conditions or oneself)
5. Phenomenal consciousness (feeling, sensing, perceiving, etc.)

There is good evidence for the existence of all these states and processes in an adult human's "*normal state*", i.e. when he is neither mentally impaired nor asleep nor influenced in his thinking and feeling by drugs or other means. While I write these lines, I am well rested, thinking about the rest of the sentence, thinking about what I know about this subject and how I reproduce it in a clear and structured way, giving it my full attention and blocking out the background noise in the room, but I am very much feeling the increasing heat of the day and an aching back. I am in a state that contains all five of the previously mentioned components. The question now arises as to which of these states and processes may be attributed to animals.

Recently, from his point of view, Wild summarised the current evidence in such a way that mammals and birds can be attested to consciousness in the sense of 1, 2, and 5, but it remains unclear whether they can do something intentionally and whether they have self-awareness in the sense of having access to their own mental states [918]. However, he refers to research results on so-called "*intentional communication*", according to which some animal species also want to communicate intentions.[7] For example, chimpanzees inform conspecifics of imminent danger [754]. It has been shown that these communicative acts are purposeful, that they are produced voluntarily, and that they serve as a means to achieve the intended goal in the receiver, i.e. to change the receiver's behaviour in a way that is conducive to the realisation of one's own goal. A comprehensive compilation of all evidence for such intentional communications in animals has shown that they seem to be present at least in apes, whales, and corvids [924], and these animal groups therefore also have consciousness in the sense of (3).

There are also approaches to formulating objective criteria for the presence of animal consciousness, which is important, especially for ethical questions and social decisions, including binding laws, and for moving away from "all-or-nothing" judgements. These approaches are particularly advisable when making decisions

[7] However, it is stated restrictively that here only a "*first-order intentionality*" is possible [919], namely, that the signaller intends to signal in order to generate a response in the receiver, but not higher-order intentionalities, which involve the supplementation of the signal intention by the attribution of mental states (by the receiver to the signaller). Mutual understanding of each other's intentions and beliefs was seen as an essential feature of human communication [920] and as an ability that presupposes a sophisticated, preexisting meta-psychological framework [921, 922]. Many psychologists and philosophers of language have followed this linguistic conception of language proposed by Grice, but recently the assumption that human communication must aim to alter mental states has been questioned [923].

about the presence of consciousness in individual animal species or larger taxonomic units. The most commonly proposed criteria are as follows [925]:

K1. Nonverbal or non-vocal behaviours
K2. The ability to use language
K3. The similarity of brain structure to that of humans
K4. The ability to learn to solve problems and be creative

Without doubt, humans fulfil all of the above criteria, and stones none. We also cannot attribute any of the abovementioned abilities to plants, fungi, bacteria, or viruses, even though these are living organisms. In the case of animals, the criteria help to draw boundaries or make gradual distinctions. Some animals fulfil two or three criteria, while others fulfil only one. We may perhaps assume that the more these criteria are fulfilled, the more likely an animal is to be conscious (even though none of the above criteria may be necessary or sufficient). However, this discussion becomes difficult when there are still gradual differences within the criteria, which is the rule rather than the exception.

The cognitive abilities of animals described in this book generally refer to the fourth criterion (K4). Therefore, I would like to focus on this in particular below. I would also like to present studies for the third criterion, as these have recently received a great boost with the rise of neuroscience. The first two criteria will be addressed briefly, as in earlier chapters of the book they were already discussed.

The first criterion assumes that reactions and behaviours that animals perform like humans are likely to have similar causes, i.e. to be the result of similar brain processes. The philosopher Rocco Gennaro from the University of Southern Indiana in Evansville (USA) [925] cites, as an example, a dog that behaves in a similar way to a human when someone steps on its leg, and this seems to indicate that the dog is conscious. Particularly in the question of the sensation of pain in animals, such analogies are invoked, but they will only bring basal bodily consciousness into view. It may make sense to start with the assumption that mental states cause bodily movements in order to deduce that the former (pain, suffering) are present when the latter are observed. However, without further evidence, for example, from the field of neurobiology (K3), one will not be able to derive any robust conclusions from this.

For the second criterion, the ability to use language, opinions differ widely. For some, the mere evidence of a form of communication is sufficient, while for others, human language is constitutive of consciousness. Undoubtedly, there are usually conscious acts of thought behind human linguistic utterances. Therefore, quite a few support the proposal of the philosopher Daniel Dennett of Tufts University in Boston (USA) [926] that consciousness requires language in the sense of "*storytelling*", and thus, only humans can be conscious. Indeed, the ability to accurately report events is routinely used for scientific and medical applications in humans. However, this places us at the other extreme end of the scale of consciousness, namely, self-awareness. The conscious "*self*" is truly just a kind of narrative, Dennett argues, that we tell ourselves. The existence of human-like language as a necessary condition for consciousness seems to be an unnecessarily high hurdle when other forms of consciousness are meant.

An explicit verbal report of a conscious experience is considered the "gold standard" in the sense that it guarantees the presence of consciousness. But responses need not be verbal; for example, eye movements are also used in cases of paralysis or "*lucid dreaming*" [927]. In young children too, pressing levers or blinking one's eyes can replace a verbal report. Analogously, this approach can also be taken with animals, as one can have the animal make a "comment" by pressing a button. This is important not only for the study of visual consciousness, which is relevant to the phenomenon of "*blindsight*"[8] and a central research topic in macaques, among others [928], but also for the well-known matching-to-sample experiments conducted in a variety of species, including primates, cetaceans, and birds such as African grey parrots and budgerigars. Such procedures of accurate behavioural reporting are used to thus demonstrate consciousness [929].

The Neurobiology of Consciousness

The third criterion, the similarity of animal brain structures with that of humans, also has its pitfalls and—as we will see—becomes more problematic the further we move away from humans in the animal family tree. However, since the question of the material causation of consciousness (not only in humans) is a fixed part of the ever-growing literature on the neurobiology of consciousness, this criterion deserves a little more space. With the increasing knowledge about the neural processes involved in conscious actions, there are also an increasing number of proposals to use comparative neurophysiological and neuroanatomical findings to determine animal consciousness. In doing so, philosophically trained neurobiologists do not claim that states of consciousness are identical to neurological states or that they are mere epiphenomena [918] but instead use the term "*neural correlate of consciousness*" [601, 927, 930].[9]

As a qualification to everything, I will subsequently write about these neural correlates; it must be said that consciousness, which is addressed here as a biological phenomenon, also contains a subjective and thus private component that is inaccessible to direct scientific observation. However, consciousness can be modified by various physical means, including drugs, sleep states, and damage to brain tissue, all of which have direct effects on neural integrity and activity. Thus, even the

[8] In human patients with lesions of the primary visual cortex, it was found that they, despite denying seeing the stimuli (phenomenal blindness), can nevertheless respond to (by forced-choice guessing) them in their phenomenally blind fields and they can learn to detect, localize, and distinguish between visual stimuli presented within their visual field defects [928].

[9] A neural correlate of consciousness is a specific pattern of neural activity that correlates specifically with conscious experience and not with unconscious perception or action. Such a pattern of brain activity, a neural footprint as it were, is a physical process and therefore amenable to empirical investigation, which initially remains neutral to questions of causality. The question of how something material becomes phenomenal experience, which has preoccupied philosophers since antiquity, is thus omitted.

subjective phenomenon of consciousness can be directly influenced by objective, quantifiable agents that affect neurons. Although the neural basis for consciousness cannot be studied directly, it is possible to study the neural basis for changes in consciousness and, from this, to formulate hypotheses about the former [931].

The attempt to prove the similarity of brain structures that may be necessary for consciousness is naturally most successful in the class of animals to which we humans also belong—mammals. Thus, a variety of theoretical and empirical arguments have been put forward in support of the thesis that consciousness is present in all mammals [932]. Human consciousness is correlated with widely distributed, relatively rapid, low-amplitude interactions in the thalamocortical nucleus of the brain, which are activated by current tasks and conditions. Indeed, such neural patterns are also found in other mammals, which suggests that consciousness is an important biological adaptation for mammalian species [933]. Moreover, many animals share several additional primal areas of the brain (such as the amygdala in the limbic system) that are responsible for emotions, such as fear. Neurobiologists thus have high hopes for discovering specific neural correlates of consciousness in humans to shed light on animal consciousness, especially given our shared evolutionary history.

However, as mentioned earlier, the more similar the relevant brain structures are, the more successful this endeavour is. This has led some neurobiologists to assume that, since animals lack some of the brain structures that are responsible for more sophisticated mental abilities in humans, they are incapable of having these kinds of mental states. This difference seems particularly evident in the lack of a prefrontal cortex, which is central to our thinking. Moreover, this region of the brain has shown the greatest growth dynamics over the course of mammalian evolution. An increase in the size of the brain was accompanied above all by an enormous increase in the cortex, which is also where we find the greatest differences. From a shrew that has a brain surface area of less than 1 cm^2 to a false killer whale, which is actually a dolphin that has one of 7,400 cm^2, we find a 10,000-fold increase in brain size [934].

If we restrict our consideration to the frontal or prefrontal cortex (PFC),[10] which in humans is the seat of working memory, action planning, control, and intelligence in general, the differences are less dramatic but still considerable. In fact, as Terrence Deacon of Harvard University (USA) has claimed in an oft-cited paper [935], humans have the largest PFC in the animal kingdom. Though its ratio to the rest of the brain is 38%, this is only slightly greater than that of our closest relatives (gorilla 37%, chimpanzee 35%, gibbon 30%). In absolute terms, the human brain is two times larger than that of the gorilla and three times larger than that of the chimpanzee [936]. The enormous increase in the size of the human PFC affects mainly the dorsal region.

[10] Although most comparative research focuses on the forebrain components most directly involved in the emergence of a state of consciousness, it is nevertheless necessary to recognise that a number of ascending systems, such as those of the *locus coeruleus*, the raphe nuclei, the reticular activation system via the intralaminar nuclei, and the cholinergic nucleus basalis, are either essential or at least contribute to maintaining a state of consciousness.

On the question of cognitive abilities, it has been argued that it is not so much the area or volume but rather the number and size of the neurons, the number of connections, and the speed of conduction that are crucial [934, 937]. While dolphins and elephants have very large pyramidal cells in the cortex, those of primates are relatively small. However, as the size of these important cortical neurons increases, their density decreases. Again, this number is high in primates, especially in small monkeys, in which 75,000 neurons per mm^2 are observed in marmosets and only 6,000–7,000 neurons are observed in dolphins and elephants. Humans reach a very high density with an average of 50,000, which can increase to up to 100,000 neurons per mm^2 in some areas of the brain.

By combining neuron size and density, the number of neurons in the cortex can be extrapolated. Here, humans are ahead, with around 16.3 billion, followed by chimpanzees (6.2 billion), elephants (5.6 billion), and gorillas (4.3 billion) and then by a large margin rhesus monkeys (1.7 billion), squirrel monkeys (1.3 billion), and capuchin monkeys (1.1 billion) [938–941].

Overall, we can now say that comparing these neuron numbers with the first large-scale quantitative studies of cognitive abilities across different mammalian species suggests that the absolute number of neurons in the mammalian cerebral cortex is a good correlate of cognitive diversity: the greater the number of neurons, regardless of brain or body size, the better the performance of a species on the same task [167, 325, 939].

However, 16.3 billion neurons in the human cortex represent only 19% of all brain neurons. This is astonishing when one considers that 82% of the human brain mass consists of the cerebral cortex (including the subcortical white matter) and 42% consists of the cerebral cortical grey matter alone. This is because the cerebral cortex houses 72% (or 61 billion) of all non-neuronal cells in the brain. In contrast, the cerebellum, which accounts for only 10% of total brain mass, contains 69 billion neurons, 80% of all neurons in the human brain. The team of Suzana Herculano-Houzel from Vanderbilt University (USA) constructed these figures using a method that allows the number of cells that make up the entire brain to be quantified quickly and reliably. Therefore, the authors emphasise that the ratios between glial cells and neurons in the human brain structures are very similar to those in other primates and that the number of cells is as expected for a primate with human proportions [938]. In fact, the human brain almost exactly follows the scaling rules of those six primate species whose brains have been fully quantified thus far: the mass and cellular composition of the human brain differ by only 10% from those expected for a primate weighing 75 kg. In other words, the human brain is a normal, isometrically scaled primate brain in terms of the number of neural and nonneural cells [940].

Certainly, the number of synapses between neurons also plays a role in information processing in the brain, but the estimates are still very vague. If you extrapolate the estimated average number of 20,000 synapses per neuron in the human brain, you arrive at the enormous figure of 300 trillion synapses. Estimates from other mammals are scarce. Finally, the processing speed of information is considered an important measure of brain performance. This depends on the distance between the neurons, the conduction speed, and the synaptic transmission speed. Interestingly, whales with an unfavourable combination of large distance between neurons and low conduction speed perform poorly here. While human brains are five times superior in this respect.

In addition to the metric factors mentioned above as indicators of brain performance, the structuring or architecture of the brain is increasingly being studied. Parcellation into functionally separate units for sensory, motor, and sensorimotor information is a readily observable trend in the evolution of the mammalian brain. Some philosophers highly value this too in the sense of modularity [942]. In contrast, insectivores have a dominant olfactory cortex, where there are only two somatosensory, two visual, one auditory, and one motor region yet hardly any integration centres [943]. Meanwhile, the human cortex is considered to have around 150 integration centres with an average of 60 connections between them, giving approximately 9,000 area-to-area connections [944]. However, this parcellation is not a general trend with the increasing of mammalian brain size, as again, the largest mammals (whales, elephants) deviate here. These mammals have only one or a few very large sensory areas—in whales, the auditory system; in elephants, the somatosensory–vibratory system—and very little parcellation. Thus, in these animals, overall, we find few cortical areas with specific functions: large limbic, parietal, and temporal areas, but no substantially emerging non-limbic prefrontal cortex areas [945].[11]

To summarise mammalian brains, while there is a trend towards larger brains in absolute terms (elephants and whales have the largest brains, weighing up to 10 kg), there is no trend in relative terms due to so-called negative brain allometry. Thus, relative to body size, elephants and whales tend to have small brains (only 0.01% in the blue whale but 10% in the shrew) [934]. Seen in this light, the human brain is large in two senses, both in weight (about 1.35 kg) and in relation to the body (2%). Add to this the enormous thickness and density of the human brain's cortex, which together, as I said, result in the highest number of neurons (approximately 15 billion) in the animal kingdom. If one then also adds its very high processing speed, due to the high conduction speed of the nerve tracts, alongside the small distance between the neurons and the high degree of parcellation, one can easily recognise the superlatives of the human brain [328]. However, it is more difficult to classify the brains of other mammals. The combination of the abovementioned characteristics gives primates, especially great apes, a very good overall brain score. However, it is difficult to classify the elephants and whales, which perform very well in terms of several important characteristics (absolute brain size and neuron number) but surprisingly poorly in others (relative brain size, cortex thickness, packing density, conduction velocity, and parcellation). Nevertheless, it can be said that the very similar, homologous architecture of mammalian brains allows the conclusion that certain types of consciousness (see above) are present in all mammals.

[11] Much of the above data on mammalian brains can be found in the major compilations by Roth [934] and Kaas [943].

However, even if direct comparability is limited, brain superlatives also exist in other animal groups. Among birds, the brains of corvids and parrots stand out. These two bird groups also have very large brains—within birds in absolute terms, compared to mammals in relative terms. Their endbrain in particular, called the cerebrum (Latin *telencephalon*), is comparatively large, accounting for 70–80% of the entire brain of these animals. Due to the combination of very small neurons and an extremely high density, these brains, weighing only a few grams (corvids 8–16 g; parrots up to 25 g), contain several 100 million neurons, reaching values found in smaller New World monkeys.

What about the architecture of the bird brain? Are there areas that correspond to the cortex, which is so important for human consciousness? The answer at the present time is a vague "yes-no". Although analogous, i.e. functionally identical, areas have been found in birds, it turns the argument on its head because it is then no longer possible to draw conclusions about identical functions, in our case consciousness, on the basis of homologous (same evolutionary origin) brain regions. The question of the similarity of brains is further complicated by the fact that it is not the large-scale anatomy alone that is decisive for the realisable functions. It is also the type of connection—the so-called "*connectome*"—and even the function of individual cells, such as neurons, interneurons, glia, etc., that has a decisive influence on cognitive output. Although there is no neocortex (in the proper sense) in birds, there are nevertheless certain types of constituent and ancient neuron cell types that are also found in the mammalian neocortex [946]. After all, the mammalian neocortex did not develop de novo but is composed of cells and their assemblies, which have existed for a long time (as has been demonstrated with genetic markers) [947] and have reorganised themselves.[12] In many cases, the microarchitecture has been preserved so that both their functions and their connections to other parts of the brain, such as the thalamus, have been preserved in these miniature networks [950].

One may no longer be surprised by the capacity of the bird brain after learning of the achievements in tool use, time travel, and social intelligence in birds, such as those described in this book. However, what is surprising is that similarities to the mammalian neocortex can also be found in the brains of fish. Their location and structure are different (as in birds), though a number of neurons exist in fish that

---

[12] However, recent research has shown that the cerebrum of birds and mammals is a mixture of homologous and analogous parts [714, 948, 949]. Even what appears to be homologous is organised very differently. Birds do not have a neocortex, but several large pallial aggregates with no obvious layering (laminar structure). These aggregations could correspond to cortical layers. Independently, birds and mammals have evolved similar brain organisations that are likely to form the neural basis of their cognitive abilities. Birds have a functional analogue to the prefrontal cortex (PFC), the nidopallium caudolaterale (NCL), which likewise has executive functions. Overall, there is a surprising number of similarities between the avian pallium and the mammalian neocortex. It is very likely that each of the avian parallels to the mammalian neocortex represents a mixture of homologous basic elements and convergent patterns. NCL and PFC may represent a spectacular case of evolutionary convergence in that both have evolved very similar—especially similar cognitive—functions over 300 million years.

have been preserved in the mammalian neocortex [951]. There may be biophysical reasons why the most anterior part of the brain, the endbrain, did not bulge as massively in fish and birds as it did in mammals; for example, it may simply be due to hydraulics that this part remained small [946].

Reptiles also lack a neocortex as well as a hippocampus. But although the latter is central to mammalian navigation, turtles, for example, can also solve complex spatial tasks. My former colleague Anna Wilkinson (now at the University of Lincoln, UK) was able to show that red-footed tortoises are similarly adept at orienting themselves in the radial maze as rats are [952, 953]. As detailed in my final Chap. 11, these animals also mastered tasks in the perceptual and social domain, such as class distinctions and social learning, despite the absence of a neocortex and the complete separation of visual, auditory, and somatosensory centres. Comparative neuroanatomists have found that there are populations of neurons in the endbrain of vertebrates that correspond to groups of cells in the mammalian neocortex in terms of connections, activations, and functions. Some of these non-laminar populations, located within the dorsal ventricular ridge (hypopallial region), are thought to represent the homologous equivalent of the mammalian (laminar) neocortex in reptiles and birds [954, 955].

In vertebrates, there were several dramatic changes, especially in the diencephalon and the endbrain, that involved significant enlargements but also reductions and simplifications. Such changes are well explained in the context of new or lost sensory systems, for example, electrosensory or echolocation, but explained worse in that of cognitive abilities. In particular, the dorsal thalamus and the closely related dorsal cerebral cortex (pallium) of the endbrain show dramatic development. Beginning with amphibians, two trends of cerebral cortex "exploitation" are discernible. First, in the six-layered cortex of mammals evolved from the dorsal pallium to a multisensory and associative centre that receives signals primarily from the formal thalamus. Second, but independently, the ventral cerebral cortex of sauropsids developed into the dorsal ventricular crest (hypopallial region) for reptiles and the mesonidopallium for birds; these regions also received sensory inputs from the dorsal thalamus, which culminated in a cortex-like integration centre in birds. One can therefore speak of analogous developments on a homologous basis: "*homoiology*" [802].

In view of these data, it is difficult to make a final judgement on the various vertebrate classes. Some similarities, whether homologous or analogous, are counterbalanced by many differences that reduce one's confidence in attributing consciousness. Without a firm metric for brain similarities and a criterion for when it is sufficient to assume consciousness, even researchers familiar with the latest findings in the behaviour and cognition of these animals do not see a solid basis for assuming that fish, amphibians, and reptiles possess the psychological and neural processes required for consciousness [956].

Finally, if we turn to invertebrates, attempts to establish neuroanatomical similarity are many times more difficult. The cephalopods are often mentioned in the consciousness literature; first and foremost, the octopus has a nervous system widely distributed throughout its body and even "thinks" with its eight arms—two-thirds of the neurons are located there [957]![13] The rope-ladder nervous system of molluscs is built completely differently from that of vertebrates. Even if some zoologists call the frontal thickening, which resulted from the fusion of large nerve nodes (ganglia), a "brain", the oesophagus, for example, runs right through it. Although the nervous system of cephalopods is not only the most efficient of all molluscs but also even that of all invertebrates, its structure is fundamentally different from that of vertebrates. This can be seen very well in the eye, which is outwardly strikingly similar to that of mammals, but in individual development, it is constructed quite differently (evers, receptors directed outwards, instead of inverses, directed inwards). Here, we are dealing with an astonishing convergence in the evolution of centralised nervous systems because the common ancestor goes back deep into the Palaeozoic era.[14] Thus, the fourth criterion for the proof of consciousness, at least in terms of phylogenetic similarity of the brain structure with humans, is not applicable.

Another group of invertebrates that has received much attention regarding consciousness beyond human brains is arthropods. Here, insects in particular, as well as some crustaceans and arachnids, have found their way into the philosophy of mind. After all, arthropods were among the first animals to develop complex active bodies and brains that they could control [958]. While, for some, the possibility of the evolution of consciousness in these animals is due to functional considerations, namely, the ability to solve problems that arise from controlling complex bodies [959], others even argue on the basis of neuroanatomical similarity to the mammalian midbrain. Considering that basic subjective experience in humans requires an integrated midbrain and basal ganglia structures, which are among the oldest vertebrate brain systems, and that the insect brain supports similar functions to the vertebrate midbrain, insects are sometimes even considered to have the capacity for subjective experience [960].

Although the brain of the honeybee contains almost a million neurons in a volume of one mm^3, it still bears relation to neither the human brain, which has many billions of neurons, nor to octopuses, birds, and many mammals, which have many millions of neurons. Above all, the insect brain has a completely different structure

[13] In fact, the nervous system of an octopus has approximately 550 million neurons, similar in number to small monkeys, but 350 million of these are in the eight arms, 160 million in the optic lobes, and a mere 40 million in the thickening of the supra-oesophageal ganglion known as the "brain".

[14] The last common ancestor of vertebrates and cephalopods lived in the Ediacarium, approximately 600 million years ago, and was probably a flat little worm. Unfortunately, no fossil evidence is available. In the groups emerging from these earliest bilaterians (bipeds), the protostomes (primordial mouths), and the deuterostomes (new mouths), more complex brains evolved in only three groups, besides the Vertebrata (vertebrates) and also in Arthropoda (arthropods) and in one group of the Mollusca (molluscs), namely, Cephalopoda (cephalopods).

[961].[15] Homologous similarities, i.e. similarities based on relationships, are very unlikely because arthropods and molluscs separated from the branch to which vertebrates belong when only very simple nervous systems (viz. diffuse nerve networks) were present.[16]

Those who look at the neural structures of animal species for the question of consciousness must therefore argue on a functional level, i.e. explain the similarities through convergent developments and, for the generation of consciousness, rely on the argument of multiple realisation [962]. Multiple realisations are also the main reason why identifying conscious states through the activity of certain parts of the nervous system seems questionable [918].

Neurobiologists therefore suggest that instead of—or at least in addition to—neuroanatomical tracing, we should focus on the detection of neurophysiological similarities. This proposal is guided by the effort to elucidate the neural conditions under which mental states, including the various types of consciousness, arise in the brains of humans and probably also in some animal species. Note that this initially speaks of necessary but not yet sufficient conditions, which is important for distinguishing between correlative and causal explanatory patterns. An essential finding, which is well established by various neurophysiological methods (EEG, MEG, fMRI), is that "higher" mental states and different types of consciousness are linked to the activity of the thalamocortical system, which is closely connected to the activities of many other brain systems, such as the reticular formation, the basal forebrain, and cortical and subcortical limbic centres. Interestingly, states of consciousness in humans (and non-human primates) are preceded by unconscious processes lasting around 200–400 ms [896, 963, 964], and in addition to a minimum intensity of cortical activity, a minimum time of 100 ms is also needed [965–968]. Ultimately, however, what is crucial is that with this knowledge and with the help of accurate brain wave measurements, one can predict conscious mental states based on specific brain processes with a probability of 60–100% and vice versa [969].

The search for the neural correlates of consciousness is not exhausted by the localisation of certain activities but rather attempts to elucidate the mechanisms

[15] The nervous system of insects has thickenings in the anterior region of a paired ventral cord. Ganglia are fused into both a superior and inferior ganglion, and both are connected to each other by pharyngeal connectives. The upper pharyngeal ganglion is again functionally equated with a brain because it reveals a very large anterior region, the protocerebrum, and two other thickenings beside it—deuterocerebrum and tritocerebrum. The protocerebrum contains the lateral visual centres, which receive their innervation from the large compound eyes, nerves extend from the deuterocerebrum into the antennae, and the tritocerebrum supplies the foregut. The protocerebrum attracted the attention of cognition researchers because it contains structures that form the neuronal substrate for learning and memory and are powerfully developed, especially in insects: the unpaired central body and, above all, the paired mushroom bodies that perform an integration of several senses. With 200,000 neurons, the mushroom bodies make up 20% of the neurons in the honeybee. Within insects, the bee's brain is very large; fruit flies are 50 times smaller with one tenth of the neurons.

[16] The idea that a tripartite, relatively complex nervous system was already present at the root of bipedal animals is discussed, but then secondary simplifications would have to have occurred very often later, which is unlikely (on this, see [934], section "Betting on Your Own Knowledge").

involved. Thus, it is assumed that during conscious states, processes of "rewiring" of existing neural networks, for example, short-term modifications of synaptic coupling, take place. This happens predominantly in the dorsolateral prefrontal cortex, the seat of working memory, in close interaction with associative parietal and temporal regions, where meaning-making information processing takes place and corresponding memories are localised. In particular, in the context of attention, evaluation of incoming information, and goal setting, rapid rewiring of synaptic coupling can be measured because these processes are metabolically demanding and lead to a significant increase in local glucose and oxygen consumption, which is accompanied by an increase in local cortical blood flow [970]. Neuromodulators, especially dopamine and acetylcholine, also play important roles.

Three decades ago, neuroscientists discovered striking oscillatory activities and synchronisations of cortical networks that are more or less directly associated with states of consciousness. This was thought to "bind together" the neural activities within distributed localised networks into meaningful units [971, 972]. But at first, there was no good evidence for a direct link between cortical oscillation synchronisation and consciousness. Rather, the only thing that was certain was that these phenomena are involved, among other things, in the orientation of attention [972], possibly also representing a nonspecific prerequisite for consciousness [896, 973].

Particularly in the visual field, it was possible to reproduce the process of becoming aware of visual perception in the primate brain with precise stimulation and measurement, for example, with the temporally precise method of event-related EEG. This resulted in the central assumption of the consistent recurring upward and downward activation of the primary and secondary sensory (and associative) cortex areas along the processing pathways of visual information [974–979]. As long as the processing of certain sensory experiences is only ascending—from the primary and secondary visual cortex (V1, V2) followed by activation of higher-order visual areas (V4) after 200–250 ms—and does not activate descending connections back to the primary areas, the process remains unconscious. Only when, after a short delay, the primary and secondary visual cortex are activated again after around 300 ms are the visual stimuli perceived consciously [963] (this finding coincides well with the occurrence of the P300 wave in event-related EEG, which is often regarded as the moment when unconscious processes become conscious).

The purpose of this repetitive upward and downward processing is to bring together details that are still without meaning and memories or interpretive schemata that lack such details. First, incoming visual stimuli are processed unconsciously in primary visual cortex areas (V1 and V2) according to basic visual properties (e.g. contrast, wavelengths, edge orientation, direction of motion, disparity). The results of this detailed processing are then sent to associative visual cortex areas, where they are further processed according to their meaning with the help of other cortical and subcortical areas and by adequate association with memory content. Only the interaction of the cortex with limbic centres enables the evaluation of the perception content—according to individual and social experience—and enables the "*creation of meaning*". This global and meaning-giving "*interpretation*" is then sent back to the primary and secondary visual cortexes and then made conscious.

Thus, the cortex in humans and other large-brained mammals is a gigantic association machine in which not only is sensory information multimodally connected but is also manipulated in the short term, abstracted and interpreted, stored in the medium and long term, and finally transformed in a superordinate way into "purely mental representations as a kind of self-description of the cortex", as the neurobiologist Gerhard Roth (University of Bremen, Germany) puts it [934].

Due to recent neurobiology research, it is now possible to narrow down the seat of human consciousness relatively well. The electrical stimulation of cortical tissue and the examination of patients after the loss of certain regions, due to injuries or diseases, were particularly informative. For example, the removal of even small regions of the posterior cortex led to a loss of entire classes of consciousness, such as recognising faces or seeing movement, colour, or space. Therefore, it seems that many sensations and experiences are generated by regions within the posterior cortex; indeed, almost all conscious experiences originate there. But even so, the exact differences between these posterior regions and much of the anterior prefrontal cortex are still unclear. Interestingly, these differences do not directly contribute to subjective experience. It is also disputed whether a correlation signifies a seat or origin of consciousness rather than a (mere) involvement.

Finally, recent results from functional imaging in humans and electrophysiology in awake monkeys indicate that there are important differences between the striate and extrastriate visual cortexes in terms of correlating neural activity with consciousness. In summary, thalamocortical signals, rapid, irregular, low-amplitude electroencephalographic (EEG) signals, and widespread cortical activity can be considered the neural correlates of human consciousness [980]. Moreover, these correlates of human consciousness can also be found in other primates, indeed in all mammals (tested thus far). A frequently used test procedure is "*binocular rivalry*", in which different visual stimuli are presented to the right and left eyes. Instead of seeing a superimposition of the two images, we first see one image for a few seconds, then the second, then the first again, and so on. When the two images alternate, the brain cannot decide on this ambiguous input. To test this phenomenon in non-human animals, rhesus macaques were trained to press a lever to report a specific stimulus [981]. Neurons in the inferotemporal cortex of macaques[17] correlated with the reported stimulus, while neurons in the V1 visual area responded to the visual signal instead. This observation is consistent with findings in humans of a critical role of this cortex in visual awareness and is thus an example of how comparisons can be made between humans and animal species in the field of consciousness research.

Is there evidence that the neural correlates of consciousness found in humans and other mammals also exist in birds and other species with complex or large brains? To date, no features typical of mammalian thalamocortical neurons, such as low-threshold calcium^{2+} spikes and slow oscillations, have been found in birds [980].

[17] The inferotemporal cortex is an area of the temporal lobe of primates that processes visual stimuli at a higher level of form perception and is primarily concerned with face recognition.

Nevertheless, similarities between the EEG patterns of awake birds and mammals, as well as the slow-wave electrical activity recorded during avian sleep, indicate [982] neural dynamics that could support the assumption of conscious states in birds. One might ask which neural features postulated to generate consciousness in mammals, particularly with respect to thalamocortical loops, are also present in birds. Like most vertebrates, birds share the so-called ascending systems with mammals, which are projection pathways running from the periphery to brain centres, but the discussion mostly focuses on the forebrain systems in question. While complex cognitive abilities in mammals and birds are clearly produced by substantially different neural structures [454, 711], there are clear similarities in the patterns of connectivity between brain areas. A comparison of neural architectures and connectivity between avian and mammalian brains can therefore shed light on the crucial neural features involved in the generation of higher-level cognition and, by association, higher-level consciousness [983].

In birds, especially corvids and parrots, the neural substrate for the generation of higher cognitive outputs includes a distinct dorsal thalamus (or comparable structure) and a distinct pallium, resulting in a relatively high brain-to-body ratio. In addition, a number of neural features of the distinct avian pallium show striking similarities to the corresponding structures of the mammalian cerebral cortex.[18] Interestingly, in this context, very similar neural activity patterns, namely, so-called "*neural bursting*" of the thalamus and the production of the 40 Hz rhythm, the gamma wave, have also been found in zebra finches [984], which is typical of neural activity in the primate thalamocortical circuitry that appears to be correlated with consciousness [985].

Comparisons of the neurodynamics of mammalian brains and the corresponding regions of the central nervous system in other animal groups are much more difficult. As previously mentioned, cephalopods have the largest accumulation of neurons in all invertebrates. Particularly revealing for the possibility of consciousness is the discovery in early electrophysiological studies of EEG patterns, including event-related potentials, that resemble those of awake vertebrates while differing from those known in other invertebrates [986]. However, no EEG patterns have yet been identified that reflect rapid irregular low-amplitude activity similar to that observed in human conscious states. This would first require narrowing the measurement to suitable areas of the brain, those being the optic, vertical, and superior lobes that are critical for learning and memory. Overall, however, neurophysiological data have not been able to prove the existence of conscious states in cephalopods.

Insects with miniature brains [329, 330, 987] are even more difficult to assess than cuttlefish. Without signs of individualised social life and mature communication—bees and ants are important exceptions—they rarely raise suspicion of greater

[18] Butler [931] has summarised these features as follows: (1) large, multipolar, glutamatergic neurons with (2) extensive dendritic trees that (3) are densely endowed with spines, (4) have extensive local and reciprocal connections, (5) have extensive regional connectivity, (6) receive GABAergic input from local interneurons, (7) receive sensory and other inputs from the dorsal thalamus, (8) participate in GABAergic circuitry with the thalamic reticular nucleus, and (9) are part of double-negative loops (involving two GABAergic synapses) through the basal ganglia.

cognitive ability and consciousness. The search for the neural correlates of consciousness is also problematic in insects: do insects have no consciousness because they have no cortex?

At this point, it is useful to try to make principled distinctions between different forms or degrees of consciousness because people can also be conscious without or with little involvement of the cortex. Although cortical damage has profound effects on the contents of consciousness, a so-called residual consciousness can persist even in patients without a cortex. In particular, the basic capacity for subjective experience may be spared despite massive damage to part of the cortex, even if these patients cannot otherwise come into contact with the world [988, 989]. Similarly, the production of unconsciousness through anaesthetics is attributed to damage to connections of subcortical structures and to the associated ability of the brain to integrate information rather than to the functions of the cortex itself [990]. Neurobiologists are therefore now inclined to assume that subcortical structures underlie the basic capacity for subjective experience in humans and that cortical structures subsequently take over the elaboration of its detailed content [991].

The capacity for subjective experience is described as the basic and, in evolutionary terms, first form of consciousness. While various classifications of consciousness have been proposed [908], all include one that consists of direct, unreflected awareness of the world as the basic form of consciousness. The defining feature of this most basic level of consciousness is the capacity for subjective experience. In humans, this capacity is not generated in the cortex but rather in the subcortical areas. The integrating structures of the midbrain seem to be particularly important here; indeed, they seem to be sufficient in all vertebrates to make possible the capacity for subjective experience [907, 959, 992, 993]. Damage to the structures of the midbrain has fatal effects on this ability [907]. It is assumed that an important function of the midbrain is the combination of interoceptive sensory information about the environment—the stimuli from inside the body about physical needs—and exteroceptive sensory information, the animal's location and movement in the environment. In humans and other vertebrates, these two strands of information are integrated in nuclei located between these stimulus sources to create a unified multimodal neural model of the spatial location of potential resources relative to the animal. The internal weighting of this model for action planning depends on the extent to which different resources are required by the animal [907]. Vertebrates organise their behaviour with reference to this integrated model of the environment rather than by responding to independent sensory inputs [994]. In mammals, the sensory topographical maps of the different sensory systems—depending on the species—vision, hearing, somatosensory, infrared, magnetic, electric, or echolocation—are overlaid in the midbrain roof,[19] providing an integrated sense of space that includes the position, state, and movement of the body [959]. The integrated model

[19] Strictly speaking, this superimposition and integration takes place in different layers of the superior colliculus of the corpora quadrigemina in the midbrain roof (*tectum mesencephali*) [995]. In birds, the anterior part is very strongly developed, because this is where the primary visual centre is located, which is why it is called the visual roof (*tectum opticum*) in birds.

also provides an efficient neural solution to the reafference problem [996], which is caused by self-motion.[20] Thus, in these parts of the brain, spatial movement control is used for directed attention, which is crucial for reaching and grasping targets.

For mobile creatures, not only spatial orientation and navigation but also the search for objects, both obvious and hidden ones, are important, including the selection of the most efficient ways to reach them. This requires a unified metric space that allows effective calculation of position in relation to the actual and hypothetical targets within that space. This goes beyond the primary sensory input but can be enabled by a simulation of the environment constructed from that input [997].

With respect to the selection of the goal itself, the ability to decide between competing behavioural motivations and to rank the required resources according to urgency and availability is highly beneficial. In humans and all mammals, this subtask of goal-directed behaviour is performed in the midbrain by hypothalamic structures and associated nuclei. Information from the nervous system and the endocrine system is integrated, and thus, information about the physiological state of the organism is gathered. However, the ultimate goal, the maintenance of homeostasis (physiological balance), also requires the active search for the required resources with the help of assessing the situation in space. Integrative structures within the midbrain and basal ganglia[21] combine the results of these inwards and outwards assessments to gather and resolve competing motivations, prioritise resources, and select goals and actions [907, 994]. This process of goal selection is very similar to volition, which I described in section "Act Willingly" (Chap. 2) as the coordination of perception, attention, cognition, emotion, motivation, activation, and movement control, but without anticipation of *future* need situations.

In decision-making, goal selection and action selection must interact. While action selection is influenced by motivational factors and their prioritisation of goals, goal selection is influenced by the localisation of these goals and thus loops back to action selection. Merker [907, 959] described this as a "core *control* system" for behaviour within which these decisions are made. Accordingly, he interprets consciousness as a biological function that evolved in mobile animals as a solution to neural logistical problems in controlling their environmental orientation. Since

[20] Physical movements cause extensive changes in sensory input, but we can easily distinguish our own movements from the movements of objects around us. Although eye and body movements produce a corresponding "movement" of the environment on the retina, the environment is not perceived as moving. One way of solving this problem is to form an efference copy, which anticipates the expected change on the retina and is sent from the command centres to the execution systems (e.g. the muscles). When efference copy and feedback from the sensory organs coincide, the environment is perceived as stable and unchanged. This model, known as the principle of reafference, was described in 1950 by Erich von Holst and Horst Mittelstaedt [996].

[21] They also include the *periaqueductal* grey (*substantia grisea periaquaeductalis*), the black matter (*substantia nigra*), the ventral thalamus, striatum, and reticular formation (*formatio reticularis*). The basal ganglia (*nuclei basales*) are core areas of the end brain (*telencephalon*) that lie below the cerebral cortex (i.e. subcortically). These subcortical basal nuclei are of great importance for important functional aspects of motor, cognitive, and limbic regulation, including initiative, willpower, drive, expectations, and motor selection.

the high complexity of sensory and motor systems arose at various points in evolution, accompanied by the diversification of such systems in the equipment of a given species, increased control problems and decision-making dilemmas arose. Problems include the coordination of multiple systems, the sharing of and competition for resources, the ranking of behavioural priorities, and decision-making. Ultimately, the result of the operation of these many components, the behaviour, must remain coherent, unified, and organised. The simplest and most primitive form of consciousness (primary [998] or core consciousness [999, 1000]) accordingly arose as an interface between the spatial senses and the motor demands of motivated behaviour [907].

The basic principle of primary consciousness is to integrate perceptual and motor events, match them with memory, and finally construct a multimodal scene. This was already postulated by William James more than 100 years ago [892]. The multimodal scene can also be related to the value system and the past learning of the individual conscious animal. On this basis, animals can adaptively change their behaviour. An animal with primary consciousness can plan to deal with the immediate scene in the remembered present [914] and thus has an advantage over an animal that lacks such a capacity for planning.

To whom can such primary consciousness be attributed? Both the abundance of highly complex behavioural decisions and the size of the centralised nervous system suggest that insects have this core control system [1001]. Not every nervous system is capable of a core control system; simple distributed nervous networks such as those of jellyfish are probably not, but centralised ones, those where one can actually speak of a brain, are. The insect nervous system contains an enlarged ganglion specialising in sensory processing and integration. Using this compaction, often referred to as the "*cerebral ganglion*", insects decide on action and target selection, process sensory information, and carry out command functions through the behavioural system [1002]. Moreover, a central complex within it specialises in processing spatial information and organising movement. Therefore, it has been argued that this area performs functions analogous to those of the tetrahedral plate of the vertebrate midbrain. Like the latter, the central complex plays a crucial role in processing spatial information originating from different senses. The central complex not only represents visual space in such a way that the insect can orient itself and navigate but also processes topographically organised mechanosensory and proprioceptive information. This information from the insect's own body is corrected using reafference so that the movements and orientations can be performed correctly [1003]. What is crucial in this motor control is that behavioural responses are not simply controlled by the primary sensory input—just from the outside, as it were—but also from the inside, by egocentric features (whether the meaning of the stimulus needs to be updated and whether this matches what the insect is momentarily focusing on). In bees, it was found that activity in the optic nerve is influenced by input from the central complex, and this input enhances the neural responses to stimuli that are sites of visual fixation [1004]. Taken together, electrophysiological studies are more convincing than those of behavioural studies alone, as Andrew B. Barron and Colin Klein of Macquarie University in Sydney (Australia) [960, 1001] emphasised. These studies are evidence for the subjective and egocentric nature of neural representations of the environment in insects. They also support the claim that these animals have the capacity for selective attention and thus the capacity for subjective experience.

However, what about the higher forms of consciousness that play a central role in philosophical debates, especially in the question of the special position of humans? Here, we will have to fall back, willy nilly, on behavioural studies, whether as a guide to understanding the underlying mechanisms or as the only way to achieve meaningful comparisons with conscious human actions. First, however, another approach will be highlighted, namely, the use of analogy as an indication of functional similarities between human and non-human brain states.

Functional Criteria for Consciousness

It is undisputed that humans are capable of more complex forms of consciousness after their complete development to reach full possession of their mental powers. For example, people can think about their own mental states and communicate these states and feelings to others. Whether the ability to communicate, which culminates in the human ability to speak in its diversity and subtlety, is a constitutive part of this highest degree of consciousness remains questionable. What is certain, however, is that linguistic reports are an invaluable means of bearing witnesses to mental states and contents of consciousness.

Without a doubt, higher forms of consciousness require more than the mere capacity for subjective experience. They are based on the ability to represent and, if necessary, reflect on one's own subjective experience and, ultimately, to have an awareness of oneself as a self. As mentioned earlier, the levels corresponding to these higher capacities are variously referred to in the literature as "*self-reflective consciousness*" [908], "*access consciousness*" [894], "*higher-order consciousness*" [998], or "*'I' consciousness*" [1005].

In psychology, in the tradition of William James [892], two forms of the self or ego have been distinguished. On the one hand, the "Me"—the self as an object of perception or self-attribution as known through its physical and mental properties—and on the other hand, the "I", the self as a knowing subject, the source of perception, action, cognition, and emotion. In cognitive neuroscience, self-related processes have also been distinguished from self-awareness processes. Self-related processing involves evaluating or judging a feature in relation to one's own perceptual image or mental concept of oneself [1006], whereas self-awareness is self-specific,[22] that is, it involves a process that specifies the self as subject and agent by

[22] Two paradigmatic, well-studied examples of self-specifying processes are sensorimotor integration and homeostatic regulation, both of which underlie the self-awareness of being a bodily agent. We then speak of *bodily awareness*, which is based on the fundamental distinction between signals that arise as a result of the organism's own outgoing (efferent) processes (self) and incoming (afferent) signals that arise as a result of events in the environment (non-self). By implementing this functional self/non-self-distinction, the integration of afferent and efferent signals implicitly specifies the self as a bodily agent [1007, 1008].

implementing a functional self/non-self-distinction [1008]. Self-referral processing is not exclusively self-specific because it also recruits a variety of other cognitive processes, such as retrieving information from memory, reasoning, and perspective taking.

However, how can one now determine whether non-human creatures can reach these higher levels? Ultimately, according to the tenor, despite all the successes of modern neuroscience, this is possible only with the help of behavioural studies that reveal cognitive performance. This is particularly evident where neuroanatomical and neurophysiological comparisons reach their limits, i.e. essentially in the entire animal kingdom apart from mammals, if not only primates.

Where to look? Which cognitive performances could provide decisive clues for this? The human ability to generate accurate verbal reports, popular both in basic science and in clinical applications, can, as we have already seen, hardly or only very conditionally be used as a criterion for proving higher-order consciousness in non-human animals. Keystroke and similar reactions can admittedly be used as a substitute for an exact report. But the ability to discriminate and generalise between classes of stimuli is widespread in the animal kingdom and can even be performed by computers in a way that resembles accurate reporting; few scientists would go as far as calling these machines conscious.

To test predictions that concern the possession of higher forms of consciousness, more than a dozen additional properties of human consciousness have been proposed [932]. However, the list contains almost exclusively characteristic properties of the human brain during the presence of conscious states, and it is questionable how far the analogies may be strained to expect a functionally equivalent connection between consciousness and certain mental properties or cognitive performances in non-human creatures. Three of these characteristics represent particularly important, if not necessary, criteria for human consciousness. First, there is a particular EEG-signature of consciousness—irregular, low-amplitude, and rapid electrical activity in the brain in the range of 12–70 Hz—which differs markedly from brain activity in unconscious states such as deep sleep, epileptic loss of consciousness, and general anaesthesia. All of these unconscious states are rather characterised by regular, high-amplitude, and slow voltages at less than 4 Hz. The other two criteria for human consciousness are its dependence on the thalamocortical complex and the widespread distribution of brain activity determined by current tasks and conditions because, in contrast, unconscious content evokes only local cortical activity.

These two criteria are less easy to test and difficult to narrow down and can be regarded only as strict definitions of human consciousness. Among other things, consciousness seems to be characterised by an extraordinary range of different content perception, rapid adaptability, and fleetingness of conscious scenes, consistency limitation, capacity limitation and seriality, binding of sensory impressions, self-attribution, accurate reporting ability, subjectivity, facilitation of learning, and stability of content [932].

With the exception of the first criterion for human consciousness, we are dealing here with a purely functional view, which, although it makes partial references to the neural correlates, can nevertheless do without them.

Recently, two information-theoretic approaches have been proposed; one by Giulio Tononi (University of Wisconsin, USA), Christof Koch (Allen Institute for Brain Science, Seattle, USA), and colleagues, which relies on purely information-theoretic quantities—the so-called integrated information [1009], the other by Stanislas Dehaene (Collège de France, Paris, France) and colleagues, which relies on the depth and nature (quality) of the processed information [1010]. Both approaches differ from other theories of consciousness in that they are *resolutely computational.*

In the second approach, two types of information-processing computations for consciousness are invoked [1010]. One type is said to be based on the selection of information for global dissemination in the brain so that this information is made flexibly available for computation and reporting (C1). The second type is said to be based on the self-monitoring of these computations, which leads to a subjective sense of certainty or error (C2). In this approach, machines that have C1 and C2 would thus also be conscious. Dehaene thinks that such a machine, like an anthropoid robot, would know that it perceives something, would trust this, would be able to communicate this to others, might suffer from hallucinations when its monitoring mechanisms breakdown, and might even have the same visual illusions as humans.

Global availability (C1) and self-monitoring (C2) are constitutive elements of human consciousness but are also possessed by other organisms that face certain cognitive challenges. A primary challenge is related to the diversity of incoming and stored information. There is a need for integration and coordination. Although the organisation of the brain into many, specialised modules or subsystems is efficient, it has a downside: the animal must perform *an* action and must decide in favour of a single path of action, often very quickly and not infrequently, for survival. The integration of all available evidence for a single decision requires global availability. A sophisticated architecture is needed for efficient aggregation (bundling) of all available information from different sources, from different senses, and from long-term memory, further the weighing of available options and selecting the best one; and ultimately, consistency is required to stick to the decision once made. While the brain of an elephant in search of a waterhole fulfils exactly these requirements, in simpler creatures there is only an unconscious competition of uncoordinated sensorimotor systems.

In fact, type C1 consciousness fulfils exactly the requirements I described in the section on volitional action (Chap. 2, section "Act Willingly"). Not only is it necessary to bridge major temporal delays, but short-term motivations must also be suppressed in favour of long-term winning strategies. C1 architecture thus enables coherent, cautious planning. Exactly when this C1 consciousness emerged during evolution remains questionable, but as an information-processing system that could solve the problem of information bundling, it seemed to have emerged very early. As soon as the first more complex brains emerged, it was necessary to break through this parallelism for the purpose of fast and targeted decision-making.

It has been hypothesised that consciousness in its first form (C1) arose in terms of an internal *global workspace* [916, 1011, 1012]. At the top of a hierarchy of specialised modules, a global neural workspace emerged with limited capacity to select information, hold it at the ready for a time, and share it with multiple modules. Only a small portion of the processed information wins the competition for access to this mental arena, which enables global dissemination and targeted decision-making. It is precisely these representations that are perceived consciously. Thus, consciousness manifests itself through the temporal dominance of these representations (over all other processes running in the brain).

One could say that C1 consciousness acts like a bottleneck, making only one piece of content consciously available at a time in the sense of an all-or-nothing decision [981, 1013]. Brain scans in humans and neural recordings in monkeys have confirmed that there is a consciousness bottleneck, which consists of a network of neurons that are distributed mainly in the cortex, especially in the higher associative centres. The enormous increase in the size of the prefrontal cortex in humans may be related to this, in that it has allowed an increase in the capacity for multimodal convergence and integration. Empirical evidence for the serial functioning of consciousness includes phenomena such as "*attentional blinks*" (a brief attention deficit) [1014] and the "*psychological refraction period*" [1014, 1015], where conscious access to content A prevents or delays the perception of content B.

However, the enlargement of the prefrontal cortex also seems to be related to the second function of higher-order consciousness. C2 consciousness is the ability to represent oneself reflexively [1016, 1017]. Unlike C1 consciousness, it is inward-looking, a kind of self-monitoring. We have become familiar with this function (in the psychological literature) with the term metacognition, which refers to something like cognition about cognition, knowledge about knowledge [1018]. In fact, the ability to look inwards to make the right decisions through correct self-assessment is one of the cognitive abilities most often associated with consciousness. This brings us back to the starting point of this book, namely, to the question of what abilities comparative cognitive research has now unearthed in non-human animals that support the assumption of higher mental states ("thinking") including consciousness.

Degrees of Consciousness

Following the neurobiological and human psychological approaches to determining consciousness in animals, I will conclude by explicitly addressing the question of what cognitive abilities have been demonstrated in non-human animals that suggest higher-order consciousness (and not merely body or core consciousness). Since we are dealing with different states, degrees, and forms of consciousness, the immediate question is how to detect and distinguish them in the widely ramified tree of life. In other words, how can we make sense of the variation of consciousness in the animal kingdom? Is it meaningful to say that some animals are more "conscious" than others? Does consciousness occur in degrees, and if so, how can it be measured and studied?

Philosopher Jonathan Birch (London School of Economics and Political Science, London, UK) and two comparative psychologists, Alexandra Schnell and Nicola Clayton, of the University of Cambridge (UK) have recently proposed a multidimensional framework for understanding variation in states of consciousness between different animal species [1019]. The framework spans five key dimensions of variation: perceptual richness, evaluative richness, integration at a time, integration across time, and self-consciousness. Using these to assess a species appropriately yields its characteristic "profile of consciousness". This approach differs from conventional one-dimensional classifications—typically a single scale on which a species could be classified as more or less conscious. And the five-dimensional approach is based on the application of established experimental paradigms, which result in a battery of convergent experimental results.

In the course of this book, we have come across all of the abilities proposed for the five-dimensional profile but have not yet associated them with consciousness of varying degrees. For the first dimension, animals are compared on the basis of the richness and fine-grainedness of their perceptions and their discriminative abilities based on them. These are primarily cognitive tasks associated with conscious perception in humans: specific forms of learning to discriminate [1020], reversal learning [1021], and trace conditioning (in which the two stimuli are widely separated) [611, 1022]. A noncognitive but neurological pathway along this first dimension is the detection of induced blindsight. It has been shown that monkeys with lesions of the primary visual cortex, V1, respond similarly to humans with blindness in that when trained to report the presence or absence of a visual stimulus, they report its absence in a region of their visual field, but at the same time use information about that stimulus to guide action in forced-choice tasks. The monkeys reached the correct location even though they indicated that they could not see the stimulus [928].

The second dimension concerns the richness of affect-based decision-making [1019]. There is thought to be wide variation in the animal kingdom in terms of affective evaluation of events and situations. Although all conscious animals are thought to have an appraisal system of one kind or another, there remains ample room for refinement. Like us, some animals can constantly evaluate small changes in their internal states and external environment to make nuanced decisions, while others respond only to more substantial changes. Rich affect-based decision-making takes into account a lot of information—external and internal—at once (evaluative bandwidth) and is sensitive to small differences in these inputs (evaluative acuity). Examples of experimental paradigms that comparatively examine such variables include motivational weighing to make flexible decisions [1023–1025]; flexible behavioural changes with the valence of a reward [126, 1026]; cognitive bias, i.e. the influence of the current emotional state on decision-making [1027, 1028]; as well as receptivity to the emotions of others [1029, 1030].

The third dimension, described by Birch and coauthors, concerns the uniformity of consciousness. However, this dimension is strongly influenced by the study of human patients and is difficult to transfer to animals. While conscious experience is very uniform in healthy adult humans, it appears to break down in patients with split-brain syndrome of whom the corpus callosum has been partially or completely severed [1031]. However, it is unclear whether animals that naturally have no connection at all in the form of a corpus callosum perceive the world inconsistently. Although birds, for example, have no equivalent connection between the two hemispheres of the pallium and the visual pathways—they are completely separate—pigeons do still show a transfer of information from one eye to the other in experiments [1032]. In the "*interocular transfer*" experiment, the animal was trained to perform a task in response to a stimulus presented to one eye while the other eye was occluded. Researchers then checked whether the task could still be performed when the stimulus was presented to the other eye (it was hardly possible). The question of unity becomes even less clear in molluscs, such as squid, where the cerebral ganglia are interconnected but have a degree of functional autonomy. It cannot be assumed that an octopus has nine conscious perspectives on the world [957].[23]

Another form of consciousness unity is temporal. Over both short and long temporal distances, events and memories appear to us humans to be so strongly integrated that we perceive time as a continuous stream rather than as a succession of snapshots [1033]. The film industry has taken advantage of this, because we no longer perceive the rapid succession of images, more than around 25 per second, as individual frames.[24]

Even over much longer periods of time, this stream of events appears to be integrated and unified. Even if we sometimes have large gaps in our memory, we can still remember a great many experiences, sometimes very vividly and in great detail. In addition, we can simulate future experiences, which allows us to span an enormous time horizon in the middle of which a comparatively short present is perceived. The seemingly arbitrary distancing in both directions thus takes on the form of mental time travel [1035].

I have already described in detail the ability of animals to travel in both directions of the arrow of time, albeit limited in comparison to humans (in Chaps. 5 and 6). Here, it may be sufficient to agree with the proponents of the five-dimensional categorisation of consciousness [1019] that this ability almost certainly requires considerable cognitive sophistication, and even if some conclusions of experiments on spontaneous planning have been criticised [535, 537], it should be considered a promising nonverbal indicator of conscious temporal integration.

---

[23] Octopuses have nine brains; in addition to the central brain, each of the eight tentacles has a mini-brain that allows them to act independently. Because each arm has a mini-brain, the central brain only sends a higher-level signal to the arm. In us humans, the brain would direct and control every movement of the arm. In an octopus, however, each arm acts almost independently while, for example, advancing into a crevice and exploring the inside.

[24] A well-studied example of psychophysics is the illusory movement, which can be illustrated even more impressively with a colour change. In the colour phi illusion, two alternating flashing dots of different colours placed next to each other are perceived as a single moving dot that changes colour halfway through the picture [1034]. Nothing is actually moving, nor is the colour changing anywhere; thus, our brain constructs a rather dramatic illusion, which has caused strong debates in the philosophy of mind [926].

However, the question remains how animals represent time, especially whether they integrate past, present, and future events in such a way that they can "travel along" it. And are the events of this journey linked to subjective experiences? The concept of mental time travel for Tulving envisages a close connection between episodic memories of past events and thoughts about personal future events. This concept of mental time travel thus includes subjective phenomena, such as consciousness evolving from a noetic to an autonoetic form. It is ultimately the introspective self that keeps our memories and personal experiences alive and enables us to imagine future events.

The close connection between memory and consciousness has long since ceased to be purely theoretical but is supported by a large number of empirical findings in humans. In particular, research into the cognitive development of children's memories has shown this connection. Gradual development is assumed, starting from a rudimentary state of anoetic consciousness with a first kind of "*self-awareness*" that draws on implicit experiential and procedural memory to "*noetic*" and "*autonotic*" consciousness based on semantic and episodic memory systems [1036].

At around 4 years of age, children can competently solve verbal false belief tasks [645], have acquired an awareness of self that persists over (subjective) time [1037], and are able to mentally construct certain future events [1038]. According to Tulving, they have an autonoetic consciousness of their experiences in the continuity of subjectively apprehended time [1039].

From this point on, children are also able to apply theory of mind to their own mental states [1040] by viewing their self from the perspective of a third person. In this way, theory of mind is associated with mental time travel; it is the "*introspective self*" that is capable of episodic memories and episodic future thinking.

The aforementioned functional and developmentally supported connections from mental time travel, theory of mind, and introspection are now widely accepted and included in textbooks of human psychology. However, for comparative cognitive researchers, theory of mind's close relationship with consciousness is problematic, especially because it complicates the study of episodic memory and future thinking in non-human animals if it relies too heavily on subjective phenomena [584]. Therefore, alternative views on the ability to mentally construct scenes to differentiate past events have been developed, which not only question temporal elements as identification features but also, more importantly, subjective personal experiences. Accordingly, episodic memory aims not to embark on a journey through time but simply to distinguish complex events [583, 584]. Non-temporal features such as memory strength are sufficient for deciding the time of occurrence of an event and making a linear chronology dispensable [585]. Moreover, such parsimonious concepts of memory would also relieve us of the obligation to provide episodic memory with evidence of conscious experience.

The fifth and final dimension described by Birch and coauthors concerns self-consciousness, which means the awareness that one is distinct from the outside world [1019]. Like with all four of the previously mentioned dimensions, gradations could be allowed here, and thus, minimal levels of self-consciousness could be present in a variety of animals [1041]. One minimal ability is the registration of a difference between self and other, which is reflected in separate experiences and representations of internal bodily events versus events in the external world. Every complex, actively moving animal needs this ability to distinguish changes in its sensory input, due to its own movements, from changes due to events in the world [959, 1042]. As noted earlier in this chapter, this ability already appears to be present in insects [960, 1001]. This simplest and most primitive form of self-consciousness, which is mostly primary [998] or core consciousness [999, 1000], is the interface between the spatial senses and the motor demands of motivated behaviour. Its basic principle is to integrate perceptual and motor events, match them with memory, and ultimately construct a multimodal scene in the present [907].

In the literature, and especially among philosophers, however, a more sophisticated self-consciousness is usually discussed, which goes beyond the perception of one's own body as a continuing object existing in the world. Central to these discussions is the question of whether a possible self-consciousness in animals is merely a bodily self-consciousness or perhaps a psychological self-consciousness after all [1043]. Before concluding on this question, I would like to describe some approaches that suggest self-consciousness on the basis of bodily self-awareness.

The Self in the Mirror: Bodily Self-Awareness

One of the most cited but also most controversial tests for self-awareness in animals is the so-called "*mirror test*". Although Darwin already addressed the reactions of children and apes in front of a mirror, this test only found its way into comparative cognition research through the experimental applications of Gordon Gallup in 1970 [1044]. It is known that some animals, including cats and budgies, show interest in moving images in the mirror or reflective surfaces, but most seem to ignore their own mirror image (let alone use the mirror as a tool to see things they would not otherwise see).

The reactions of animals in front of a mirror can be interpreted very differently—some animals react to their own reflection not only with interest but also in a way that suggests that they suspect unknown conspecifics in the reflection; as they show fear or even attack behaviour, others seem to recognise themselves. Therefore, Gallup devised an objective procedure, focusing on the spontaneous manifestations of mirror-directed self-exploration and on the spontaneous use of the mirror to examine markings on otherwise invisible parts of the body. Since he was interested in comparing apes and other monkeys, he examined both chimpanzees and macaques. At first, individuals of both species reacted as if they were in the presence of an unknown conspecific. However, macaques reacted "socially" to the reflection for a while and then got used to the presence of the "other monkey", while the chimpanzees soon began to use the reflection to carefully explore parts of their bodies they could not normally see. The individuals opened their mouths while staring at their heads in the mirror, removed the granules from the corner of their eyes, or examined their back anogenital area while looking in the mirror. Note, the chimpanzees never showed this behaviour in the absence of the mirror [1045].

This human-like behaviour of the chimpanzees prompted Gallup to test this phenomenon further. The basic idea was that if the animals know what they look like and recognise that the mirror represents their visual body image, they would recognise a discrepancy between their expected image and an actual mirror image. This should lead them to examine inconspicuous markings secretly placed on their bodies by the experimenter [1046].

After 10 days of gaining experience with mirrors, the chimpanzees and monkeys were anaesthetised and marked with a dab of paint on the forehead and one ear. Importantly, the paint neither irritated the skin nor could it be smelled. This was checked with a control without a mirror: after waking up from the anaesthesia and still in the absence of the mirror, not a single individual made any effort to touch the markings. When the mirror came into play, however, the chimpanzees, but not the macaques,[25] touched the markings on their head and examined them. Apparently, the chimpanzees understood that they saw themselves in the mirror.

The oft-cited Gallup paper in *Science*, succinctly titled "Chimpanzees: Self-Recognition" [1044] not only generated a great deal of immediate attention but also ushered in two important developments in cognitive research. First, there was a flurry of interest in species comparisons because Gallup's study suggested a cognitive gap between great apes (and humans) and all other animal species (which he continued to assert 45 years later [1045]). In recent decades, a considerable number of species have undergone mirror tests. After other great apes and other Old World monkeys, as well as New World monkeys, were confronted with the mirror, it was soon the turn of dolphins and elephants, some bird species, and finally fish. The tests, which are, after all, considered by many to be a measure of the "*consciousness of the self*" or the "*idea of me*" [1048] triggered a fierce philosophical and cognitive debate. At its core, this dispute revolves around the argument between a lean, soft interpretation, and a broad, hard interpretation [1049]. Must one become the object of one's attention in front of a mirror and thus become aware of one's own existence [1050], or is it enough to have the ability to distinguish feedback from other perceptual stimuli to pass the marking test [1051, 1052]?

Before I try to answer this difficult question, let me first look at humans or, more precisely, children in marking tests. Children of different ages were also marked with paint or stickers on their faces or heads as well as on additional, normally invisible parts of the body. By comparing reactions in the test and control sessions with and without the mirror, respectively, different types of behaviour were observed in response to the mirror: mirror-directed behaviour (e.g. touching, playing), contingent movements (moving the head or body while watching the movements in the mirror), self-directed behaviour (e.g. touching or visually inspecting parts of the body), and marking-related behaviour. Mirror self-recognition was exhibited by the

[25] In the meantime, however, there is evidence that macaques also recognise themselves in the mirror [1047].

majority of infants at 24 months of age who promptly inspected the markings on their face or stickers on their legs [1053, 1054]. Before that, they are interested in their reflections but do not touch the mark. So, in humans, mirror self-recognition first appears between 18 and 21 months of age. At the age of 2, a quick glance in the mirror is usually enough for them to notice changes in their body or clothing [1055]. Since other cognitive abilities such as intentional deception, perspective taking, and empathy are not yet present at this age, the results at this age tend to speak for a "weak" interpretation in the sense of the ability to use the mirror as a perceptual tool. Thus, children at this age can understand mirror images of objects but not of themselves. However, passing the marking test implies the ability to perform visual–kinesthetic matching [1056], which means knowing one's own appearance and being able to compare it with the mirror image.

The chimpanzees in Gallup's pioneering study first had to spend several hours or days in front of the mirror to form memory images of their appearance, which later enabled them to detect unusual markings as deviating from their normal appearance. Interestingly, chimpanzees exhibit a very similar time course for passing the marking test as children do [1057]. This could also argue for the "weak" interpretation, as children develop the ability to use the mirror as a perceptual tool before they reach the age where they begin to use it for self-recognition, as undoubtedly shown by naming the self [1048].

The tendency to overestimate the marking test has since been challenged by ambiguous findings in other species. While bonobos and orangutans passed the marking test,[26] gorillas regularly failed it. Only very few of them showed signs of reliable responses to the marks [1058], but overall, the evidence for mirror image recognition in gorillas has remained controversial among experts until recently [1045]. The situation is similarly unclear for lesser apes, such as gibbons and siamangs [1055]. Other reports, in turn, suffer from a lack of repeatability. In the case of cotton-top tamarins [1059], the attempt to replicate the original positive results failed completely [1060]. A remarkably successful elephant study [1061], in which one out of three Asian elephants increasingly touched its marked head with its trunk during or immediately after the mirror experience, contrasts with two failed attempts [1062, 1063]. However, recent studies have shown that elephants recognise themselves as separate from other individuals and objects and, for example, understand their body weight as an obstacle to reaching objects [1064].

The ambiguity of the findings is again surpassed in studies with animal species that do not have hands or other body parts to touch a mark. Among mammals, these are mainly whales. So far, only two bottlenose dolphins have been tested for mirror self-recognition, and according to the experimenters, they have shown responses to reflective surfaces that are consistent with using a mirror to examine marked body parts. In fact, they examined marked body parts with "clearly self-oriented, marker-oriented behaviour" [1065; p. 5942]. Another study was able to confirm this, and

[26]A good overview of all primate mirror recognition studies conducted until 2015 can be found in [1045].

even more, body-oriented behaviour was shown in front of a mirror by very young animals [1066]. Both young dolphins exhibited the behaviour of the older animals from the first mirror studies and were even younger than the human and chimpanzee children. Interestingly, this change occurred in parallel with advanced sensorimotor development and the emergence of complex and reciprocal social interactions. This result is also remarkable because dolphins, unlike chimpanzees that groom each other, do not pay attention to markings on conspecifics.

While large-brained, cunning dolphins may be easily credited with passing the mirror test, bony fish remarkably also exhibited similar behaviours to the chimpanzees in front of the mirror—they had social responses to reflection, repeated idiosyncratic behaviour towards the mirror, and frequent observation of their reflection. Strikingly similar to dolphins, cleaner wrasses attempt to remove a coloured mark by scraping their bodies in front of the mirror (they showed no response to transparent marks nor coloured marks in the absence of a mirror) [1067]. The phylogenetic distance of fish from apes has led comparative cognitive biologists to doubt the usefulness of the mirror test as a "litmus test" for self-awareness and to assume a gradualist evolution of self-awareness [1068].

The findings from birds are particularly interesting. They also lack long body parts such as arms and trunks to examine hidden areas of the body, but they could partially compensate for this handicap with a long beak or even with their feet. But, another possible handicap remains, or at least has not yet been solved with appropriate technical modifications. Recognition in a mirror is not only a cognitive problem but also a perceptual problem. Generally, animals have little experience with mirrors, such as those hanging in our front rooms and bathrooms. Here, the reflecting surface is made of glass, which partially or completely absorbs wavelengths in the UVA spectrum (315–400 nm). This is a serious obstacle for animals for which UV light is an important source of information, as is the case for many birds because these wavelengths can be important for visual signalling [1069]. For example, corvids of the genus *Corvus* have some ultraviolet (UV) sensitivity. Therefore, the reflection seen by a bird may not be an accurate representation of its true appearance. Furthermore, clear reflection could be rather confusing or disturbing if only dull or blurred reflections are present in the visual environment of birds.

The use of mirrors thus examines not only self-recognition but also how animals react to reflections and process information about spatially displaced objects and actions. Therefore, different performances that are retrieved in the mirror test need to be distinguished. Although it was designed for the question of self-recognition, other perceptual and cognitive abilities, or their deficits, have also come to light in comparative cognition research.

In principle, mirrors can also be used for nonsocial tasks, such as finding hidden objects. When an animal sets out to find an object that cannot be seen without a mirror (because, e.g. the object is under a table on which the animal is sitting in front of a mirror), the animal has understood the correspondence between the location of the object and its reflection in the mirror [1062].

Another ability is mirror-mediated object discrimination, where the animal uses the mirror to choose between objects. Two grey parrots were tested for these different types of mirror use and passed most of them. They were able to locate hidden objects and discriminate objects in the mirror [1070].

As far as social responses based on conspecific recognition are concerned, the parrots showed behaviour similar to that already found in pygmy marmosets, various apes, dolphins, and young children (<18 months) and in the first spontaneous responses of orangutans and young chimpanzees [reviewed in 1071, 1072]. However, there was no evidence of self-recognition in the mirror. This was also confirmed in a recent study with two other parrot species, kea and Goffin's cockatoos [1073]. After several weeks of habituation to the mirrors, both species were subjected to the marking test using different markings and marker placements in front of elongated, horizontally, and vertically arranged mirrors. The cockatoos showed fewer social responses to their horizontal mirror image than to their vertical mirror image, suggesting that they may have interpreted them differently (similar to various corvids, as I will discuss). This finding is evidence for how strongly perceptual or technical changes influence perception and how much the humanised perspective can mislead. The kea were therefore subjected to an additional control test in which their reaction to a marked or unmarked conspecific behind a transparent partition was compared with that to their own reflection. They reacted very similarly to both, confirming that they interpreted the reflection as a conspecific and not as a reflection of themselves. Overall, the findings of these parrots thus fit into a picture already drawn by earlier, but much less well controlled, work with other bird species.[27]

As shown several times in this book, corvids, along with parrots, have achieved primate-like intelligence, including tool use, time travelling, perspective taking, and metacognition. Thus, they are also good candidates for mirror tests. However, the findings of this group of birds do not reveal a clear trend. Four captive large-billed crows frequently exhibited pecking and wing flapping while looking at a mirror during 2.5 hours of exposure. This was interpreted as aggressive behaviour towards a conspecific [1074]. Self-referential or even contingent behaviour, on the other hand, was completely absent. Similarly, ten wild-caught New Caledonian crows showed rather unfriendly, agonistic behaviour towards the mirror image in front of a large vertical mirror, as well as exploration of the mirror from the front and back; moreover, there were no signs of self-recognition [1075]. However, four animals—one immediately and the others after some time—understood the use of a smaller, horizontal mirror to find otherwise invisible food.

Carrion and hooded crows behaved quite differently, as they hardly showed any aggressive behaviour towards their mirror image [1076] but immediately began examining the mirror, which raised hopes of self-recognition. However, not a single crow showed significant self-referential behaviour during the subsequent marking test, let alone attempts to remove the red mark on their throat. These findings, as

[27] Simple mirror experiments were carried out with budgerigars, house sparrows, kea, black-capped chickadees, zebra finches, glaucous-winged gulls, dusky grouse, cedar waxwings, juncos, and rosy-faced lovebirds (reviewed in [1070]).

well as other experiments with jackdaws, confirmed the problematic nature of applying the marking test developed in primates to birds [1077].

The very few positive findings in the marking tests of birds are therefore all the more surprising. In a test with Clark's nutcrackers, one bird showed more behaviour towards an area of the laryngeal feathers marked with a red dot than under control conditions [1078]. However, the fact that only one out of ten birds showed this strange behaviour and that the reactions could also be explained by tactile irritation of the larynx left the researchers in doubt. So, they then devised an alternative test that is not only methodologically but above all ecologically sound. Since nutcrackers are known for hiding food as well as defending their hiding places, the authors chose a caching task that would better show whether the jays could recognise themselves in the mirror. They allowed the birds to hide food either alone, in the presence of a conspecific, or in the presence of a mirror. The birds did not hide food in front of a normal mirror, a typical behavioural response to prevent theft by conspecifics. Apparently, they saw a conspecific in the mirror image. Surprisingly, on the other hand, they hid it in a cloudy mirror, as if they could perceive themselves—even if only in a blurred way. However, why should self-recognition be made possible with a blurred reflection of all things? Wouldn't it be much more obvious to think that they could no longer perceive a conspecific in the cloudy mirror?

A look at the visual environment of animals may help to clarify this. In a natural environment, self-generated images to which an animal is exposed rarely contain clear identity information (the recognizable visual characteristics of an individual), while self-contingent *movements* (self-triggered synchronous movements of the body and image) are much more common, such as in shadows or distorted reflections in water. Therefore, the two different mirrors could contribute to self-recognition of different degrees depending on an animal's learning history and a blurred image that provides only contingent-movement information could facilitate self-recognition in nutcrackers [1078].

These arguments for birds are not entirely convincing, especially given that scrub jays neither showed self-recognition in the marking test nor the one for hiding food in front of a mirror [1079]. This is despite the fact that scrub jays are known not only for their strategies in hiding and defending food but also for their impressive performance in many other cognitive tasks. The scrub jays' behaviour in front of the mirror was not significantly different from behaviour in the absence of a mirror (and conspecifics), which could mean that they actually recognised themselves in the mirror. However, the differences between both the conditions with conspecifics and mirrors were small and not statistically significant. Moreover, there are alternative explanations, such as that the birds interpreted the mirror image as a "strange" conspecific [1080], which could have caused them to display food-hiding behaviour atypical of the presence of another individual. More importantly, however, the scrub jays failed the classical marking test because they showed no signs of removing a surreptitiously placed red mark when in front of the clear or the cloudy mirror.

In summary, the experiments with birds in front of a mirror can still be interpreted as positive in the sense that the mirror was used as a tool for object discrimination and localisation but negative in the sense that there was missing evidence for mirror self-recognition. This sobering summary applies particularly if one takes the behaviour of humans and apes in front of the mirror into account. But even so, it should not be overlooked that only a rather modest proportion of apes exhibited typical self-referential behaviour. Even in chimpanzees, the most studied species with the most convincing results, clear evidence of self-recognition could by no means be obtained for all the individuals tested. Povinelli and coauthors tested 105 individuals [1081], yet only 21 showed clear evidence of self-recognition. Similarly, in another study of 73 chimpanzees, 71% of the individuals failed to show compelling evidence of self-recognition [1082]. Even in 18-month-old infants, only approximately 65% passed the marking test [1083].

Against this background, it was not surprising that a certain report of magpies caused a considerable stir. A group of researchers at the Ruhr University of Bochum, Germany, led by Onur Güntürkün investigated the mirror self-recognition of magpies, a species rather inconspicuous in the cognition literature (apart from reaching the highest level of object permanence according to Piaget). The magpies showed clear mirror-induced behaviour [1084]. Three out of five of the corvids showed interest in the mirror, and two avoided it. After that, two birds performed contingent behaviour—systematic back-and-forth movement with the head or the whole body in front of the mirror. Finally, in the three birds that showed interest in the mirror, actions directed at their own bodies could be observed, and of these, two tried to reach markings on their chest or back with their beak or a foot. For the authors and quite a few behavioural scientists, these results provided the first solid evidence of mirror self-recognition in a non-mammalian species. However, these results have not yet been replicated and must therefore be interpreted with due caution.

I will conclude my review of studies on mirror self-recognition with this magpie study, although several other species have also been investigated. A full report would go beyond the scope and contribute little more to our fundamental question: "do animals have consciousness?".[28] For now, I will summarise the problems that the mirror test poses for the detection of self-awareness in animals. First, as already mentioned, there are perceptual problems. The mirror is a tool that works well for humans and their vision, and it works comparably well for primates, which have been shown to have very similar visual systems. This does not apply to animals that see quite differently or have been tested under water. In addition, animals have a lack of experience with reflections.

Another problem is a lack of motivation to check the mark. Dolphins, for example, are not interested in the greatly altered appearance of conspecifics due to markings [1086]. This suggests that appearance may not be as important to these animals as it is to primates. But even if one animal is interested in appearance, this does not necessarily mean that the animal wants to modify it. Even examining parts of one's body that one does not usually see does not necessarily have to be intentional. The mere fact that even many chimpanzees—in one study, as many as 10 out of 11—did

[28] Mirror self-recognition has also been studied—without regard to completeness—in ants, cichlids, giant pandas, domestic horses, giant manta rays, killer whales, sea lions, pigeons, and squid (large-finned reef squid) [for a summary, see 1076].

not remove the markings on their heads points to a motivational, not cognitive failure [1087].

Taken together, these problems call for a rethinking of the mirror test, at least as the preferred method of demonstrating self-awareness in non-human animals. At the very least, the marking test would still need to be combined with other mirror test variants—for example, those that require distinguishing one's own body or its identity from conspecifics, as in the hiding food task. However, even this approach might be unsatisfactory and therefore requires more ecologically sound procedures that are better suited to the sensory characteristics of animal species. In any case, the absence of "human" reactions in front of the mirror is not yet strong evidence for the absence of self-awareness.

Even more serious than the methodological problems of the mirror test are the conclusions drawn from it because the test has promoted a binary view of self-awareness, according to which a few species possess this ability while others do not. Moreover, this approach often leads to claims about a cognitive gap between great apes, including humans, and the rest of the animal kingdom [1088]. But, if the test favours organisms that are comparable to humans, then its exclusive use easily leads to false conclusions [1049].

The evolution of cognitive abilities is rarely erratic but mostly gradual, with new abilities building on old ones—similar to the development of self-awareness in human ontogeny [1089]. Even less likely is a development that conforms to the "*scala naturae*" model, the Aristotelian concept of progressive complexity with humans at the top [1090]. In this respect, we should try to clarify which framework conditions are necessary for the construction of a self and which stages or formations of self-representations in different animal species would correspond to these. Thus, at least an intermediate stage between the total absence of a self-concept and a fully matured one seems possible. Human children also express curiosity about their mirror image long before they pass the mirror test [1089]. Intermediate or preliminary stages to fully, mature self-awareness are indicated when animals correctly assess the properties of their bodies with their weight when they want to pick up a mat to stand on [1064, 1091], their size when they walk with their antlers between trees standing close together [84], or their shade when they do not want to be discovered by a predator [1092].

Body concepts are, in this sense, precursors of ego concepts, where an organism recognises its identity over time. This low form of self-consciousness was described by Ulric Neisser (Cornell University, USA) [917] as the "*ecological self*": the processing of self-referential (visual, auditory, kinaesthetic, and other proprioceptive) information from the body. This process is at the lowest level of his five-level model of consciousness [1093].[29] The ecological self conveys a direct awareness of the self

[29] Special receptors in muscles, tendons, ligaments, and joints, which react to pressure or deformation, and in the organ of equilibrium (vestibular apparatus in the inner ear of vertebrates) convey information about movements, posture, and position of the body or individual limbs. Taken together, proprioception is the perception of one's own body—its position in space, the position of the head, torso, and limbs in relation to each other, as well as their changes as movements,

in relation to the physical environment, especially position and movement. Since this is a basic form of body consciousness or primitive knowledge of one's body in relation to the environment, the ecological self is presumably the cause of some form of subjective perspectivity [1095]. The ecological self thus represents a "lower" form of (self-)consciousness [908].

The question now is to what extent the ecological self enables self-recognition in the mirror. The psychologist Robert Mitchell of Eastern Kentucky University (USA) provides a plausible answer to this question [1056, 1096]. He is convinced that self-recognition, as evidenced by the mirror test, does not have to be understood as strongly as Gallup assumed, for example, but that it already begins as a self-concept that does not need to be reflected upon [1097]. This self-concept is the result of proprioceptive information about the position of one's own body and one's own body feeling as well as of visual images of one's own and other bodies. He calls this theory the "*kinesthetic–visual matching theory*" because the organism matches its kinaesthetic sensations to its visual image in the mirror, and if these match, it can thereby predict that the mark on the nose on the face in the mirror indicates a mark on its own nose [1056].

If self-awareness in front of the mirror, as measured by passing the marking test, results merely from discovering the similarity between the feeling of extension and movement of one's body and one's appearance, the self-awareness required for this would not involve complex mental states. For Mitchell, self-awareness in the marking test is merely a behavioural skill that is equivalent to other associative processes. The disturbing fact for this assumption that only a few species have passed the test thus far could be explained by the methodological problems of the test described above, along with its sensory and motivational obstacles. However, one could also object that recognising oneself in the mirror is not merely based on a purely physical self-perception but also on a social classification, where the "*Self in the Mirror*" [637, 1098] is constructed in a reflecting object, as described by Wolfgang Prinz of the Max Planck Institute for Human Cognitive Research in Leipzig, Germany.

On the Awareness of Being Conscious

Phylogenetically later, in humans but perhaps already in other species of animals that pass the mirror test, self-awareness is extended so that the individual has knowledge or understanding of his or her personal history or motives, actions, and hopes. At the highest level, recursive perception enables the individual to be aware that he or she is aware [1099, 1100].

---

including the sensation of heaviness, tension, force, and speed. In sensory physiology, a distinction is made between three sensory modalities of proprioception: the sense of movement (kinaesthesia), the sense of position, and the sense of force. In humans, this is also referred to synonymously as depth perception [1094].

Some authors distinguish between two qualities of self-consciousness: objective self-consciousness, which refers to an organism's ability to distinguish itself from the rest of the world and to attend to the stimuli that correspond to itself, and subjective self-consciousness, by means of which the organism identifies with these stimuli [1101]. This identification was described by Lewis [1048] as "the idea of me".

In accordance with the subdivision made for human consciousness, I now refer to the highest level, where the organism not only focuses and reacts to the stimuli of the external environment but also builds a self-model that enables planning, evaluation, and self-monitoring. It processes and evaluates new information from the perspective of a personal history, remembers the past, and projects into the future. In this respect, this self-awareness is "*meta-representational*" [911].

With this kind of self-awareness, the organism becomes a reflective observer, knowing not only its body and appearance (self-awareness in the mirror) but also its own knowledge (metacognitive) and memories. This includes episodic (and thus autobiographical) memory content and contents of self-assessment. Subsequently, the organism can also put itself in future scenarios, assess its future needs, and plan accordingly. This, in turn, is crucial in making and using appropriate tools and in solving problems with multiple courses of action. Finally, perspective-taking and mind-reading skills are important byproducts of self-reflection, because an individual harnesses self-assessment to project onto others. We can also describe these skills using Tulving's [499] distinction between *noetic* and *autonoetic* consciousness and ask whether non-human animals can develop self-consciousness with autobiographical content in addition to a consciousness of (semantic) perceptual, conceptual, or memory representations.

As I mentioned at the beginning of the chapter on mind reading, it has been argued that awareness of oneself also implies awareness in others [636] because the self is, at its core, a social being that is largely determined by its position in the community [637]. Indeed, the connection between mind reading, introspection, and consciousness is illustrated by the fact that people who have problems imagining the thoughts and desires of others also have difficulty looking inside themselves and anticipating the consequences of their own actions [638]. It therefore seems to be important for human consciousness to think or feel its way into others, to know its own desires and needs well in a wide variety of situations and to project this knowledge onto others. The interplay of introspection and projection allows us to exchange ourselves with others, to coordinate our own actions with those of others, and to work together in a closely coordinated way.

As already stated at the beginning of the chapter on metacognition, René Descartes, with his "I think, therefore I am", already saw this form of reflection on one's own thinking as a prerequisite for a concept of self and fully reflective consciousness. This reflection is accompanied by a sense of confidence in one's own knowledge and the rightness of one's own decisions. People can report this, but it is also possible to assess from the outside how much confidence people have in their decisions by measuring (retrospectively) whether they stick permanently to their original decision or by testing (prospectively) whether they "get out" before attempting to solve a difficult task. These are precisely the two procedures used to try to determine the capacity for metacognition in non-human creatures [623]. Neurobiologists have located specific networks in the human brain for the ability to explicitly extract and manipulate trust signals [603]. However, such prefrontal cortex areas have also been demonstrated in monkeys and rats [1102, 1103].

An indirect connection between (self-)consciousness and metacognition is also visible in certain neurological procedures. In particular, neural activity in the prefrontal cortex is important for conscious perception, exactly where metacognition occurs [1104]. When an activity in this part of the brain is manipulated with transcranial magnetic stimulation (TMS), explicit (conscious) metacognition is also affected [1105]. Apparently, the brain mechanisms responsible for higher-order conscious perception and explicit metacognition are shared. Either the two are causally linked or the older capacity for metacognition has been "recycled" to enable the younger capacity for consciousness [899].

It is also important to note the difference between introspection and metacognition. While introspection refers to an act of deliberate and effortful self-observation and evaluation of ongoing conscious experiences, metacognition simply means that one cognitive process relates to another process. Although most theorists assume that explicit judgements are conscious. These are judgements about one aspect of another process and self-confidence about how well one can perform a cognitive or memory task. There is also the possibility that metacognition can occur at a subpersonal level—for example, when we spontaneously notice that we have made a mistake without being asked.

Recently, there has been an increasing expression of the view that even the higher-order theory of thought (HOT) needs not to be so sophisticated that all the processes involved must be explicit and verbalised (leaving HOT consciousness incompatible with animal consciousness [904] or even leaving animals without any conscious states at all [624]). For instance, Rocco Gennaro has defended the view that HOT is compatible with animal consciousness in the vast majority of cases [925, 1106, 1107]. In his opinion, some recent experiments strongly suggest that many animals can exhibit metacognitive states (see Chap. 7).

A similar change of mind has been made with regard to "*ego-minds*". Here, too, the increasingly sophisticated experiments on metacognition have led to rethinking among some philosophers, who now grant animals mental states about mental states or "*cognitions*" about other mental representations [3, 1041, 1108]. However, we know from recent studies in humans that control over cognitive processes is not necessarily metacognitive [625] or uses metacognitive representations, for example, of decisions about inconsistent actions [627, 628]. According to the dual-system theory [630], the first control process acts quickly, often operates autonomously and in parallel, and is implicit, i.e. non-conscious, while the second process is carried out by executive functions at a higher level, serially and with greater time expenditure, and is explicit/conscious [629]. In humans, much of the cognitive control that falls under a broad notion of metacognition has been shown to occur outside of System 2 [1109]. Subjects exhibit decision uncertainty, inhibit responses to stimuli they cannot know about because of visual masking, and slow their decisions after making mistakes they do not know that they have made [1110].

Are non-human animals, which are similarly able to use metacognitive information, restricted to System 1? I have covered in this book a number of convincing pieces of evidence that under "uncertain" conditions, animals are more likely to search for additional information, reject decisions, and place lower stakes on difficult choices. However, sceptics claim that these remarkable performances could also be due to simple reinforcement learning [614, 1111]. Even in the most convincing experiments with chimpanzees or rats, critics consider it possible that unconsciously ongoing internal System 1 processes automatically trigger reward-seeking behaviour. In any case, critics do not see enough evidence for the additional use of the metacognitive capacity of System 2.

But why is System 1 insufficient for humans, and what is the added value of System 2? I have elaborated on this in the chapter on metacognition, but briefly summarised, the hypothesis is that System 1 concerns only control processes within an individual, while System 2 enables control processes that exist between two or more individuals [629]. Only here, and this is the crucial point, is higher consciousness involved because only this ensures the coordination of sensorimotor representations between two or more individuals and uses the metacognitive information of the interacting individuals [629]. Only these metacognitive representations of System 2 are available for verbal reports, and this is precisely what some theorists consider to be the decisive characteristic of consciousness and thus the unique selling point of humans. Verbal reports are important because they ensure optimal control when two or more people cooperate [632]. It is undeniable that joint decision-making about perceptions and mental states is much more effective through linguistic communication than merely through gestures or other nonverbal means and that verbal communication is an advantage, especially in the case of metacognitive content [632, 633]. Who can be trusted, who speaks the truth, who is uncertain? All of these factors characterise the suprapersonal or interpersonal coordination of metacognitive content and connect it to the theory of mind [634], especially because metacognition enables an explicit understanding of false assumptions or beliefs.

Chapter 11
How Do Animals Think?: A Seven-Point Summary

In this book, I have discussed and proposed answers to (most of) the big and, at the same time, most difficult questions in comparative cognitive research. Starting with the question of whether we can ascribe rationality to animals and, if so, to which species and to what degree, I have highlighted recent research findings that either challenge preferred interpretations or even require entirely new ones. It was particularly important for me to critically question hasty categorisations, often derived from human and everyday psychological notions, and to look at animal (mis)performance from the perspective of species-specific, natural requirements. The most important questions that I have addressed in this book can be briefly summarised as follows[1]:

Can We Attribute Mind to Animals?

Yes, animals undoubtedly have mental abilities that go beyond rigid, instinct-bound action control and thus enable flexible behaviour and, above all, the solving of novel problems. Many animals modify their behaviour through learning, are not bound to the here and now, gain experience, store it in memory, adjust behaviour according to future events and needs, communicate with each other, learn from each other, and act together. Some animal species—both apes and several other mammals, as well as birds and other vertebrates, and in some cases even invertebrates (e.g. certain hymenoptera and cephalopods)—are capable of the highest forms of nonlinguistic rationality summarised in this book.

[1]An older version of this chapter appeared in my review of a book by Frans de Waal with the significant title *Are we smart enough to know how smart animals are*? [1112].

L. Huber, *The Rational Animal*, https://doi.org/10.1007/978-3-031-60803-2_11

How Is It Different from That of Humans?

Like every animal species, the species *Homo sapiens* also possesses species-specific characteristics and abilities. For this reason, biologists avoid the term "special status", which is popular among some philosophers and anthropologists, and replace it with terms such as "unique traits" or "peculiarity", which are used to describe the characteristics of humans and their species-specific expression in comparison to that of other animals in terms of evolutionary biology [890, 1113]. Depending on the scale, fine-grainedness, and terminology, differences between human and analogous characteristics in other animal species tend to be only quantitative or yet qualitative. Findings in comparative cognition research have recently triggered much fascination, partly due to the fact that unexpectedly complex, *human-like* performances have been found in not only closely related animals such as chimpanzees but also in birds, fish, reptiles, frogs, bees, and octopuses. However, we must not miss the forest for the trees; the human mind, human culture, or human morality are certainly unique, even if there are similarities between their individual components and those of other animals.

Philosophers, anthropologists, and linguists often mention two abilities in particular as special, outstanding features of human beings: language and consciousness. However, are these truly unique features of humans? Do they really aid in the elevation of human cognition above that of the rest of the animal kingdom? In the case of language, I have admitted that it is unique to our species in its full human form, mainly because it gives us the ability to express everything we can think of. But while this ability relies on a wide range of biological mechanisms that are shared with other animals, many other species can actually communicate both emotions and knowledge. Thus, it is perfectly arguable whether the limitless expressive capacity, qualitatively and categorically, is what sets human language apart from all other communication systems.

As far as consciousness is concerned, today's debate among cognitive researchers runs along the lines of whether non-human animals can be said to have a "metarepresentational" sense of self. Recent evidence of planned action, episodic memory, metacognition, and mind reading in some animal species suggests that they do construct a, at least modest, kind of self-model that enables planning and self-monitoring.

The aforementioned abilities may have been combined with others in human evolution that were particularly advantageous for group coordination. In sophisticated cooperative actions such as joint hunting, higher forms of metacognitive representations can be exchanged between individuals with the help of increasingly powerful language, thus enabling a higher, supra-individual form of control. A recently proposed hypothesis [629] goes as far as suggesting that the human System 2 of metacognition provides thoughts, ideas, and knowledge for oral reports and is thus thought to be the defining characteristic of human consciousness. With this kind of self-awareness, the human being becomes a reflective observer who knows not only his own body and appearance but also his knowledge ("metacognitive") and memories, and he is able to communicate these to others. System 2 enables rapid and effective decision-making about perceptions and sensitivities in the collective and, in turn, improved control within cooperative activities [632]. If this hypothesis is correct, in the evolution of the genus *Homo*—possibly to some extent even before it—there was a synergy in the greater development of several cognitive abilities, which positively influenced each other to form together the human mind in an emergent way. One could understand this mind as a conglomerate whose elements—at least in a more modest form—also occur in other living beings, but which, taken as a whole, can now be found in only one species, *Homo sapiens*.

How Can We Determine the Mental Abilities of Animals?

By attempting to answer all four of Tinbergen's questions [837] with a holistic approach using great knowledge of the species, its living conditions, its evolutionary and functional traits, added methods and findings from all other biological and cognitive science disciplines, a combination of (sophisticated, controlled, repeated, and above all experimental) laboratory research and field observation, and progressive elucidation of the neural bases and mechanisms of cognitive performance, we are making progress in this matter [1114]. I agree with Frans de Waal that we always have to ask ourselves whether we are clever enough to understand the cleverness of animals [1115], but the zeal and creativity of the scientists working on this matter make me very optimistic.

Can the Human (Anthropomorphic) Perspective Be Abandoned or Its (Epistemological) Limitations Overcome?

The sooner we advance the methods described in Point 3, the sooner we can crack the hard nuts of cognitive research. Above all, we need to clarify terms, question everyday concepts, and learn to appreciate the great diversity of evolution, according to which animals did not get stuck halfway to humans but rather developed in completely different directions. It is important to find a neutral language or a language that does justice to animals, to dismantle old prejudices and wishful thinking, and to develop species-independent models that have great predictive power [1116]. In addition to refining the methods of one's own subject, the addition of findings from other biological disciplines (ecology, physiology, neurobiology, genetics, etc.) and findings from the large subject canon of the cognitive sciences (anthropology, psychology, social and cultural sciences, computer science, linguistics, etc.) is necessary for a comprehensive understanding of cognitive processes in animals.

How Can Different Non-human Animal Species Be Compared?

There is no silver bullet, but species-appropriate task settings and devising convergent task series, which have systematic variation in as many non-cognitive influences on solution finding as possible, facilitate intrinsically difficult species comparisons [1117, 1118]. It is indispensable to find a task that is adequate for the animal species and to exclude alternative explanations for the behaviour shown by using suitable control and comparison conditions. To do this, it is important to know the animal species, its everyday problems, and its coping strategies used. For example, it does not make sense to apply tasks in which handedness is an advantage to hoofed or cloven-hoofed animals. Even dogs and wolves are at a disadvantage in object manipulation tasks compared to primates. Even within birds, there are ones that use both their feet and their beaks, while others are limited to only their beaks [1119, 1120].

However, caution is also needed when using species-specific tasks: if animals are required to perform only in tasks tailored to their needs, demands, adaptations, etc., instead of abstract tasks, species comparisons are made more difficult and undermined. Any difference in problem-solving ability could then be explained away as a methodological flaw in the task. What proves suitable for one species is then too abstract and ecologically uninformed for the other.

In the cognitive sciences, the association between lifestyle, evolution, and cognition is addressed in various theories—be it with regard to the distinction between "e-rationality" (economic, ecological) and "PP-rationality" (that of philosophers and psychologists)—for the debate about the psychometric factors *g* and *G* or for the theory of embodied mind [254]. The question of whether animal species possess *domain-general* rather than *domain-specific* intelligence has been the subject of frequent debate [1121]. Should evolution not equip species with tailor-made toolboxes instead of Swiss army knives [1122]? High flexibility and solvability of problems in all possible environments are, after all, costly. According to the slogan "simple heuristics make us smart" [82], ecological rationality should prevail in the animal kingdom.

In recent years, the integration of comparative cognition research and evolutionary biology has made it possible to answer questions about the phylogenetic distribution of cognitive traits and the evolutionary processes that drove their development [1123]. Methods have been developed to quantitatively test whether specific cognitive abilities are correlated with life history (e.g. lifespan), morphology (e.g. brain size) or socio-ecological variables (e.g. social system). Sophisticated statistical analyses can even be used to measure how strongly relatedness predicts the distribution of cognitive abilities across species. Finally, even the evolutionary sources of a particular cognitive trait can be narrowed down through experimental determination of an extant species' cognitive performance [1124].

What Are the Benefits of Cognitive Skills?

Crucial for understanding the cognitive abilities of animals is answering the question of what benefit or use these abilities have for the respective animal species, not whether they possess human-like qualities. From an evolutionary point of view, animals should be able to do or know what is beneficial for the survival of the individual and its reproduction. This inevitably raises the question of whether animals are thus functionally shackled to the extent that "higher" or "complex" abilities emerge only when they are necessary for survival. But isn't it precisely the nature of cognition to master newly emerging problems? Doesn't some degree of functional surplus need to be present for this to happen? This leads us to a difficult cost–benefit calculation. As we know, the cost of high intelligence is great. Our brains consume an enormous amount of energy relative to other parts of the body (which is necessary for a brain with 86 billion neurons chasing electrical impulses at 400 km per hour). Although our brain makes up only 2% of our body weight, it consumes 20% of the available oxygen. In return, no less than 1000 litres of blood are pumped through our brain every day.

How Common Are Higher Levels of Intelligence in the Animal Kingdom?

Evolutionary psychology [1125] offers no satisfactory answer to this question [1124]. Since it is essentially short-sightedly limited to hominid evolution, the major trends in the evolution of cognitive abilities cannot be determined. Even if one considers all primates or even all mammals, this branch of psychology remains extremely selective. The great rest of the animal kingdom must also master life problems and have a certain amount of cognition to do so. We therefore need "comparative cognition research", which, following "comparative behavioural research" (i.e. ethology), attempts to do justice to a comprehensive, balanced, and thus less anthropocentric view of the great family tree of life by considering the various "fruits on the tree of knowledge" [1126]. Even within vertebrates, mainstream research is one-eyed since two major classes are virtually left out: amphibians and reptiles. For evolutionary reasons such as the origin of avian or mammalian cognition, however, these large animal groups are indispensable, and "cold-blooded cognition research" is timely [1127, 1128]. Although research in this area is lagging behind that in mammals and birds, the first signs of a race to catch up are already visible. For example, my colleague Anna Wilkinson, now at the University of Lincoln (UK), was able to show in sensational experiments that solitary red-footed tortoise [1129] and central bearded dragons [1130] are capable of imitation in the sense of true imitation. This turned the previously prevailing opinion that imitation was possible only for socially living animals on its head.

Less surprising, meanwhile, are the reports of extraordinary reasoning powers in fish. These include the special abilities of cleaner fish to recognise their clients, to cooperate, or to deceive them [1131, 1132], and also the evidence found in various other fish species for social strategies, social learning and tradition building, cooperative hunting, special foraging abilities, use of tools, cognitive maps, behaviour towards predators, and manipulation of the environment. Together, all these abilities have revealed astonishing evidence for "underwater cognition" [1133]. Seen in this light, it is not all that startling that cleaner fish learn faster than pigeons, rats, capuchin monkeys, and even orangutans and chimpanzees when choosing between permanent and ephemeral food [1134].

Among invertebrates we also find "intelligent beasts" under water. The cephalopods, especially the octopus, have gained the greatest notoriety in this respect [957, 1135, 1136]. Their intelligence is not limited to special hunting techniques but extends—as recently demonstrated in the common octopus—to future planning [582].

The question of the evolution of cognitive abilities was literally flipped by the sensational findings of honeybees, bumblebees, and other *Hymenoptera* made in recent years by teams led by Martin Giurfa (Université de Toulouse, FR) and Lars Chittka (Queen Mary University of London, GB) [reviewed in 331, 1137, 1138–1140]. Although their neuron condensations—rather than brains in the true sense of the word—are among the largest among insects, with a maximum of 960,000 neurons, they are many times smaller than those of vertebrates. Even if the brain size of honeybees is normalised to their body size, the relative brain size is in the lowest range of that of vertebrates. For a long time, it was assumed that a small brain provided an animal with less computational capacity, and bees were therefore considered to be purely instinctly driven. Their referential communication using the waggle dance was also long considered an example of an innate, less flexible behaviour, but this has now been disproved—as described at the beginning of my chapter on language. Moreover, recent experimental research has painted an astonishing picture of other cognitive performances. It has been shown that internal representations do not consist exclusively of chains of associations for stimulus-response links but are much more complex, flexible, and integrative in structure. Honeybees exhibit configural forms of conditioning, the capacity for biconditional discrimination, context-dependent learning and remembering, and even some forms of concept formation, such as that for abstract rule or the relationship between same and different [332, 1141]. Recently, bees have also joined the groups of animal species capable of social learning and even tool use [1142], and bumblebees can represent objects intermodally (tactile and visual) [1143].

Research into higher cognitive abilities (such as the sextet of animal intelligence described in this book) is still in its infancy in invertebrates, fish, amphibians, and reptiles, but comparative cognition research—like any other empirical science—is open-ended. Further spectacular discoveries like those described in this book are to be expected. However, at the same time as the "search for clues in cognitive biology" is being blazed, the field of comparative cognition research must reform because it must not be biased towards confirming more extraordinary abilities in animals, which is unfortunately too often the case. Critics suggest that there is likely to be a high rate of false-positive discoveries and a certain bias against exceptional abilities, even in the presence of strong methodological criticism [1144]. In addition, there is a lack of formal methods for evaluating evidence for the *absence of* a cognitive ability, ambiguous definitions used for claims, and, most importantly, a "replication crisis" [1145, 1146]. However, there is potential for reformability, and a growing number of (self-)critical cognitive biologists have proposed initiatives that can improve replicability and reproducibility in comparative cognition research, which often suffers from the use of small and idiosyncratic samples [1147]. This includes increasing transparency and implementing good practices in all areas of experimental research, such as increasing both homogeneity and heterogeneity in experimental design; statistically modelling variation; increasing the use of preregistration (the disclosure of research questions, variables, experimental methods, and planned analyses prior to data collection); and especially promoting replication experiments [1148]. Another constructive way forward for comparative cognitive research (and many other empirical sciences) lies in the consistent application of "open science" principles [1149]. One recent way to address replication problems has been through large-scale collaborations using open science platforms. "Big Team Science" [1118, 1150, 1151] involves research groups working across national borders on a joint project, for example, in the field of primate cognition [1152], bird cognition [1153], and research on human infants [1154]. My laboratory has also joined such an initiative and has collaborated with many other laboratories to study the understanding of human pointing in dogs. This project helped to determine both the influence of differences between populations and breeds and the influence of testing environments on the robustness of results [1155]. All of these collaborations will increase the reliability and validity of published research results and provide a bright future for comparative cognition research.

Epilogue: What Can We Learn from This?

First, everything that is generally true for empirical (basic) sciences applies to comparative cognitive research: knowledge satisfies human curiosity. Both by nature and by culture, humans strive to know and understand themselves and the world. We want to be able to explain natural phenomena to uncover their causal mechanisms because we hate uncertainty. Without the constant pursuit of knowledge, human culture would not exist. The more knowledge we have, the more we change the world. This is particularly evident at the moment with climate change, but it is generally important to consider the consequences of one's actions, both small and large. We are not only able but also challenged to think about the consequences of our actions, because acting without knowledge is negligent.

Knowledge gain does not always lead immediately and directly to practical knowledge. All too often, practical applications became possible on the basis of previously established general knowledge. Therefore, even the basic researcher must justify her or his actions: she or he must question her or his knowledge because knowledge without action is irresponsible. The knowledge provided by scientific findings can be decisive for our future.

However, we must not overlook the fact that basic science is fundamentally different from applied science. Basic science is a source of inspiration and innovation; its goal is not to support economic growth or the availability of consumer goods but to create new knowledge and challenge existing knowledge. Driven by curiosity, it ceaselessly challenges the limits of current knowledge, overcomes the limitations of traditional concepts, and generates new hypotheses, new theories, new materials, new reactions, and new methods. Truly innovative research is different from technical application research aimed at quick success. Although basic science is promoted by a different mindset, it is nevertheless the driving force behind societal development in all fields [1156].

Basic science has a special value for society in its function to create the fact- and evidence-based prerequisites for societal developments while at the same time being able to foresee and understand the effects of societal decisions. It is of special interest to society and political leaders to make science-based decisions rather than

L. Huber, *The Rational Animal*, https://doi.org/10.1007/978-3-031-60803-2

emotional ones, since the latter can lead to arbitrary outcomes with potentially harmful consequences.

In the field of animal research, too, we are in the midst of a dramatic social change that is accompanied by particularly heated and emotional debates. There is no doubt that this change is also fuelled by the new findings from comparative cognitive research and that the resulting knowledge about animals makes its way into central questions of animal ethics, namely, on how we should treat animals.

To illustrate this with an example: While I was still studying biology, it was a standard practice to kill countless rats to dissect them (still warm to the touch) in comparative morphological exercises. Today, animals are (largely) replaced by models, sometimes even in virtual reality (3D goggles). At my current place of work, the University of Veterinary Medicine Vienna, the so-called VetSIM laboratory was established in 2012, where students learn and practise practical medical skills on models before applying them to animal patients. A particularly popular and effective model is a dog dummy, on which the appropriate grips and actions are practised before, during, and after veterinary treatment.

Animal experiments[1] in basic science are increasingly criticised because our attitude towards the benefit–harm ratio has changed. The greater the intervention and the more serious the consequences are, the greater the need for justification. Meanwhile, experiments on great apes are banned throughout the EU, however great their benefit. Unfortunately, this still does not apply to all other primates. Their evolutionary proximity to humans makes them particularly suitable for research into human diseases. They are used, for the most part, for final safety tests of future drugs before they are used on humans. They are also used for important fundamental questions about the function of healthy organisms. In neuroscience, monkeys are still being researched for nonmedical reasons, such as the significance of individual brain areas or even individual neuron classes (e.g. so-called mirror neurons). Invasive methods are used—with the insertion of electrodes or the removal (lesions) or silencing of entire brain parts. Although academic research on monkeys accounts for just under a fifth of all animal experiments with primates,[2] the lesser necessity for the welfare of humans places this research under a particularly high obligation to justify and provide reasons for it. After all, animal experiments are permitted not only in the EU but also in many non-European countries if the stress on the animals is ethically justifiable with regard to the purpose of the experiment. Committees specially appointed for this purpose must examine whether the necessity and

[1] "Animal experiments are all experimental interventions on or treatment of living vertebrates which are stressful for the animal, in particular involving fear, pain, suffering or permanent harm, and which [...] have the aim of testing a scientific assumption, obtaining information, obtaining or testing a substance or determining the effect of a specific measure on the animal" (§ 2 of the Austrian Animal Experiments Act, Federal Law Gazette No. 501/1989).

[2] In Germany in 2015, 17% of monkeys were used for academic research and 83% by industry, for example, for mandatory toxicity and safety tests. These included mainly Javanese monkeys (85.6%), along with tamarins and marmosets (9.7%), prosimians (3.2%), rhesus monkeys (1.3%), and baboons (0.2%). Source: Experimental animal figures 2016. Statistics of the Federal Ministry of Food, Agriculture and Consumer Protection (BMEL) Germany.

appropriateness of the planned animal experiment outweigh the expected stress. The general rule is that the greater the suffering is, the greater the (scientific) benefit must be.

Researchers who want to conduct an animal experiment must already weigh whether their project is ethically justifiable in the application for approval. The problem here is, on the one hand, that it is very difficult to predict the future benefit and, on the other hand, that one is caught in an almost unresolvable ethical tension between the desire to conduct research for the benefit of humanity and the problem of consciously accepting the suffering and death of animals for this purpose. As long as animal experiments cannot be completely replaced by alternative methods,[3] this ethical dilemma will persist.

In purely quantitative terms, however, monkeys are not the predominant animals sacrificed for medical and scientific progress. Worldwide, more than 100 million animals are killed in animal experiments every year,[4] predominantly (approx. 80%) mice and rats. The reason for this is that—apart from the practical advantages of keeping and breeding them—it is possible to alter the genetic material of rodents in such a way that cancer; metabolic, infectious, or nervous diseases, and other serious ailments can be easily targeted. In the chapter on episodic memory, I described a series of rat studies that impressively demonstrate the enormous mental powers of which these animals are capable. It has been known for much longer that rats also have a superb spatial memory and an extraordinary ability to orient themselves [1158]. Of particular relevance to the ethical question of how to deal with rats, which are the most frequently used laboratory animals after mice, are the torturous experiments that are the subject of loud protests and not only from animal protection organisations.[5] Life in cramped cages and death, which usually occurs during or after the experiment, is still the lesser evils compared to stress and fear and agony and pain. If one considers that rats not only are particularly clever creatures but are also particularly social, even prosocial (they free their conspecifics from cages and even give up popular sweets in return, as a series of sensational experiments has shown [1159, 1160]), then, the ethical discrepancy becomes obvious.

[3] There is widespread agreement in the community of scientists who use animal experiments to keep them to a necessary minimum or to replace them with alternatives. The guiding principle is the "3Rs principle": Replace, Reduce, and Refine. The three terms were coined by two British researchers, the zoologist William Russell and the microbiologist Rex Burch [1157].

[4] However, the number of unreported cases is probably many times higher, as many animals are not recorded in the animal experiment statistics.

[5] There are countless reports of experiments in which their stomachs are cut open while they are still alive, their brains are wired with microelectrodes, or they are infected with diseases, poison, or their own faeces. Particularly cruel is the "Forced Swim Test", in which rats are placed in a tank of water from which there is no escape. In panic, they try to climb up the walls and pedal like mad to keep their heads above water. Eventually, their strength wears off. Some pharmaceutical companies and even renowned British universities used this cruel experiment to develop treatments for depression.

It becomes even more disturbing, however, when one looks at farm animals. In relation to the 750 million living creatures that are fattened and slaughtered for meat consumption every year in Germany alone, laboratory animals make up only 0.5%. Although the number of vegetarians and vegans is continuously increasing, the consumption of animals is much more socially accepted than the use of these animals for scientific knowledge or medical progress. In this respect, the analysis of the German philosopher Richard David Precht [1161] holds true, according to which the gap between what people think is right in dealing with animals and what is actually practised has never been as wide as it is today.

How can this irrational attitude be explained? After all, living without meat is possible without any problems, but testing vaccines, medicines, or chemicals without animals is risky and endangers people's health.

In scientific discourse, it is ethicists who point out such discrepancies, denounce inconsistencies and contradictions, question traditions and social practices, and often call for rethinking. Animal ethics and, more broadly, bioethics are disciplines of moral philosophy that have been concerned with these questions since Aristotle (with reference to ideas of Socrates, Plato, and the Sophists). In the case of animal ethics, these are questions of ethical appropriateness and the consequences of our behaviour for animals. Setting ethical boundaries, transforming them into binding laws, and implementing them through the legislative and executive branches are central tasks of society, but these boundaries must be appropriately prepared and justified. The German philosopher Otfried Höffe from the University of Tübingen described this process as follows (my translation):

> Where traditional ways of life and institutions lose their self-evident validity, philosophical ethics, guided by the idea of a meaningful human life, seeks methodically [...] and without final appeal to political and religious authorities [...] or to what has been customary and proven from time immemorial, to make universally valid statements about good and just action. [1162, p. 66]

Almost all approaches in animal ethics today agree that humans have moral obligations towards sentient animals. What is less clear, however, is how this moral obligation towards animals can be justified and what conclusions are to be drawn from it. For example, one can ask whether, in addition to the negative obligation not to inflict (unnecessary) pain, suffering, or harm on animals,[6] there are also positive obligations, particularly a duty to care for and promote the welfare of sentient animals. In most cases, however, the moral principles of *non-maleficence* and *beneficence* complement each other.

To achieve improvements here, more than technical and spatial changes are needed; one must understand why animals behave the way they do. The so-called *animal welfare sciences* have made great progress in this regard, for example, in researching not only the vital needs (for survival) but also the emotional and social

[6]The second sentence of the preamble to the German Animal Welfare Act requires: "No one shall cause pain, suffering or harm to an animal without reason".

ones of animals. This has required a better understanding of both their behavioural repertoire and of their inner "mental life".

For a long time, the guidelines and regulations developed to protect the welfare of farm animals emphasised the importance of providing an environment that ensures good health and a normal physiological and physical condition. Mental needs were at most marginal. Today, however, there is a growing recognition that animal welfare is very much, if not predominantly, dependent on their "cognitive needs" [1163]. When these needs are met, the physical needs of the animals are usually met as well. Accordingly, approaches to assessing farm animal welfare have evolved from concepts such as the "Five Freedoms"[7] to more animal-centred approaches that also incorporate the individual differences of farm animals [1164].

Thus, an important insight of welfare sciences is that animal welfare is highly dependent on the mental, psychological, and cognitive needs of the animals concerned [1163]. For example, learning and memory skills are thought to have an impact on an animal's ability to cope with housing conditions [1165] and to cope with changing social conditions, such as separation and regrouping. However, although welfare concepts emphasise the importance of detailed knowledge of the cognitive abilities of farm animals—their ability to mentally grasp, process, store and use information—the much more important question for our subject remains whether knowledge of higher cognitive abilities, as I have described them in this book, can also contribute to better animal welfare, as this knowledge may not initially facilitate farmers' efforts to adapt husbandry systems and design elements to the needs and preferences of their farm animals.

I believe that knowledge of the sextet of animal intelligence, together with new evidence from the fields of language and consciousness, is of ethical and social interest for at least two main reasons. First, it contributes significantly to the conviction that animals can feel pain as well as suffering and other morally relevant psychological states.[8]

The very sensation of pain leads directly to the question of the moral consideration of animals. For a long time, it was very controversial whether fish feel pain, but in recent years, there has been increasing recognition that they do [1167, 1168] and that their overall sentience is probably much greater than what has long been suspected [1169]. Without the growing evidence for extraordinary cognitive

---

[7] The concept of the Five Freedoms arose from the report of the Technical Committee to Inquire into the Welfare of Animals Kept in Intensive Animal Husbandry Systems, the Brambell Report, December 1965 (HMSO London). This stated that farm animals should have the freedom to "stand up, lie down, turn around, groom themselves and stretch their limbs". The *Farm Animal Welfare Council* (FAWC) developed these into the Five Freedoms, which provide a framework for analysing animal welfare. FAWC online: http://webarchive.nationalarchives.gov.uk/20121007104210/http:/www.fawc.org.uk/freedoms.htm, last accessed 20.7.2021.

[8] The research of physiologists, ethologists, and animal welfare scientists on the sensation of pain in animals was preceded by the ethical demand to base the moral attitude towards animals on these very conditions and capacities. This view culminated in the famous statement of the English jurist and philosopher Jeremy Bentham (1789), which summarises the animal-ethical position of pathocentrism: "The Question is not, Can they reason? Nor, Can they talk? But, Can they suffer?" [1166].

abilities (such as those described in Point 7), this paradigm shift would probably have been much slower and more laborious. Overall, it can now be said that vertebrates can feel pain, and even the evidence for octopuses, squid, crabs, or shrimps has increased [1170, 1171].

Pain is a clear example of phenomenal consciousness or basal body consciousness, but beyond that, animal species that have extended consciousness, that can plan into the future and metacognitively control their own sensations, and that are emphatic about the emotional lives of their conspecifics and act prosocially might also be capable of suffering to a certain extent. Suffering cannot be precisely delimited since it refers to a broad spectrum of unpleasant emotional states, such as fear, anxiety, pain, or frustration; these states are long-lasting and/or severe but, in any case, closely related to consciousness [1172]. Much animal welfare science has therefore been concerned with the capacity of animals to suffer and, more recently, with their capacity to experience pleasure and other positive emotions [1173].

It becomes even more difficult when examining more subtle emotions, feelings, and mental states. For example, animals with higher states of consciousness might suffer from loneliness and boredom, the restriction of not being able to pursue their interests, perhaps even witnessing the killing of conspecifics or feeling completely at the mercy of others. Having longer-term interests and being able to pursue them are particularly often suggested to be criteria for special protection and promotion, especially in so-called interest-based moral conceptions. For example, the Australian philosopher Peter Singer (Monash University, Australia, later Princeton University, USA), who is considered one of the founders of modern animal ethics,[9] demands that equal weight be given to the similar interests of all those affected by an action [1174]. However, for Singer, *sentience* is sufficient to be able to have an interest. For others, only those beings who can have beliefs have interests or even preferences. This brings us to the debate I outlined at the very beginning of the book in the section "To Err Is Human?". No matter how far one may concede these abilities to non-human creatures, the demand to protect animals with increasing degrees of cognitive abilities and competences to communicate, feel pain, and be aware has now found its way into textbooks and encyclopaedias [1175].

The second reason for the relevance of comparative cognitive research for animal ethics issues and social change is due to the fact that the way we treat animals very much depends on how we see and judge animals. Put fully, one could say, "Those who think morally divide the world into two parts: that which they respect and that which they condemn" [1161, p. 285]. Since respect is related to esteem, the attribution of values increases both esteem and respect. Among the values we esteem is intelligence. Inventors, Nobel Prize winners, and other intellectual greats are usually highly regarded in society. This connection between intelligence, esteem, and the worthiness of protection can also be seen in the evaluation of animals.

---

[9]The principle of equal consideration of interests is a centrepiece of his preferential utilitarian approach, a variant of utilitarianism. In classical utilitarianism (Bentham; Mill), the ultimate goal is to provide for "the greatest happiness of the greatest number". Utilitarianism does not have the welfare of a particular group in mind but that of all those affected by an action.

An example of this connection—albeit in the opposite direction—is the assessment of animals that serve as food. The knowledge that the animals we eat have minds makes them similar to us in a morally important way, but this knowledge is at odds with their use or exploitation as food. Humans are granted moral rights on the basis that they have minds (without drawing the reverse conclusion that would affect the treatment of newborns and the mentally ill) [1176]. Being reminded that animals have minds, but at the same time serve as food, can cause moral conflict for meat eaters. In fact, humans are reluctant to harm things that have minds [1177]. Moreover, as a study by Australian and British scientists has shown, this dissonance motivates people to deny sanity to animals they like to eat or at least to grant them only a diminished mental capacity [1178].

Under these circumstances, the use of the popular terms such as "stupid pig" is hardly surprising. In terms of volume, pork is the most consumed type of meat worldwide, with Austrians and Germans in particular consuming a disproportionate amount of pork.[10] The pig is thus valued as a staple food but not as a creature of reason. For this reason, I decided in 2013 to contribute to the elucidation of the cognitive performance and emotional sensitivity of this farm animal species and to investigate its abilities and characteristics. However, since most previous research on pigs had been undertaken with ones bred for rapid growth, kept in unnatural barn environments, and slaughtered after just a few months, I decided to establish my own seminatural pig farm. With generous financial support from the Swiss Messerli Foundation, this project succeeded. After extensive research, I chose a New Zealand domestic pig breed known for its slow growth, year-round outdoor housing, and friendly nature: the Kune-Kune pig.[11] These rather stocky animals, with strong, multicoloured hair, have proven very useful for research into the natural behaviours of domestic pigs, about which far too little is still known [1180].

My incredibly dedicated colleague Marianne Wondrak, many eager students (especially my PhD student Ariane Veit), and I began to adapt various standard comparative cognition tests to suit pigs and thus tested a group of 40 animals that grew up on our free-range farm. These pigs were able to shape their own family life

[10] To meet the demand for meat and other animal products, livestock farming has been greatly expanded worldwide. Global meat production has more than quadrupled since 1961. This trend has a strong east–west divide. While Europe's meat output has approximately doubled and North American output has increased 2.5-fold, in Asia, meat production has increased 15-fold since 1961. Globally, ca. 360 million tonnes of meat were produced in 2022, of which around 26 billion chickens, 1.5 billion cattle, 1.3 billion sheep, and 1 billion pigs were kept globally. Source: FAO ("Production"/"Crops and livestock products"/"Live Animals"); https://www.fao.org/faostat.

[11] Kune-Kune pigs are known as the Maori pigs of New Zealand because they were kept by the Maori as free-range pigs. They were allowed to roam freely around and inside the houses in their settlements in search of food. The long and strong contact with humans is probably the reason why these pigs have a very friendly, good-natured, and calm temperament. They also have an eye-catching appearance. They are rather small, long-haired pigs with short legs, a crowded and rounded body, and a blunt, somewhat open, nose. Their origin is not fully understood; they probably arrived in New Zealand with Spanish and Portuguese sailors in the early nineteenth century [1179].

for several years. Already the first results confirmed our suspicion that these animals were criminally underestimated. Although pigs have a relatively modest sense of sight compared to their outstanding sense of smell and excellent sense of hearing, they could quickly manage to distinguish human faces on a computer screen. Moreover, they were able to apply what they had learned to new and sometimes alienated images [1181] and also use their sense of sight effectively in an observational learning task. They learned from their mother or aunt how to open a box to reach pieces of apple hidden inside, remembering the opening technique even the next day [1182]. After this, we were particularly convinced of their socio-cognitive abilities, especially because we had observed the animals foraging together and doing other social activities for several years. We then decided to use a test that was originally developed for chimpanzees and had already yielded the first surprising results in pigs a decade ago in England.

The *informed forager* test checks whether pigs can outsmart other pigs when foraging, thus checks if they act strategically to a certain extent. Susanne Held from the University of Bristol (UK) was indeed able to show that pigs can learn to outwit their competitors when foraging in pairs in an arena. First, it was shown that the informed animals remembered where to find food and that the uninformed animals were able to exploit the informed animal's knowledge by following them [1183]. Subordinate domestic pigs engaged in behaviours similar to those described as tactical deception in apes and corvids. When individuals were aware of both more and less abundant foraging sites and were tested with a dominant uninformed animal, they were less likely to visit the good foraging site first. They also increased their search speed compared to that of individuals searching with inferior animals or when they were alone. Thus, they responded flexibly and effectively to exploitation by scrounging dominant animals in competitive feeding situations with multiple feeding sites [1184].

These British results inspired us to confront Kune-Kune pigs in analogous situations within large arenas on our pasture. Together with Ariane, Marianne tested ten informed animals of medium rank, with 10 uninformed animals each of higher and lower rank over a whole year; ultimately, more than 800 videos of such tests were available for evaluation. In each test, the informed animal was first allowed into the arena to check its memory of where a bowl of maize was hidden. Immediately afterwards, they were allowed to enter the arena again, but this time, they entered the arena simultaneously with a partner who was either higher or lower in rank. By and large, Held's original results could be confirmed. Not only did the informed animals adapt their behaviour to the rank of the competitor, in individual cases we even found evidence that they deceived their uninformed dominant partner by leading them to the wrong feeding place before quickly running to the food on the other side of the arena.

In recent decades, increasing importance has been given to understanding the cognitive abilities and emotional life of farm animals [1185]. This field of research is still underdeveloped, but it is already becoming apparent that the cognitive abilities of animals that are used economically by humans—whether as food suppliers, as fur animals, as riding and carrying animals, or as attractions in circuses and show fights—are often underestimated. Recent findings on the ability to discriminate between quantities, to mentally represent objects that are out of sight (object permanence), and to cooperate and empathise are largely similar to those we know from the non-human animal species I have described in this book [1186]. In addition, scientists are increasingly focusing on complex social interactions (such as cooperation and prosocial behaviour) and deriving recommendations from them to promote animal welfare [1187].

It is becoming apparent that the increase in well-documented evidence—not only of higher intelligence but also of the social and prosocial traits of animals (cooperation, alliance building, helping, altruistic action, reconciliation [1030, 1188–1191]), which can be elicited only from species-appropriate, natural group living—increases the psychological pressure on humans to give them their preferred life. Evidence of these so-called moral analogue capacities ultimately forces us to ask whether freedom from pain and viability is good enough for them. As my esteemed colleagues Lesley Rogers and Gisela Kaplan of the University of New England in Brisbane (Australia) once expressed so beautifully:

> the ultimate aim must surely be that we do not just want animals to survive but want them to have a quality of life commensurate with their needs and dignity: physical, psychological, social, and cultural. [1192, p. 196]

In humans, such needs are usually protected by inalienable rights. In the future, the growing understanding of animals' capabilities and needs will increasingly lead us to question the very systems we have set up to use them for our purposes. Such developments are indeed likely to seriously challenge human self-understanding, as they tend to dissolve the boundaries between humans and animals [12].

In contrast to Richard David Precht, who proposed an "ethics of not knowing" [1161]. I prefer an "ethics of knowing". It may well be that, despite all the modern research, we cannot judge with absolute certainty about the inner life of animals. However, as natural scientists, we are aware that we cannot know and judge anything with absolute certainty, but we can know and judge many things with a high degree of probability. Precht is right in saying that philosophy, in order to be able to think about animals adequately, must "reduce its presumptuous claim to knowledge in this respect" [1161, p. 303]. But he himself admitted that our attitude towards animals depends strongly on what we know or do not know about them:

> Those who know a lot about octopuses, about their high sensitivity, about their childlike play instinct and about their learning ability will think twice about whether they think they have to eat an octopus. [1161, p. 303; my translation]

Intuitions will not be enough to make us treat animals differently. Research, reflection, disclosure, awareness, knowledge, and education are powerful tools on the way to appropriate animal welfare. According to Goethe's saying "With knowledge grows doubt", should our increasing knowledge about the cognitive abilities of animals feed our doubts as to whether we are treating them properly. Knowledge and morality should go hand in hand, and we should promote moral progress alongside scientific and technical progress, in line with a sentiment attributed to Mahatma

Gandhi: "The greatness and moral progress of a nation can be measured by how it treats animals".

Role models also help. As an "Honorary Ambassador" of the Austrian Jane Goodall Institute, I had the opportunity to talk to Jane several times about her own views and the motivations behind her tireless commitment to animals, to people, and to the environment. Jane still commits to tirelessly travelling around the world at well over 80 years of age to give lectures, attend sponsoring events, and beat the advertising drum for her projects. Her personal commitment has encouraged and motivated millions of people to rethink and take action themselves. Children and young people around the world are becoming involved in Jane's "Roots and Shoots" projects for animals, people, and the environment. After all, the great persuasive power of her arguments nourishes the intuition that a harmonious coexistence of humans and animals on earth is possible. Together with the biologist Marc Bekoff from the University of Colorado in Boulder (USA), she has proposed ten concrete steps on how to recognise the interdependence between humans and animals in the knowledge of our close evolutionary connections to them, and how to respect animals in their own right to live [1193]. Above all, we should have the courage to fight for our beliefs, recognise that our actions make a difference, and act in the knowledge that we are not alone in doing so. There is reason for hope [1194]:

> Only if we understand, can we care.
> Only if we care, will we help.
> Only if we help, shall they be saved. [1195]

References

1. Heyes, C.M., Huber, L. (eds.): The Evolution of Cognition. MIT Press, Cambridge, MA (2000)
2. Darwin, C.: On the Origin of Species by Means of Natural Selection, or the Preservation of Favoured Races in the Struggle for Life. John Murray, London (1859)
3. Hurley, S., Nudds, M. (eds.): Rational Animals? Oxford University Press, Oxford (2006)
4. Davidson, D.: Rational animals. Dialectica. **46**(4), 317–327 (1982)
5. Griffin, D.R.: The Question of Animal Awareness. Evolutionary Continuity of Mental Experience. Rockefeller University Press, New York (1976)
6. Griffin, D.R.: Animal Thinking, ix, 237 p. Harvard University Press, Cambridge, MA (1984)
7. Benz-Schwarzburg, J.: Cognitive Kin, Moral Strangers? Linking Animal Cognition, Animal Ethics & Animal Welfare Human-Animal Studies. Brill, Leiden, London (2020)
8. Malcolm, N.: Thoughtless brutes. Proc. Addresses Am. Philos. Assoc. **46**, 5–20 (1972)
9. Wild, M.: Die anthropologische Differenz: der Geist der Tiere in der frühen Neuzeit bei Montaigne, Descartes und Hume. Walter de Gruyter, Berlin (2006)
10. Oelze, A.: Animal Minds in Medieval Latin Philosophy: A Sourcebook from Augustine to Wodeham, vol. 27. Springer Nature, Cham (2021)
11. Riedl, R.: Biology of Knowledge. The Evolutionary Basis of Reason, p. 252. Wiley, Chichester, New York (1984)
12. Wild, M.: Tierphilosophie. Zur Einführung, 232 S. Junius, Hamburg (2008)
13. Brandt, R.: Können Tiere denken? Ein Beitrag zur Tierphilosophie. Suhrkamp, Frankfurt am Main (2009)
14. Damasio, A.: The Feeling of What Happens. Body and Emotion in the Making of Consciousness. Harcourt, Inc, San Diego (1999)
15. Huber, L.: Gestalten, Konzepte und nonverbale Begriffe. Eine vergleichende Untersuchung täubischer und menschlicher Kategorisierung. Dissertation, Universität Wien (1991)
16. Premack, D.: Intelligence in Ape and Man. Erlbaum, Hillsdale (1976)
17. Herrnstein, R.J.: Objects, categories, and discriminative stimuli. In: Roitblat, H.L., Bever, T.G., Terrace, H.S. (eds.) Animal Cognition, pp. 233–261. Lawrence Erlbaum Associates, Hillsdale (1984)
18. Huber, L.: Categories and concepts: language-related competences in non-linguistic species. In: Breed, M.D., Moore, J. (eds.) Encyclopedia of Animal Behavior, pp. 261–266. Academic Press, Oxford (2010)
19. Allen, C.: Animal concepts revisited. The use of self-monitoring as an empirical approach. Erkenntnis. **51**, 33–40 (1999)

L. Huber, *The Rational Animal*, https://doi.org/10.1007/978-3-031-60803-2

20. Glock, H.-J.: Begriffliche Probleme und das Problem des Begrifflichen. In: Perler, D., Wild, M. (eds.) Der Geist der Tiere. Philosophische Texte zu einer aktuellen Diskussion, pp. 153–187. Suhrkamp, Frankfurt am Main (2005)
21. Dretske, F.: Explaining Behavior. Reasons in a World of Causes. MIT Press, Cambridge, MA (1988)
22. Bermúdez, J.L.: Thinking Without Words. Oxford University Press, Oxford (2003)
23. Millikan, R.G.: Varieties of Meaning. MIT Press, Cambridge, MA (2004)
24. Fodor, J.: A Theory of Content and Other Essays. MIT Press, Cambridge, MA (1990)
25. Ewert, J.P., Buxbaum-Conradi, H., Dreisvogt, F., Glagow, M., Merkel-Harff, C., Rottgen, A., Schurg-Pfeiffer, E., Schwippert, W.W.: Neural modulation of visuomotor functions underlying prey-catching behaviour in anurans: perception, attention, motor performance, learning. Comp. Biochem. Physiol. A Mol. Integr. Physiol. **128**(3), 417–461 (2001)
26. Huber, L.: On the biology of perceptual categorization. Evol. Cogn. **1**(2), 121–138 (1995)
27. Huber, L.: Visual categorization in pigeons. In: Cook, R.G. (ed.) Avian Visual Cognition. Comparative Cognition Press (1999) On-line: http://www.pigeon.psy.tufts.edu/avc/
28. Huber, L., Aust, U.: A modified feature theory as an account of pigeon visual categorization. In: Zentall, T.R., Wasserman, E.A. (eds.) The Oxford Handbook of Comparative Cognition, pp. 497–512. Oxford University Press, New York (2011)
29. Huber, L.: Perceptual categorization as the groundwork of animal cognition. In: Taddei-Ferretti, C., Musio, C. (eds.) Downward Processes in the Perception Representation Mechanisms, pp. 287–293. World Scientific, Singapore, New Jersey, London, Hong Kong (1998)
30. Huber, L.: Generic perception: open-ended categorization of natural classes. In: Fagot, J. (ed.) Picture Perception in Animals, pp. 219–261. Psychology Press, Hove (2000)
31. Harnad, S. (ed.): Categorical Perception. The Groundwork of Cognition. Cambridge University Press, Cambridge (1987)
32. Herrnstein, R.J.: Riddles of natural categorization. In: Weiskrantz, L. (ed.) Animal Intelligence, pp. 129–144. Clarendon Press, Oxford (1985)
33. Herrnstein, R.J.: Levels of stimulus control: a functional approach. Cognition. **37**, 133–166 (1990)
34. Lazareva, O.F., Shimizu, T., Wasserman, E.A.: How Animals See the World: Comparative Behavior, Biology, and Evolution of Vision. Oxford University Press, New York (2012)
35. Thompson, R.K., Oden, D.L., Boysen, S.T.: Language-naive chimpanzees (Pan troglodytes) judge relations between relations in a conceptual matching-to-sample task. J. Exp. Psychol. Anim. Behav. Process. **23**(1), 31 (1997)
36. Zentall, T.R., Wasserman, E.A., Urcuioli, P.J.: Associative concept learning in animals. J. Exp. Anal. Behav. **101**(1), 130–151 (2014)
37. Pinker, S.: The cognitive niche: coevolution of intelligence, sociality, and language. Proc. Natl. Acad. Sci. USA. **107**, 8993–8999 (2010)
38. Pinker, S.: The Stuff of Thought: Language as a Window into Human Nature. Penguin Group (Viking Press), New York (2007)
39. Sapir, E.: Language. Harcourt, Brace, and World, New York (1921)
40. Carroll, J.B. (ed.): Language, Thought, and Reality. Selected Writings of Benjamin Lee Whorf. MIT Press, Cambridge, MA (1956)
41. Whorf, B.L.: Language, Thought and Reality. MIT Press, Cambridge, MA (1956)
42. Humboldt, W.V.: Schriften zur Sprache. Zweitausendeins, Frankfurt am Main (2008)
43. Soboleva, M.: Animal rationale und animal symbolicum: was heißt es rational zu sein? In: Andreev, J., Lensky, E., Angelova, P. (eds.) Das Interpretative Universum, pp. 397–410. Konigshausen & Neumann, Würzburg (2017)
44. Frege, G.: Logische Untersuchungen. Dritter Teil: Gedankengefüge. Beitr. Philos. Dtsch. Idealismus. **3**, 36–51 (1923)
45. Tugendhat, E., Wolf, U.: Logisch-semantische Propädeutik Durchges. Ausg. 1993 ed. Reclams Universal-Bibliothek, 268 S. Reclam, Stuttgart (2014)

46. Wittgenstein, L.: Philosophical Investigations. Macmillan, New York (1953)
47. Bates, E., Benigni, L., Bretherton, I., Camaioni, L., Volterra, V.: The Emergence of Symbols: Cognition and Communication in Infancy. Academic Press, New York (1979)
48. Liszkowski, U., Brown, P., Callaghan, T., Takada, A., de Vos, C.: A prelinguistic gestural universal of human communication. Cogn. Sci. **36**(4), 698–713 (2012)
49. Leavens, D.A., Hopkins, W.D., Bard, K.A.: Indexical and referential pointing in chimpanzees (Pan troglodytes). J. Comp. Psychol. **110**(4), 346–353 (1996)
50. Pika, S., Mitani, J.: Referential gestural communication in wild chimpanzees (Pan troglodytes). Curr. Biol. **16**(6), R191–R192 (2006)
51. Vea, J., Sabater-Pi, J.: Spontaneous pointing behaviour in the wild pygmy chimpanzee (Pan paniscus). Folia Primatol. **69**(5), 289–290 (1998)
52. Pika, S., Bugnyar, T.: The use of referential gestures in ravens (Corvus corax) in the wild. Nat. Commun. **2**, 560 (2011)
53. Vail, A.L., Manica, A., Bshary, R.: Referential gestures in fish collaborative hunting. Nat. Commun. **4**, 1765 (2013)
54. Tomasello, M., Carpenter, M.: Shared intentionality. Dev. Sci. **10**(1), 121–125 (2007)
55. Tomasello, M., Carpenter, M., Call, J., Behne, T., Moll, H.: Understanding and sharing intentions: the origins of cultural cognition. Behav. Brain Sci. **28**(5), 675–691 (2005).; discussion 691–735
56. Byrne, R.W., Whiten, A.: Machiavellian Intelligence. Oxford University Press, Oxford (1988)
57. Campos, J., Sternberg, C.: Perception, appraisal, and emotion: the onset of social referencing. In: Lamb, M., Sherrod, L. (eds.) Infant Social Cognition: Empirical and Theoretical Considerations. Erlbaum, Hillsdale (1981)
58. Striano, T., Rochat, P.: Emergence of selective social referencing in infancy. Infancy. **1**(2), 253–264 (2000)
59. Boesch, C.: Joint cooperative hunting among wild chimpanzees: taking natural observations seriously. Behav. Brain Sci. **28**, 692–693 (2005)
60. Carpenter, M.: Instrumental, social, and shared goals and intentions in imitation. In: Rogers, S.J., Williams, J. (eds.) Imitation and the Development of the Social Mind: Lessons from Typical Development and Autism, pp. 48–70. Guilford, New York (2006)
61. Whiten, A., McGuigan, N., Marshall-Pescini, S., Hopper, L.M.: Emulation, imitation, overimitation and the scope of culture for child and chimpanzee. Philos. Trans. R. Soc. Lond. Ser. B Biol. Sci. **364**(1528), 2417–2428 (2009)
62. Huber, L., Salobir, K., Mundry, R., Cimarelli, G.: Selective overimitation in dogs. Learn. Behav. **48**(1), 113–123 (2020)
63. Huber, L., Popovová, N., Riener, S., Salobir, K., Cimarelli, G.: Would dogs copy irrelevant actions from their human caregiver? Learn. Behav. **46**(4 (Special issue: Canine Cognition)), 387–397 (2018)
64. Searle, J.R.: The Construction of Social Reality. Free Press, New York (1985)
65. Gilbert, M.: On Social Facts. Routledge, New York (1989)
66. Tuomela, R.: The Importance of Us. Stanford University Press, Stanford (1995)
67. Vygotskiĭ, L.S., Rieber, R.W., Carton, A.S.: The Collected Works of L.S. Vygotsky: Volume 1: Problems of General Psychology, Including the Volume Thinking and Speech. Plenum, New York (1987)
68. Vygotsky, L.S.: In: Cole, M., et al. (eds.) Mind in Society. Harvard University Press, Cambridge, MA (1978)
69. Csibra, G., Gergely, G.: Natural pedagogy. Trends Cogn. Sci. **13**(4), 148–153 (2009)
70. Turing, A.M.: Computing machinery and intelligence. Mind. **59**(236), 433 (1950)
71. Searle, J.R.: Minds, Brains and Science. Harvard University Press, Cambridge, MA (1986)
72. Cheney, D.L., Seyfarth, R.M.: How Monkeys See the World: Inside the Mind of Another Species. University of Chicago Press, Chicago (1990)

73. Perler, D., Wild, M.: Der Geist der Tiere – eine Einführung. In: Perler, D., Wild, M. (eds.) Geist der Tiere. Philosophische Texte zu einer aktuellen Diskussion, pp. 10–74. Suhrkamp, Frankfurt am Main (2005)
74. Glock, H.-J.: Animal minds: conceptual problems. Evol. Cogn. **5**, 177–186 (1999)
75. Bermúdez, J.L.: Animal reasoning and proto-logic. In: Hurley, S., Nudds, M. (eds.) Rational Animals? pp. 127–137. Oxford University Press, Oxford (2006)
76. Köhler, O.: Vom unbenannten Denken. Zool. Anz. **16**, 202–211 (1952)
77. Dretske, F.I.: Minimal rationality. In: Hurley, S., Nudds, M. (eds.) Rational Animals? pp. 107–115. Oxford University Press, Oxford (2006)
78. Zadeh, L.A., Fu, K.-S., Tanaka, K., Shimura, M.: Fuzzy Sets and Their Applications to Cognitive and Decision Processes. Academic Press, New York (1975)
79. Tversky, A., Kahneman, D.: Judgment under uncertainty – heuristics and biases. Science. **185**(4157), 1124–1131 (1974)
80. Kahneman, D.: Thinking, Fast and Slow. Farrar, Straus and Grioux, New York (2011)
81. Simon, H.: Models of Bounded Rationality, vol. 2. MIT Press, Cambridge, MA (1982)
82. Hutchinson, J.M., Gigerenzer, G.: Simple heuristics and rules of thumb: where psychologists and behavioural biologists might meet. Behav. Process. **69**(2), 97–124 (2005)
83. Brunswik, E.: "Ratiomorphic" models of perception and thinking. Acta Psychol. **11**, 108–109 (1955)
84. Lorenz, K.: Behind the Mirror. A Search for a Natural History of Human Knowledge. Methuen & Co Ltd, London (1977)
85. Riedl, R., Huber, L., Ackermann, G.: Rational versus ratiomorphic strategies in human cognition. Evol. Cogn. **1**(1), 71–88 (1991)
86. Riedl, R., Ackermann, G., Huber, L.: A ratiomorphic problem solving strategy. Evol. Cogn. **2**(1), 23–61 (1992)
87. Skinner, B.F.: Operant behavior. Am. Psychol. **18**(8), 503–515 (1963)
88. Dickinson, A.: Contemporary Animal Learning Theory. Cambridge University Press, Cambridge (1980)
89. Piaget, J.: The Origins of Intelligence in Children. International University Press, New York (1952)
90. Nickerson, R.S.: Aspects of Rationality. Reflections on What it Means to Be Rational and Whether We Are. Psychology Press, New York (2007)
91. McGonigle, B.O., Chalmers, M.: Are monkeys logical? Nature. **267**(5613), 694 (1977)
92. McGonigle, B.O., Chalmers, M.: Monkeys are rational! Q. J. Exp. Psychol. **45B**, 198–228 (1992)
93. Zentall, T.R.: The case for a cognitive approach to animal learning and behavior. Behav. Process. **54**(1–3), 65–78 (2001)
94. Chalmers, M., McGonigle, B.: Are children any more logical than monkeys on the five-term series problem? J. Exp. Child Psychol. **37**(2), 355–377 (1984)
95. Lazareva, O.F.: Transitive inference in non-human animals. In: Wasserman, E., Zentall, T. (eds.) Handbook of Comparative Cognition, pp. 718–735. Oxford University Press, New York (2012)
96. Allen, C.: Transitive inference in animals: reasoning or conditioned associations. In: Rational Animals, pp. 175–185. Oxford University Press, Oxford (2006)
97. Seyfarth, R.M., Cheney, D.L.: The structure of social knowledge in monkeys. In: de Waal, F.B.M., Tyack, P.L. (eds.) Animal Social Complexity. Intelligence, Culture, and Individualized Societies, pp. 207–229. Harvard University Press, Cambridge, MA (2003)
98. Paz-y-Mino, G., Bond, A.B., Kamil, A.C., Balda, R.P.: Pinyon jays use transitive inference to predict social dominance. Nature. **430**(7001), 778–781 (2004)
99. Thorndike, E.L.: Animal intelligence: an experimental study of the associative processes in animals. Psychol. Rev. Monogr. Suppl. **2**, i–109 (1898)
100. Köhler, W.: The Mentality of Apes. Harcourt, Brace & World, New York (1925)
101. Thorpe, W.H.: Learning and Instinct in Animals. Methuen, London (1956)

102. Epstein, R., Lanza, R.P., Skinner, B.F.: "Self-awareness" in the pigeon. Science. **212**(4495), 695–696 (1981)
103. Sloman, S.A.: The empirical case for two systems of reasoning. Psychol. Bull. **119**(1), 3–22 (1996)
104. Neisser, U.: The multiplicity of thought. Br. J. Psychol. **54**, 1–14 (1963)
105. Premack, D.: Cause/induced motion; intention/spontaneous motion. In: Changeux, J.P., Chavaillon, J. (eds.) Origins of the Human Brain, pp. 286–308. Oxford University Press, Oxford (1995)
106. Premack, D., Premack, A.J.: Levels of causal understanding in chimpanzees and children. Cognition. **50**, 347–362 (1994)
107. Call, J., Carpenter, M.: Do apes and children know what they have seen? Anim. Cogn. **4**, 207–220 (2001)
108. Call, J.: Inferences about the location of food in the great apes (Pan paniscus, Pan troglodytes, Gorilla gorilla, and Pongo pygmaeus). J. Comp. Psychol. **118**(2), 232–241 (2004)
109. Kaminski, J., Call, J., Fischer, J.: Word learning in a domestic dog: evidence for "fast mapping". Science. **304**(5677), 1682–1683 (2004)
110. Carey, S., Bartlett, E.: Acquiring a single new word. Papers Rep. Child Lang. Dev. **15**, 17–29 (1978)
111. Schloegl, C., Bugnyar, T., Aust, U.: Exclusion performances in non-human animals: from pigeons to chimpanzees and back again. In: Watanabe, S., et al. (eds.) Rational Animals, Irrational Humans, pp. 217–234. Keio University Press, Tokyo (2009)
112. Pepperberg, I.M.: The Alex Studies: Cognitive and Communicative Abilities of Grey Parrots. Harvard University Press, Cambridge, MA (1999)
113. Schloegl, C., Dierks, A., Gajdon, G.K., Huber, L., Kotrschal, K., Bugnyar, T.: What you see is what you get? Exclusion performances in ravens and keas. PLoS One. **4**(8), e6368 (2009)
114. Diamond, J., Bond, A.B.: Kea, Bird of Paradox: The Evolution and Behaviour of a New Zealand Parrot. University of California Press, Berkley, Los Angeles, London (1999)
115. Heinrich, B.: Mind of the Raven. Harper, New York (1999)
116. Aust, U., Range, F., Steurer, M., Huber, L.: Inferential reasoning by exclusion in pigeons, dogs, and humans. Anim. Cogn. **11**, 587–597 (2008)
117. O'Hara, M., Auersperg, A.M.I., Bugnyar, T., Huber, L.: Inference by exclusion in Goffin cockatoos (Cacatua goffini). PLoS One. **10**(8), e0134894 (2015)
118. O'Hara, M., Schwing, R., Federspiel, I., Gajdon, G.K., Huber, L.: Reasoning by exclusion in the kea (Nestor notabilis). Anim. Cogn. **19**(5), 965–975 (2016)
119. Goschke, T.: Volition und kognitive Kontrolle. In: Müsseler, J., Prinz, W. (eds.) Allgemeine Psychologie, pp. 232–293. Spektrum Akademischer Verlag, Heidelberg (2007)
120. Posner, M.I., Snyder, C.R.R.: Attention and cognitive control. In: Solso, R. (ed.) Information Processing and Cognition: The Loyola Symposium, pp. 55–85. Erlbaum, Hillsdale (1975)
121. Norman, D., Shallice, T.: Attention to action: willed and automatic control of behavior. In: Davidson, R., Schwartz, R., Shapiro, D. (eds.) Consciousness and Self-Regulation Advances in Research and Theory, pp. 1–18. Plenum Press, New York (1986)
122. Lorenz, K.: Innate bases of learning. In: Pribram, K.H. (ed.) On the Biology of Learning, pp. 13–93. Harcourt, Brace & World, Inc, New York, Chicago, San Francisco, Atlanta (1969)
123. Lorenz, K.Z.: The Foundations of Ethology. Springer, New York, Wien (1981)
124. Ewert, J.-P.: Neuroethology of releasing mechanisms: prey-catching in toads. Behav. Brain Sci. **10**(3), 337–405 (1987)
125. Giurfa, M.: Cognitive neuroethology: dissecting non-elemental learning in a honeybee brain. Curr. Opin. Neurobiol. **13**(6), 726–735 (2003)
126. Dickinson, A.: Associative learning and animal cognition. Philos. Trans. R. Soc. Lond. Ser. B Biol. Sci. **367**(1603), 2733–2742 (2012)
127. Ach, N.: Über den Willensakt und die Willenshandlung. In: Die Motivation menschlichen Handelns, pp. 147–154. Kiepenheuer & Witsch, Köln (1969)
128. Gazzaniga, M.S.: Who's in Charge? Free Will and the Science of the Brain. Ecco, New York (2011)

129. Haggard, P.: Human volition: towards a neuroscience of will. Nat. Rev. Neurosci. **9**(12), 934–946 (2008)
130. Haggard, P., Clark, S., Kalogeras, J.: Voluntary action and conscious awareness. Nat. Neurosci. **5**(4), 382–385 (2002)
131. Kuhl, J.: Wille und Freiheitserleben. Formen der Selbststeuerung. In: Kuhl, J., Heckhausen, H. (eds.) Enzyklopädie der Psychologie Serie IV, Band 4: Motivation, Volition und Handlung, pp. 665–765. Hogrefe, Göttingen (1996)
132. Goschke, T.: Voluntary action and cognitive control from a cognitive neuroscience perspective. In: Maasen, S., Prinz, W., Roth, G. (eds.) Voluntary Action: Brains, Minds, and Sociality, pp. 49–85. Oxford University Press, New York (2003)
133. Prinz, W.: Die Reaktion als Willenshandlung (Responses considered as voluntary actions). Psychol. Rundsch. **49**(1), 10–20 (1998)
134. Elsner, B., Hommel, B.: Effect anticipation and action control. J. Exp. Psychol. Hum. Percept. Perform. **27**(1), 229–240 (2001)
135. Gold, J.I., Shadlen, M.N.: The neural basis of decision making. Annu. Rev. Neurosci. **30**, 535–574 (2007)
136. Dörner, D.: Denken und Wollen: Ein systemtheoretischer Ansatz. In: Heckhausen, H., Gollwitzer, P.M., Weinert, F.E. (eds.) Jenseits des Rubikon. Der Wille in den Humanwissenschaften, pp. 238–250. Springer, Berlin (1987)
137. Miller, E.K.: The prefrontal cortex and cognitive control. Nat. Rev. Neurosci. **1**, 59–65 (2000)
138. Heekeren, H.R., Marrett, S., Bandettini, P.A., Ungerleider, L.G.: A general mechanism for perceptual decision-making in the human brain. Nature. **431**(7010), 859–862 (2004)
139. Heekeren, H.R., Marrett, S., Ungerleider, L.G.: The neural systems that mediate human perceptual decision making. Nat. Rev. Neurosci. **9**(6), 467–479 (2008)
140. Frith, C.D.: Making up the Mind: How the Brain Creates our Mental World, xii, 234 p., [7] leaves of plates. Blackwell, Oxford (2007)
141. Nachev, P., Rees, G., Parton, A., Kennard, C., Husain, M.: Volition and conflict in human medial frontal cortex. Curr. Biol. **15**(2), 122–128 (2005)
142. Nachev, P., Wydell, H., O'Neill, K., Husain, M., Kennard, C.: The role of the pre-supplementary motor area in the control of action. NeuroImage. **36**(Suppl 2), T155–T163 (2007)
143. Mischel, W., Shoda, Y., Rodriguez, M.L.: Delay of gratification in children. Science. **244**, 933–938 (1989)
144. Mischel, W.: The Marshmallow Test: Mastering Self-Control, 1st edn. Little, Brown and Company, New York (2014)
145. Duckworth, A.L., Tsukayama, E., Kirby, T.A.: Is it really self-control? Examining the predictive power of the delay of gratification task. Personal. Soc. Psychol. Bull. **39**, 843–855 (2013)
146. Watts, T.W., Duncan, G.J., Quan, H.: Revisiting the marshmallow test: a conceptual replication investigating links between early delay of gratification and later outcomes. Psychol. Sci. **29**(7), 1159 (2018)
147. Stevens, J.R., Stephens, D.W.: The adaptive nature of impulsivity. In: Madden, G., Bickel, W.K. (eds.) Impulsivity: The Behavioral and Neurological Science of Discounting, pp. 361–387. APA Press, Washington, DC (2009)
148. Dufour, V., Pele, M., Sterck, E.H.M., Thierry, B.: Chimpanzee (Pan troglodytes) anticipation of food return: coping with waiting time in an exchange task. J. Comp. Psychol. **121**(2), 145–155 (2007)
149. Rosati, A.G., Stevens, J.R., Hare, B., Hauser, M.D.: The evolutionary origins of human patience: temporal preferences in chimpanzees, bonobos, and human adults. Curr. Biol. **17**(19), 1663–1668 (2007)
150. Beran, M.J., Evans, T.A.: Delay of gratification by chimpanzees (Pan troglodytes) in working and waiting situations. Behav. Process. **80**(2), 177–181 (2009)
151. Frederick, S., Loewenstein, G., O'Donoghue, T.: Time discounting and time preference: a critical review. J. Econ. Lit. **40**, 351–401 (2002)

152. Kacelnik, A.: The evolution of patience. In: Loewenstein, G., Read, D., Baumeister, R.F. (eds.) Time and Decision: Economic and Psychological Perspectives on Intertemporal Choice, pp. 115–138. Russell Sage, New York (2003)
153. Hayden, B.Y., Parikh, P.C., Deaner, R.O., Platt, M.L.: Economic principles motivating social attention in humans. Proc. Biol. Sci. **274**(1619), 1751–1756 (2007)
154. Hayden, B.Y.: Time discounting and time preference in animals: a critical review. Psychon. Bull. Rev. **23**(1), 39–53 (2016)
155. Bramlett, J.L., Perdue, B.M., Evans, T.A., Beran, M.J.: Capuchin monkeys (Cebus apella) let lesser rewards pass them by to get better rewards. Anim. Cogn. **15**(5), 963–969 (2012)
156. Toner, I.J., Smith, R.A.: Age and overt verbalization in delay-maintenance behavior in children. J. Exp. Child Psychol. **24**(1), 123–128 (1977)
157. Evans, T.A., Westergaard, G.C.: Self-control and tool use in tufted capuchin monkeys (Cebus apella). J. Comp. Psychol. **120**(2), 163–166 (2006)
158. Pelé, M., Dufour, V., Micheletta, J., Thierry, B.: Long-tailed macaques display unexpected waiting abilities in exchange tasks. Anim. Cogn. **13**(2), 263–271 (2010)
159. Osvath, M., Persson, T.: Great apes can defer exchange: a replication with different results suggesting future oriented behavior. Front. Psychol. **4**, 698 (2013)
160. Dufour, V., Wascher, C.A., Braun, A., Miller, R., Bugnyar, T.: Corvids can decide if a future exchange is worth waiting for. Biol. Lett. **8**(2), 201–204 (2012)
161. Auersperg, A.M.I., Laumer, I.B., Bugnyar, T.: Goffin cockatoos wait for qualitative and quantitative gains but prefer "better" to "more". Biol. Lett. **9**(3), 20121092 (2013)
162. Laumer, I.B., Bugnyar, T., Auersperg, A.M.I.: Flexible decision-making relative to reward quality and tool functionality in Goffin cockatoos (Cacatua goffiniana). Sci. Rep. **6**, 28380 (2016)
163. Leonardi, R.J., Vick, S.J., Dufour, V.: Waiting for more: the performance of domestic dogs (Canis familiaris) on exchange tasks. Anim. Cogn. **15**(1), 107–120 (2012)
164. Beran, M.J.: Self-Control in Animals and People. Academic Press, London (2018)
165. Miller, R., Boeckle, M., Jelbert, S.A., Frohnwieser, A., Wascher, C.A.F., Clayton, N.S.: Self-control in crows, parrots and nonhuman primates. Wiley Interdiscip. Rev. Cogn. Sci. **10**, e1504 (2019)
166. Beran, M.J., Hopkins, W.D.: Self-control in chimpanzees relates to general intelligence. Curr. Biol. **28**(4), 574–579 (2018)
167. MacLean, E.L., et al.: The evolution of self-control. Proc. Natl. Acad. Sci. USA. **111**(20), E2140–E2148 (2014)
168. Duque, J.F., Stevens, J.R.: Cylinder Task. Springer, New York (2017)
169. Kabadayi, C., Taylor, L.A., von Bayern, A.M., Osvath, M.: Ravens, New Caledonian crows and jackdaws parallel great apes in motor self-regulation despite smaller brains. R. Soc. Open Sci. **3**(4), 160104 (2016)
170. Loewenstein, G., Read, D., Baumeister, R.F.: Time and Decision: Economic and Psychological Perspectives on Intertemporal Choice. Russell Sage Foundation Press, New York (2003)
171. Fawcett, T.W., McNamara, J.M., Houston, A.I.: When is it adaptive to be patient? A general framework for evaluating delayed rewards. Behav. Process. **89**(2), 128–136 (2012)
172. Diamond, A.: Executive functions. Annu. Rev. Psychol. **64**, 135–168 (2013)
173. Nigg, J.T.: Annual research review: on the relations among self-regulation, self-control, executive functioning, effortful control, cognitive control, impulsivity, risk-taking, and inhibition for developmental psychopathology. J. Child Psychol. Psychiatry. **58**, 361–383 (2017)
174. Völter, C.J., Tinklenberg, B., Call, J., Seed, A.M.: Comparative psychometrics: establishing what differs is central to understanding what evolves. Philos. Trans. R. Soc. Lond. Ser. B Biol. Sci. **373**(1756), 20170283 (2018)
175. Mackintosh, N.J.: Animal Learning and Cognition, xviii, 379 p. Academic Press, San Diego, London (1994)
176. Spearman, C.: The Abilities of Man; their Nature and Measurement, vi, 415, xxxiii p. Macmillan, New York (1927)

177. Duncan, J., Emslie, H., Williams, P., Johnson, R., Freer, C.: Intelligence and the frontal lobe: the organization of goal-directed behavior. Cogn. Psychol. **30**(3), 257–303 (1996)
178. Burkart, J.M., Schubiger, M.N., van Schaik, C.P.: The evolution of general intelligence. Behav. Brain Sci. **40**, e195 (2017)
179. Locurto, C., Benoit, A., Crowley, C., Miele, A.: The structure of individual differences in batteries of rapid acquisition tasks in mice. J. Comp. Psychol. **120**(4), 378–388 (2006)
180. Anderson, B.: Evidence from the rat for a general factor that underlies cognitive performance and that relates to brain size: intelligence? Neurosci. Lett. **153**, 98–112 (1993)
181. Wass, C., Denman-Brice, A., Rios, C., Light, K.R., Kolata, S., Smith, A.M., Matzel, L.D.: Covariation of learning and "reasoning" abilities in mice: evolutionary conservation of the operations of intelligence. J. Exp. Psychol. Anim. Behav. Process. **38**(2), 109–124 (2012)
182. Herrmann, E., Call, J., Hernandez-Lloreda, M.V., Hare, B., Tomasello, M.: Humans have evolved specialized skills of social cognition: the cultural intelligence hypothesis. Science. **317**(5843), 1360–1366 (2007)
183. Herrmann, E., Call, J.: Are there geniuses among the apes? Philos. Trans. R. Soc. Lond. Ser. B Biol. Sci. **367**(1603), 2753–2761 (2012)
184. Gergely, G., Bekkering, H., Kiraly, I.: Rational imitation in preverbal infants. Nature. **415**(6873), 755 (2002)
185. Range, F., Viranyi, Z., Huber, L.: Selective imitation in domestic dogs. Curr. Biol. **17**(10), 868–872 (2007)
186. Buttelmann, D., Carpenter, M., Call, J., Tomasello, M.: Enculturated chimpanzees imitate rationally. Dev. Sci. **10**(4), F31–F38 (2007)
187. Kaminski, J., Nitzschner, M., Wobber, V., Tennie, C., Bräuer, J., Call, J., Tomasello, M.: Do dogs distinguish rational from irrational acts? Anim. Behav. **81**(1), 195–203 (2011)
188. Huber, L., Range, F., Viranyi, Z.: Dogs imitate selectively, not necessarily rationally: reply to Kaminski et al. (2011). Anim. Behav. **83**(6), e1–e3 (2012)
189. Condillac, E.B.D., Kayling, V.Ü.: Traité des animaux. Königshausen & Neumann, Würzburg (1755 (2019))
190. Oakley, K.P.: Man the Tool-Maker. British Museum, London (1949)
191. Gibson, K.R., Ingold, T.: Tools, Language and Cognition in Human Evolution, xii, 483 p. Cambridge University Press, Cambridge (1993)
192. Gamble, C.S., Gowlett, J.A.J., Dunbar, R.: Thinking Big: How the Evolution of Social Life Shaped the Human Mind. Thames & Hudson, London (2014)
193. Leakey, L.S.B., Tobias, P.V., Napier, J.R.: A new species of the genus Homo from Olduvai Gorge. Nature. **202**, 7–9 (1964)
194. Semaw, S., Renne, P., Harris, J.W.K., Feibel, C.S., Bernor, R.L., Fesseha, N., Mowbray, K.: 2.5-million-year-old stone tools from Gona, Ethiopia. Nature. **385**(6614), 333–336 (1997)
195. Harmand, S., et al.: 3.3-million-year-old stone tools from Lomekwi 3, West Turkana, Kenya. Nature. **521**(7552), 310–315 (2015)
196. McPherron, S.P., Alemseged, Z., Marean, C.W., Wynn, J.G., Reed, D., Geraads, D., Bobe, R., Bearat, H.A.: Evidence for stone-tool-assisted consumption of animal tissues before 3.39 million years ago at Dikika, Ethiopia. Nature. **466**(7308), 857–860 (2010)
197. Lefebvre, L., Nicolakakis, N., Boire, D.: Tools and brains in birds. Behaviour. **139**, 939–973 (2002)
198. Reader, S.M., Laland, K.N.: Social intelligence, innovation, and enhanced brain size in primates. Proc. Natl. Acad. Sci. USA. **99**(7), 4436–4441 (2002)
199. Shumaker, R.W., Walkup, K.R., Beck, B.B.: Animal Tool Behavior: The Use and Manufacture of Tools by Animals, Rev. and updated edn, 282 S. Johns Hopkins University Press, Baltimore (2011)
200. Bentley-Condit, V., Smith, E.O.: Animal tool use: current definitions and an updated comprehensive catalog. Behaviour. **147**(2), 185–221 (2010)., A1–A32
201. Goodall, J.: Tool-using and aimed throwing in a community of free-living chimpanzees. Nature. **201**, 1264–1266 (1964)

202. Root-Bernstein, M., Narayan, T., Cornier, L., Bourgeois, A.: Context-specific tool use by Sus cebifrons. Mamm. Biol. **98**(1), 102–110 (2019)
203. Fayet, A.L., Hansen, E.S., Biro, D.: Evidence of tool use in a seabird. Proc. Natl. Acad. Sci. USA. **117**, 201918060 (2019)
204. Meyerriecks, A.J.: Tool-using by a double-crested cormorant. Wilson Bull. **84**(4), 482–483 (1972)
205. Farrar, B.: Evidence of tool use in a seabird? PsyArXiv, 463hk (2020) https://doi.org/10.31234/osf.io/463hk
206. Auersperg, A., Schwing, R., Mioduszewska, B., O'Hara, M., Huber, L.: Do puffins use tools? Proc. Natl. Acad. Sci. USA. **117**, 11859 (2020)
207. Brooks, J., Yamamoto, S.: Dog stick chewing: an overlooked instance of tool use? Front. Psychol. **11**(3753), 577100 (2021)
208. Van Lawick-Goodall, J.: Tool-using in primates and other vertebrates. Adv. Study Behav. **3**, 195–249 (1970)
209. Meulman, E.J., Sanz, C.M., Visalberghi, E., van Schaik, C.P.: The role of terrestriality in promoting primate technology. Evol. Anthropol. **21**(2), 58–68 (2012)
210. McCullough, M.: Abstracting Craft: The Practiced Digital Hand. MIT Press, Cambridge, MA (1998)
211. Baber, C.: Cognition and Tool Use: Forms of Engagement in Human and Animal Use of Tools, xi, 172 p. Taylor & Francis, London (2003)
212. Gibson, J.J.: The Senses Considered as Perceptual Systems. Houghton Mifflin, Boston (1966)
213. St Amant, R., Horton, T.E.: Revisiting the definition of animal tool use. Anim. Behav. **75**(4), 1199–1208 (2008)
214. Breuer, T., Ndoundou-Hockemba, M., Fishlock, V.: First observation of tool use in wild gorillas. PLoS Biol. **3**(11), e380 (2005)
215. Krützen, M., Mann, J., Heithaus, M.R., Connor, R.C., Bejder, L., Sherwin, W.B.: Cultural transmission of tool use in bottlenose dolphins. Proc. Natl. Acad. Sci. USA. **102**(25), 8939–8943 (2005)
216. Beck, B.B.: Animal Tool Behavior the Use and Manufacture of Tools by Animals Garland Series in Ethology, 307 S. Garland STPM Press, New York (1980)
217. Sanz, C., Call, J., Boesch, C.: Tool Use in Animals Cognition and Ecology. Cambridge University Press, Cambridge (2013) Online-Ressource
218. Van Schaik, C.P., Ancrenaz, M., Borgen, G., Galdikas, B., Knott, C.D., Singleton, I., Suzuki, A., Utami, S.S., Merrill, M.: Orangutan cultures and the evolution of material culture. Science. **299**(5603), 102–105 (2003)
219. Fragaszy, D.M., Fedigan, L.M., Visalberghi, E.: The Complete Capuchin: the Biology of the Genus Cebus, xi, 339 p. Cambridge University Press, Cambridge (2004)
220. Gumert, M.D., Fuentes, A., Jones-Engel, L.: Monkeys on the Edge: Ecology and Management of Long-Tailed Macaques and Their Interface with Humans. Cambridge University Press, Cambridge, MA (2020)
221. Haslam, M., Fujii, J., Espinosa, S., Mayer, K., Ralls, K., Tinker, M.T., Uomini, N.: Wild sea otter mussel pounding leaves archaeological traces. Sci. Rep. **9**(1), 4417 (2019)
222. Poole, J.: Coming of Age With Elephants: A Memoir. Hyperion, New York (1996)
223. van Lawick-Goodall, J., van Lawick-Goodall, H.: Use of tools by the Egyptian vulture, Neophron percnopterus. Nature. **212**(5069), 1468–1469 (1966)
224. Higuchi, H.: Bait-fishing by the Green-backed Heron Ardeola striata in Japan. Ibis. **128**(2), 285–290 (1986)
225. Levey, D.J., Duncan, R.S., Levins, C.F.: Animal behaviour: use of dung as a tool by burrowing owls. Nature. **431**(7004), 39 (2004)
226. Shettleworth, S.: Cognition, Evolution and Behavior, 2nd edn. Oxford University Press, Oxford (2010)
227. Sugiyama, Y., Koman, J.: Tool-using and-making behavior in wild chimpanzees at Bossou, Guinea. Primates. **20**(4), 513–524 (1979)

228. Boesch, C., Boesch, H.: Optimization of nut cracking with natural hammers by wild chimpanzees. Behaviour. **83**(3/4), 265–286 (1983)
229. Hannah, A.C., McGrew, W.: Chimpanzees using stones to crack open oil palm nuts in Liberia. Primates. **28**(1), 31–46 (1987)
230. Matsuzawa, T.: Nesting cups and metatools in chimpanzees. Behav. Brain Sci. **14**, 570–571 (1991)
231. Matsuzawa, T.: Field experiments on use of stone tools in the wild. In: Wrangham, R.W., et al. (eds.) Chimpanzee cultures, pp. 351–370. Harvard University Press, Cambridge, MA (1994)
232. Matsuzawa, T.: Chimpanzee intelligence in nature and in captivity: isomorphism of symbol use and tool use. In: McGrew, W.C., Marchant, L., Nisida, T. (eds.) Great Ape Societies, pp. 196–212. Cambridge University Press, Cambridge (1996)
233. Carvalho, S., Cunha, E., Sousa, C., Matsuzawa, T.: Chaines operatoires and resource-exploitation strategies in chimpanzee (Pan troglodytes) nut cracking. J. Hum. Evol. **55**(1), 148–163 (2008)
234. Clayton, N.S.: Animal cognition: crows spontaneously solve a metatool task. Curr. Biol. **17**(20), R894–R895 (2007)
235. Bird, C.D., Emery, N.J.: Insightful problem solving and creative tool modification by captive nontool-using rooks. Proc. Natl. Acad. Sci. USA. **106**(25), 10370–10375 (2009)
236. Wimpenny, J.H., Weir, A.A., Clayton, L., Rutz, C., Kacelnik, A.: Cognitive processes associated with sequential tool use in New Caledonian crows. PLoS One. **4**(8), e6471 (2009)
237. Taylor, A.H., Elliffe, D., Hunt, G.R., Gray, R.D.: Complex cognition and behavioural innovation in New Caledonian crows. Proc. Biol. Sci. **277**(1694), 2637–2643 (2010)
238. Taylor, A.H., Hunt, G.R., Holzhaider, J.C., Gray, R.D.: Spontaneous metatool use by New Caledonian crows. Curr. Biol. **17**(17), 1504–1507 (2007)
239. Mulcahy, N.J., Call, J., Dunbar, R.I.: Gorillas (Gorilla gorilla) and orangutans (Pongo pygmaeus) encode relevant problem features in a tool-using task. J. Comp. Psychol. **119**(1), 23–32 (2005)
240. Merrick, H., De Heinzelin, J., Haesaerts, P., Howell, F.: Archaeological occurrences of early Pleistocene age from the Shungura Formation, lower Omo valley, Ethiopia. Nature. **242**(5400), 572–575 (1973)
241. Sumita, K., Kitahara-Frisch, J., Norikoshi, K.: The acquisition of stone-tool use in captive chimpanzees. Primates. **26**(2), 168–181 (1985)
242. Sugiyama, Y.: Social tradition and the use of tool-composites by wild chimpanzees. Evol. Anthropol. **6**(1), 23–27 (1997)
243. Wright, R.V.: Imitative learning of a flaked stone technology-the case of an orangutan. Aust. J. Anthropol. **8**(4), 296 (1972)
244. Mannu, M., Ottoni, E.B.: The enhanced tool-kit of two groups of wild bearded capuchin monkeys in the Caatinga: tool making, associative use, and secondary tools. Am. J. Primatol. **71**(3), 242–251 (2009)
245. Engels, F.: Anteil der Arbeit an der Menschwerdung des Affen, vol. 20. Dietz, Berlin (1962)
246. McGrew, W.C.: Chimpanzee technology. Science. **328**(5978), 579–580 (2010)
247. Sousa, C., Biro, D., Matsuzawa, T.: Leaf-tool use for drinking water by wild chimpanzees (Pan troglodytes): acquisition patterns and handedness. Anim. Cogn. **12**(1), 115–125 (2009)
248. Stewart, F.A., Pruetz, J.D., Hansell, M.H.: Do chimpanzees build comfortable nests? Am. J. Primatol. **69**(8), 930–939 (2007)
249. Bermejo, M., Illera, G.: Tool-set for termite-fishing and honey extraction by wild chimpanzees in the Lossi Forest, Congo. Primates. **40**(4), 619 (1999)
250. Boesch, C., Heada, J., Robbins, M.M.: Complex tool sets for honey extraction among chimpanzees in Loango National Park, Gabon. J. Hum. Evol. **56**(6), 560–569 (2009)
251. Sanz, C.M., Morgan, D.B.: Flexible and persistent tool-using strategies in honey-gathering by wild chimpanzees. Int. J. Primatol. **30**(3), 411–427 (2009)
252. Wimpenny, J.H., Weir, A.A., Kacelnik, A.: New Caledonian crows use tools for non-foraging activities. Anim. Cogn. **14**(3), 459–464 (2011)

253. Fragaszy, D.M., Mangalam, M.: Tooling. Adv. Study Behav. **50**, 177–241 (2018)
254. Varela, F.J., Thompson, E., Rosch, E.: The Embodied Mind, p. 308. MIT Press, Cambridge, MA (1991)
255. Thelen, E., Smith, L.: A Dynamic Systems Approach to the Development of Cognition and Action. MIT Press, Cambridge, MA (1994)
256. Calvin, W.H.: A stone's throw and its launch window: timing precision and its implications for language and hominid brains. J. Theor. Biol. **104**(1), 121–135 (1983)
257. Westergaard, G., Liv, C., Haynie, M., Suomi, S.: A comparative study of aimed throwing by monkeys and humans. Neuropsychologia. **38**(11), 1511–1517 (2000)
258. Lombardo, M.P., Deaner, R.O.: Born to throw: the ecological causes that shaped the evolution of throwing in humans. Q. Rev. Biol. **93**(2), 1–16 (2018)
259. Colbourne, J.A.D., Auersperg, A.M.I., Lambert, M.L., Huber, L., Völter, C.J.: Extending the reach of tooling theory: a neurocognitive and phylogenetic perspective. Top. Cogn. Sci. **13**(4), 548–572 (2021)
260. Maravita, A., Iriki, A.: Tools for the body (schema). Trends Cogn. Sci. **8**(2), 79–86 (2004)
261. Arbib, M.A., Bonaiuto, J.B., Jacobs, S., Frey, S.H.: Tool use and the distalization of the end-effector. Psychol. Res. **73**(4), 441–462 (2009)
262. Johnson-Frey, S.H.: The neural bases of complex tool use in humans. Trends Cogn. Sci. **8**(2), 71–78 (2004)
263. Frey, S.H.: Tool use, communicative gesture and cerebral asymmetries in the modern human brain. Philos. Trans. R. Soc. Lond. Ser. B Biol. Sci. **363**(1499), 1951–1957 (2008)
264. Johnson-Frey, S.H.: What's so special about human tool use? Neuron. **39**(2), 201–204 (2003)
265. Goodall, J.: Essays on science and society: learning from the chimpanzees: a message humans can understand. Science. **282**(5397), 2184–2185 (1998)
266. Matsuzawa, T.: Primate Origins of Human Cognition and Behavior. Springer Science & Business Media (2009)
267. Inoue-Nakamura, N., Matsuzawa, T.: Development of stone tool use by wild chimpanzees (Pan troglodytes). J. Comp. Psychol. **111**(2), 159–173 (1997)
268. Mercader, J., Panger, M., Boesch, C.: Excavation of a chimpanzee stone tool site in the African rainforest. Science. **296**(5572), 1452–1455 (2002)
269. Mercader, J., Barton, H., Gillespie, J., Harris, J., Kuhn, S., Tyler, R., Boesch, C.: 4,300-year-old chimpanzee sites and the origins of percussive stone technology. Proc. Natl. Acad. Sci. USA. **104**(9), 3043–3048 (2007)
270. Estienne, V., Robira, B., Mundry, R., Deschner, T., Boesch, C.: Acquisition of a complex extractive technique by the immature chimpanzees of Loango National Park, Gabon. Anim. Behav. **147**, 61–76 (2019)
271. Hernandez-Aguilar, R.A., Moore, J., Pickering, T.R.: Savanna chimpanzees use tools to harvest the underground storage organs of plants. Proc. Natl. Acad. Sci. USA. **104**(49), 19210–19213 (2007)
272. Motes-Rodrigo, A., Majlesi, P., Pickering, T.R., Laska, M., Axelsen, H., Minchin, T.C., Tennie, C., Hernandez-Aguilar, R.A.: Chimpanzee extractive foraging with excavating tools: experimental modeling of the origins of human technology. PLoS One. **14**(5), e0215644 (2019)
273. Koops, K., McGrew, W.C., Matsuzawa, T.: Do chimpanzees (Pan troglodytes) use cleavers and anvils to fracture Treculia africana fruits? Preliminary data on a new form of percussive technology. Primates. **51**(2), 175–178 (2010)
274. Pruetz, J.D., Bertolani, P.: Savanna chimpanzees, Pan troglodytes verus, hunt with tools. Curr. Biol. **17**(5), 412–417 (2007)
275. Roffman, I., Savage-Rumbaugh, S., Rubert-Pugh, E., Stadler, A., Ronen, A., Nevo, E.: Preparation and use of varied natural tools for extractive foraging by bonobos (Pan paniscus). Am. J. Phys. Anthropol. **158**(1), 78–91 (2015)
276. Russon, A.E., Galdikas, B.M.F.: Imitation in free-ranging rehabilitant orangutans (Pongo pygmaeus). J. Comp. Psychol. **107**(2), 147–161 (1993)

277. Russon, A.E., Galdikas, B.M.F.: Imitation and tool use in rehabilitant orangutans. In: Nadler, R.D.E.A. (ed.) The Neglected Ape, pp. 191–197. Plenum Press, New York (1995)
278. Galdikas, B.M.: Orangutan tool use. Science. **243**(4888), 152 (1989)
279. Russon, A.E., van Schaik, C.P., Kuncoro, P., Ferisa, A., Handayani, D.P., van Noordwijk, M.A.: Innovation and intelligence in orangutans. In: Wich, S.A., et al. (eds.) Orangutans: Geographic Variation in Behavioral Ecology and Conservation, pp. 279–298. Oxford University Press, Oxford (2009)
280. Meulman, E.J., van Schaik, C.P.: Orangutan tool use and the evolution of technology. In: Tool Use in Animals: Cognition and Ecology, pp. 176–202. Cambridge University Press, Cambridge (2013)
281. Koops, K., Sanz, C.: Progress and prospects in primate tool use and cognition. In: Schwartz, B.L., Beran, M.J. (eds.) Primate Cognitive Studies, pp. 238–259. Cambridge University Press, Cambridge (2022)
282. Malaivijitnond, S., Lekprayoon, C., Tandavanittj, N., Panha, S., Cheewatham, C., Hamada, Y.: Stone-tool usage by Thai long-tailed macaques (Macaca fascicularis). Am. J. Primatol. **69**(2), 227–233 (2007)
283. Luncz, L.V., Svensson, M.S., Haslam, M., Malaivijitnond, S., Proffitt, T., Gumert, M.: Technological response of wild macaques (Macaca fascicularis) to anthropogenic change. Int. J. Primatol. **38**(5), 872–880 (2017)
284. Haslam, M., et al.: Primate archaeology evolves. Nat. Ecol. Evol. **1**(10), 1431–1437 (2017)
285. Visalberghi, E., Addessi, E., Truppa, V., Spagnoletti, N., Ottoni, E., Izar, P., Fragaszy, D.: Selection of effective stone tools by wild bearded capuchin monkeys. Curr. Biol. **19**(3), 213–217 (2009)
286. Visalberghi, E., Spagnoletti, N., Ramos da Silva, E.D., Andrade, F.R., Ottoni, E., Izar, P., Fragaszy, D.: Distribution of potential suitable hammers and transport of hammer tools and nuts by wild capuchin monkeys. Primates. **50**(2), 95–104 (2009)
287. Schrauf, C., Huber, L., Visalberghi, E.: Do capuchin monkeys use weight to select hammer tools? Anim. Cogn. **11**(3), 413–422 (2008)
288. Ottoni, E.B., Izar, P.: Capuchin monkey tool use: overview and implications. Evol. Anthropol. **17**(4), 171–178 (2008)
289. Visalberghi, E., Fragaszy, D.M., Izar, P., Ottoni, E.B.: Terrestriality and tool use. Science. **308**(5724), 951–952 (2005)
290. Haslam, M., Luncz, L.V., Staff, R.A., Bradshaw, F., Ottoni, E.B., Falotico, T.: Pre-Columbian monkey tools. Curr. Biol. **26**(13), R521–R522 (2016)
291. Cnotka, J., Güntürkün, O., Rehkämper, G., Gray, R.D., Hunt, G.R.: Extraordinary large brains in tool-using New Caledonian crows (Corvus moneduloides). Neurosci. Lett. **433**(3), 241–245 (2008)
292. Mehlhorn, J., Hunt, G.R., Gray, R.D., Rehkämper, G., Güntürkün, O.: Tool-making New Caledonian crows have large associative brain areas. Brain Behav. Evol. **75**(1), 63–70 (2010)
293. Rutz, C., St Clair, J.J.: The evolutionary origins and ecological context of tool use in New Caledonian crows. Behav. Process. **89**(2), 153–165 (2012)
294. Rutz, C., Bluff, L.A., Reed, N., Troscianko, J., Newton, J., Inger, R., Kacelnik, A., Bearhop, S.: The ecological significance of tool use in New Caledonian crows. Science. **329**(5998), 1523–1526 (2010)
295. Hunt, G.R.: Manufacture and use of hook-tools by New Caledonian crows. Nature. **379**, 249–251 (1996)
296. Bluff, L.A., Troscianko, J., Weir, A.A., Kacelnik, A., Rutz, C.: Tool use by wild New Caledonian crows Corvus moneduloides at natural foraging sites. Proc. Biol. Sci. **277**(1686), 1377–1385 (2010)
297. Chappell, J., Kacelnik, A.: Tool selectivity in a non-primate, the New Caledonian crow (Corvus moneduloides). Anim. Cogn. **5**(2), 71–78 (2002)
298. Weir, A.S., Chappel, J., Kacelnik, A.: Shaping of hooks in New Caledonian crows. Science. **297**, 981 (2002)

299. Troscianko, J., Von Bayern, A.M., Chappell, J., Rutz, C., Martin, G.R.: Extreme binocular vision and a straight bill facilitate tool use in New Caledonian crows. Nat. Commun. **3**(1), 1–7 (2012)
300. Martinho III, A., Burns, Z.T., von Bayern, A.M., Kacelnik, A.: Monocular tool control, eye dominance, and laterality in New Caledonian crows. Curr. Biol. **24**(24), 2930–2934 (2014)
301. Rutledge, R., Hunt, G.R.: Lateralized tool use in wild New Caledonian crows. Anim. Behav. **67**(2), 327–332 (2004)
302. Weir, A.A., Kenward, B., Chappell, J., Kacelnik, A.: Lateralization of tool use in New Caledonian crows (Corvus moneduloides). Proc. Biol. Sci. **271**(Suppl 5), S344–S346 (2004)
303. Kenward, B., Schloegl, C., Rutz, C., Weir, A.A., Bugnyar, T., Kacelnik, A.: On the evolutionary and ontogenetic origins of tool-oriented behaviour in New Caledonian crows (Corvus moneduloides). Biol. J. Linn. Soc. Lond. **102**(4), 870–877 (2011)
304. Hunt, G.R., Gray, R.D.: The crafting of hook tools by wild New Caledonian crows. Proc. R. Soc. B Biol. Sci. **271**, S88–S90 (2004)
305. Hunt, G.R.: Human-like, population-level specialization in the manufacture of pandanus tools by New Caledonian crows Corvus moneduloides. Proc. R. Soc. Lond. B. **267**, 403–413 (2000)
306. Hunt, G.R., Gray, R.D.: Diversification and cumulative evolution in New Caledonian crow tool manufacture. Proc. R. Soc. B Biol. Sci. **270**(1517), 867–874 (2003)
307. Hunt, G.R., Corballis, M.C., Gray, R.D.: Design complexity and strength of laterality are correlated in New Caledonian crows' pandanus tool manufacture. Proc. R. Soc. B Biol. Sci. **273**(1590), 1127–1133 (2006)
308. Hunt, G.R., Gray, R.D.: Parallel tool industries in New Caledonian crows. Biol. Lett. **3**(2), 173–175 (2007)
309. Whiten, A., Goodall, J., McGrew, W.C., Nishida, T., Reynolds, V., Sugiyama, Y., Tutin, C.E.G., Wrangham, R.W., Boesch, C.: Cultures in chimpanzees. Nature. **399**, 682–685 (1999)
310. Bluff, L.A., Kacelnik, A., Rutz, C.: Vocal culture in New Caledonian crows Corvus moneduloides. Biol. J. Linn. Soc. **101**(4), 767–776 (2010)
311. Kenward, B., Weir, A.A., Rutz, C., Kacelnik, A.: Behavioural ecology: tool manufacture by naive juvenile crows. Nature. **433**(7022), 121 (2005)
312. Jelbert, S.A., Hosking, R.J., Taylor, A.H., Gray, R.D.: Mental template matching is a potential cultural transmission mechanism for New Caledonian crow tool manufacturing traditions. Sci. Rep. **8**(1), 1–8 (2018)
313. Gruber, R., Schiest, M., Boeckle, M., Frohnwieser, A., Miller, R., Gray, R.D., Clayton, N.S., Taylor, A.H.: New Caledonian crows use mental representations to solve metatool problems. Curr. Biol. **29**(4), 686–692 (2019)
314. Rutz, C., Klump, B.C., Komarczyk, L., Leighton, R., Kramer, J., Wischnewski, S., Sugasawa, S., Morrissey, M.B., James, R., St Clair, J.J.: Discovery of species-wide tool use in the Hawaiian crow. Nature. **537**(7620), 403 (2016)
315. Curio, E., Kramer, P.: Vom Mangrovefinken (Cactospiza heliobates Snodgrass und Heller) 1: Ergebnisse der Deutschen Galapagos-Expedition 1962/63 I. Z. Tierpsychol. **21**(2), 223–234 (1964)
316. Fessl, B., Loaiza, A.D., Tebbich, S., Young, H.G.: Feeding and nesting requirements of the critically endangered Mangrove Finch Camarhynchus heliobates. J. Ornithol. **152**(2), 453–460 (2011)
317. Hundley, M.H.: Notes on methods of feeding and the use of tools in the Geospizinae. Auk. **80**(3), 372–373 (1963)
318. Tebbich, S., Taborsky, M., Fessl, B., Dvorak, M.: The ecology of tool-use in the woodpecker finch (Cactospiza pallida). Ecol. Lett. **5**(5), 656–664 (2002)
319. Tebbich, S., Taborsky, M., Fessl, B., Blomqvist, D.: Do woodpecker finches acquire tool-use by social learning? Proc. R. Soc. Lond. B. **268**, 1–5 (2001)
320. Tebbich, S., Teschke, I.: Why do woodpecker finches use tools? In: Tool Use in Animals: Cognition and Ecology, pp. 134–157. Cambridge University Press, Cambridge (2013)

321. Tebbich, S., Stankewitz, S., Teschke, I.: The relationship between foraging, learning abilities and neophobia in two species of Darwin's finches. Ethology. **118**(2), 135–146 (2012)
322. Jerison, H.J.: Evolution of the Brain and Intelligence. Academic Press, New York (1973)
323. van Schaik, C.P., Triki, Z., Bshary, R., Heldstab, S.A.: A farewell to the encephalization quotient: a new brain size measure for comparative primate cognition. Brain Behav. Evol. **96**(1), 1–12 (2021)
324. Healy, S.D., Rowe, C.: A critique of comparative studies of brain size. Proc. Biol. Sci. **274**(1609), 453–464 (2007)
325. Deaner, R.O., Isler, K., Burkart, J., Van Schaik, C.: Overall brain size, and not encephalization quotient, best predicts cognitive ability across non-human primates. Brain Behav. Evol. **70**(2), 115–124 (2007)
326. Benson-Amram, S., Weldele, M.L., Holekamp, K.E.: A comparison of innovative problem-solving abilities between wild and captive spotted hyaenas, Crocuta crocuta. Anim. Behav. **85**, 349–536 (2013)
327. Horschler, D.J., Hare, B., Call, J., Kaminski, J., Miklosi, A., MacLean, E.L.: Absolute brain size predicts dog breed differences in executive function. Anim. Cogn. **22**(2), 187–198 (2019)
328. Stout, D.: Human brain evolution: history or science. In: Schwartz, J.H. (ed.) Rethinking Human Evolution, pp. 297–317. MIT Press, Cambridge, MA (2018)
329. Menzel, R., Giurfa, M.: Cognition by a mini brain. Nature. **400**(6746), 718–719 (1999)
330. Menzel, R., Giurfa, M.: Cognitive architecture of a mini-brain: the honeybee. Trends Cogn. Sci. **5**(2), 62–71 (2001)
331. Chittka, L., Niven, J.: Are Bigger Brains Better? Curr. Biol. **19**(21), R995–R1008 (2009)
332. Niven, J.E., Chittka, L.: Evolving understanding of nervous system evolution. Curr. Biol. **26**(20), R937–R941 (2016)
333. Chittka, L.: The Mind of a Bee. Princeton University Press, Princeton (2023)
334. Schuster, S., Wohl, S., Griebsch, M., Klostermeier, I.: Animal cognition: how archer fish learn to down rapidly moving targets. Curr. Biol. **16**(4), 378–383 (2006)
335. Pollirer, M.: Eine Analyse der Gruppendynamik im Beutefangverhalten von Schützenfischen (Toxotes chatareus) im Labor. Dissertation, Universität Wien (2006)
336. Fodor, J.A.: The Modularity of Mind: An Essay on Faculty Psychology. MIT Press, Cambridge, MA (1983)
337. Carey, S., Spelke, E.: Domain-specific knowledge and conceptual change. In: Hirschfeld, L.A., Gelman, S.A. (eds.) Mapping the Mind: Domain Specificity in Cognition and Culture. Cambridge University Press, New York (1994)
338. Geary, D.C., Huffman, K.J.: Brain and cognitive evolution: forms of modularity and functions of mind. Psychol. Bull. **128**(5), 667–698 (2002)
339. Kanazawa, S.: General intelligence as a domain-specific adaptation. Psychol. Rev. **111**(2), 512–523 (2004)
340. Macphail, E.M.: Brain and Intelligence in Vertebrates. Clarendon, Oxford (1982)
341. Macphail, E.M.: The comparative psychology of intelligence. Behav. Brain Sci. **10**(4), 645–695 (1987)
342. Huber, L.: Psychophylogenesis: innovations and limitations in the evolution of cognition. In: Heyes, C., Huber, L. (eds.) The Evolution of Cognition, pp. 23–41. MIT Press, Cambridge, MA (2000)
343. Reader, S.M., Laland, K.N.: Animal Innovation, vol. 10. Oxford University Press, Oxford (2003)
344. Walker, S., Mikheenko, Y., Argyle, L., Robbins, T., Roberts, A.: Selective prefrontal serotonin depletion impairs acquisition of a detour-reaching task. Eur. J. Neurosci. **23**(11), 3119–3123 (2006)
345. Roby-Brami, A., Hermsdörfer, J., Roy, A.C., Jacobs, S.: A neuropsychological perspective on the link between language and praxis in modern humans. Philos. Trans. R. Soc. Lond. Ser. B Biol. Sci. **367**(1585), 144–160 (2012)

346. Tebbich, S., Sterelny, K., Teschke, I.: The tale of the finch: adaptive radiation and behavioural flexibility. Philos. Trans. R. Soc. Lond. Ser. B Biol. Sci. **365**(1543), 1099–1109 (2010)
347. Teschke, I., Cartmill, E.A., Stankewitz, S., Tebbich, S.: Sometimes tool use is not the key: no evidence for cognitive adaptive specializations in tool-using woodpecker finches. Anim. Behav. **82**(5), 945–956 (2011)
348. Visalberghi, E., Limongelli, L.: Lack of comprehension of cause-effect relationships in tool using capuchin monkeys (Cebus apella). J. Comp. Psychol. **108**, 15–22 (1994)
349. Tebbich, S., Bshary, R.: Cognitive abilities related to tool use in the woodpecker finch, Cactospiza pallida. Anim. Behav. **67**, 689–697 (2004)
350. Teschke, I., Tebbich, S.: Physical cognition and tool-use: performance of Darwin's finches in the two-trap tube task. Anim. Cogn. **14**(4), 555–563 (2011)
351. Call, J.: Three ingredients for becoming a creative tool user. In: Sanz, C., Call, J., Boesch, C. (eds.) Tool Use in Animals Cognition and Ecology, pp. 3–20. Cambridge University Press, Cambridge (2013)
352. Mendes, N., Hanus, D., Call, J.: Raising the level: orangutans use water as a tool. Biol. Lett. **3**(5), 453–455 (2007)
353. Reader, S.M., Laland, K.N.: Primate innovation: sex, age and social rank differences. Int. J. Primatol. **22**(5), 787–805 (2001)
354. Rössler, T., Mioduszewska, B., O'Hara, M., Huber, L., Prawiradilaga, D.M., Auersperg, A.M.I.: Using an Innovation Arena to compare wild-caught and laboratory Goffin's cockatoos. Sci. Rep. **10**(1), 1–12 (2020)
355. Manrique, H.M., Völter, C.J., Call, J.: Repeated innovation in great apes. Anim. Behav. **85**(1), 195–202 (2013)
356. Manrique, H.M., Gross, A.N., Call, J.: Great apes select tools on the basis of their rigidity. J. Exp. Psychol. Anim. Behav. Process. **36**(4), 409–422 (2010)
357. Manrique, H.M., Call, J.: Spontaneous use of tools as straws in great apes. Anim. Cogn. **14**(2), 213–226 (2011)
358. Manrique, H.M., Sabbatini, G., Call, J., Visalberghi, E.: Tool choice on the basis of rigidity in capuchin monkeys. Anim. Cogn. **14**(6), 775–786 (2011)
359. Maier, N.R.F., Schneirla, T.C.: Principles of Animal Psychology. Dover, New York (1935)
360. Osthaus, B., Lea, S.E., Slater, A.M.: Dogs (Canis lupus familiaris) fail to show understanding of means-end connections in a string-pulling task. Anim. Cogn. **8**(1), 37–47 (2005)
361. Range, F., Hentrup, M., Viranyi, Z.: Dogs are able to solve a means-end task. Anim. Cogn. **14**(4), 575–583 (2011)
362. Range, F., Moslinger, H., Viranyi, Z.: Domestication has not affected the understanding of means-end connections in dogs. Anim. Cogn. **15**(4), 597–607 (2012)
363. Riemer, S., Müller, C., Range, F., Huber, L.: Dogs (Canis familiaris) can learn to attend to connectivity in string pulling tasks. J. Comp. Psychol. **128**(1), 31–39 (2014)
364. Muller, C.A., Riemer, S., Viranyi, Z., Huber, L., Range, F.: Inhibitory control, but not prolonged object-related experience appears to affect physical problem-solving performance of pet dogs. PLoS One. **11**(2), e0147753 (2016)
365. Müller, C.A., Riemer, S., Virányi, Z., Huber, L., Range, F.: Dogs learn to solve the support problem based on perceptual cues. Anim. Cogn. **17**, 1071–1080 (2014)
366. Katona, G.: Organizing and Memorizing: Studies in the Psychology of Learning and Teaching. Columbia University Press, New York (1940)
367. Wertheimer, M.: Productive Thinking. Harper & Row, New York (1945)
368. Maltzman, I.: Thinking: from a behavioristic point of view. Psychol. Rev. **62**(4), 275–286 (1955)
369. Epstein, R., Kirshnit, C., Lanza, R., Rubin, L.: "Insight" in the pigeon: antecedents and determinants of an intelligent performance. Nature. **308**, 61–62 (1984)
370. Weisberg, R.: Creativity: Genius and Other Myths. WH Freeman/Times Books/Henry Holt & Co, New York (1986)
371. Epstein, R.: The spontaneous interconnection of four repertoires of behavior in a pigeon (Columba livia). J. Comp. Psychol. **101**(2), 197 (1987)

372. Epstein, R.: Bringing cognition and creativity into the behavioral laboratory. In: Approaches to Cognition: Contrasts and Controversies, pp. 91–109. Lawrence Erlbaum Associates, Hillsdale (1986)
373. Epstein, R.: Skinner, creativity, and the problem of spontaneous behavior. Psychol. Sci. **2**(6), 362–370 (1991)
374. Birch, H.G.: The relation of previous experience to insightful problem-solving. J. Comp. Psychol. **38**(6), 367–383 (1945)
375. Cook, R.G., Fowler, C.: "Insight" in pigeons: absence of means–end processing in displacement tests. Anim. Cogn. **17**(2), 207–220 (2014)
376. Duncker, K.: On Problem Solving. American Psychological Association, Washington, DC (1945)
377. Köhler, W.: The Task of Gestalt Psychology. Princeton University Press, Princeton (1969)
378. Olton, D.S., Collison, C., Werz, M.A.: Spatial memory and radial arm maze performance of rats. Learn. Motiv. **8**(3), 289–314 (1977)
379. Bruner, J.S.: Nature and uses of immaturity. Am. Psychol. **27**, 1–22 (1972)
380. Whiten, A., Ham, R.: On the nature and evolution of imitation in the animal kingdom: reappraisal of a century of research. In: Slater, P.J.B., et al. (eds.) Advances in the Study of Behavior, pp. 239–283. Academic Press, New York (1992)
381. Tomasello, M.: Emulation learning and cultural learning. Behav. Brain Sci. **21**, 703–704 (1998)
382. Whiten, A., Horner, V., Litchfield, C.A., Marshall-Pescini, S.: How do apes ape? Learn. Behav. **32**(1), 36–52 (2004)
383. Call, J., Carpenter, M., Tomasello, M.: Copying results and copying actions in the process of social learning: chimpanzees (Pan troglodytes) and human children (Homo sapiens). Anim. Cogn. **8**(3), 151–163 (2005)
384. Zentall, T.: Imitation: definitions, evidence, and mechanisms. Anim. Cogn. **9**, 335–353 (2006)
385. Huber, L.: Emulation learning: the integration of technical and social cognition. In: Dautenhahn, K., Nehaniv, C.L. (eds.) Imitation and Social Learning in Robots, Humans and Animals: Behavioural, Social and Communicative Dimensions, pp. 427–439. Cambridge University Press, Cambridge (2007)
386. Tennie, C., Call, J., Tomasello, M.: Evidence for emulation in chimpanzees in social settings using the floating peanut task. PLoS One. **5**(5), e10544 (2010)
387. Haun, D.B., Call, J., Janzen, G., Levinson, S.C.: Evolutionary psychology of spatial representations in the hominidae. Curr. Biol. **16**(17), 1736–1740 (2006)
388. Seed, A., Call, J.: Causal knowledge for events and objects in animals. In: Watanabe, S., et al. (eds.) Rational Animals, Irrational Humans, pp. 173–188. Keio University Press, Tokyo (2009)
389. Seed, A., Hanus, D., Call, J.: Causal knowledge in corvids, primates and children: more than meets the eye? In: McCormack, T. (ed.) Tool Use and Causal Cognition, pp. 89–110. Oxford University Press, Oxford (2011)
390. Bräuer, J., Kaminski, J., Riedel, J., Call, J., Tomasello, M.: Making inferences about the location of hidden food: social dog, causal ape. J. Comp. Psychol. **120**(1), 38–47 (2006)
391. Call, J.: Descartes' two errors: reason and reflection in the great apes. In: Hurley, S., Nudds, M. (eds.) Rational Animals? pp. 219–234. Oxford University Press, Oxford (2006)
392. Davidson, J.E.: The role of insight in giftedness. In: Sternberg, R.J., Davidson, J.E. (eds.) Conceptions of Giftedness, pp. 201–223. Cambridge University Press, Cambridge (1986)
393. Davidson, J.E.: The suddenness of insight. In: Sternberg, R.J., Davidson, J.E. (eds.) The Nature of Insight, pp. 125–156. Cambridge University Press, Cambridge (1995)
394. Seifert, C.M., Meyer, D.E., Davidson, N., Patalano, A.L., Yaniv, I.: Demystification of cognitive insight: opportunistic assimilation and the prepared-mind hypothesis. In: Sternberg, R.J., Davidson, J.E. (eds.) The Nature of Insight, pp. 65–124. MIT Press, Cambridge, MA (1994)
395. Piaget, J.: The Child's Conception of the World. Routledge and Kegan Paul, London (1928)

396. Mayer, R.E.: The search for insight: grappling with Gestalt psychology's unanswered questions. In: Sternberg, R.J., Davidson, J.E. (eds.) The Nature of Insight, pp. 3–32. Cambridge University Press, Cambridge (1995)
397. Selz, O.: Über die Gesetze des geordneten Denkverlaufs. W. Spemann, Stuttgart (1913)
398. Maier, N.R.F.: Reasoning in humans II: the solution of a problem and its appearance in consciousness. J. Comp. Psychol. **12**, 181–194 (1931)
399. Seed, A.M., Boogert, N.J.: Animal cognition: an end to insight? Curr. Biol. **23**(2), R67–R69 (2013)
400. Palmer, S.E.: Vision Science. Photons to Phenomenology. The MIT Press, Cambridge, MA (1999)
401. Hanus, D., Mendes, N., Tennie, C., Call, J.: Comparing the performances of apes (Gorilla gorilla, Pan troglodytes, Pongo pygmaeus) and human children (Homo sapiens) in the floating peanut task. PLoS One. **6**(6), e19555 (2011)
402. Bourne, L.E.J., Ekstrand, E.R., Dominowski, R.L.: The Psychology of Thinking. Prentice Hall, Englewood Cliffs (1971)
403. Dominowski, R.L., Dallob, P.: Insight and problem solving. In: Sternberg, R.J., Davidson, J.E. (eds.) The Nature of Insight, pp. 33–62. MIT Press, Cambridge, MA (1995)
404. Bühler, K.: Tatsachen und Probleme zu einer Psychologie der Denkvorgänge. W. Engelmann, Leipzig (1907)
405. Hammerschmidt, K., Fischer, J.: Constraints in primate vocal production. In: Oller, K., Griebel, U. (eds.) The Evolution of Communicative Creativity: From Fixed Signals to Contextual Flexibility, pp. 93–120. MIT Press, Cambridge, MA (2008)
406. Parker, S.T., Gibson, K.R.: A developmental model for the evolution of language and intelligence in early hominids. Behav. Brain Sci. **2**, 367–408 (1979)
407. Torigoe, T.: Comparison of object manipulation among 74 species of non-human primates. Primates. **26**, 182–194 (1985)
408. Brentano, F.: Psychologie vom empirischen Standpunkte (Psychology from an Empirical Standpoint). Duncker & Humblot, Leipzig (1874)
409. Seed, A.M., Byrne, R.: Animal Tool-Use. Curr. Biol. **20**(23), R1032–R1039 (2010)
410. Limongelli, L., Boysen, S.T., Visalberghi, E.: Comprehension of cause-effect relations in a tool-using task by chimpanzees (Pan troglodytes). J. Comp. Psychol. **109**(1), 18–26 (1995)
411. Kacelnik, A., Chappell, J., Weir, A., Kenward, B.: Cognitive adaptations for tool-related behaviour in New Caledonian crows. In: Wasserman, E.A., Zentall, T.R. (eds.) Comparative Cognition: Experimental Explorations of Animal Intelligence, pp. 515–528. Oxford University Press, New York (2006)
412. Povinelli, D.J.: Folk Physics for Apes. The chimpanzee's Theory of how the World Works. Oxford University Press, Oxford (2000)
413. Silva, F.J., Page, D.M., Silva, K.M.: Methodological-conceptual problems in the study of chimpanzees' folk physics: how studies with adult humans can help. Anim. Learn. Behav. **33**(1), 47–58 (2005)
414. Gigerenzer, G., Goldstein, D.G.: Reasoning the fast and frugal way: models of bounded rationality. Psychol. Rev. **103**(4), 650–669 (1996)
415. Gigerenzer, G., Brighton, H.: Homo heuristicus: why biased minds make better inferences. Top. Cogn. Sci. **1**(1), 107–143 (2009)
416. Seed, A.M., Call, J., Emery, N.J., Clayton, N.S.: Chimpanzees solve the trap problem when the confound of tool-use is removed. J. Exp. Psychol. Anim. Behav. Process. **35**(1), 23–34 (2009)
417. Seed, A.M., Tebbich, S., Emery, N.J., Clayton, N.S.: Investigating physical cognition in rooks, Corvus frugilegus. Curr. Biol. **16**(7), 697–701 (2006)
418. Taylor, A.H., Hunt, G.R., Medina, F.S., Gray, R.D.: Do New Caledonian crows solve physical problems through causal reasoning? Proc. R. Soc. B Biol. Sci. **276**(1655), 247–254 (2009)
419. Bierens de Haan, J.A.: Der Stieglitz als Schöpfer. J. Ornithol. **81**, 1–22 (1933)

420. Altevogt, R.: Über das "Schöpfen" einiger Vogelarten. Behaviour. **6**, 147–152 (1953)
421. Vince, M.A.: "String-pulling" in birds. III. The successful response in greenfinches and canaries. Behaviour. **17**, 103–129 (1961)
422. Dücker, G., Rensch, B.: The solution of patterned string problems by birds. Behaviour. **62**, 164–173 (1977)
423. Seibt, U., Wickler, W.: Individuality in problem solving: string pulling in two Carduelis species (Aves: Passeriformes). Ethology. **112**(5), 493–502 (2006)
424. Huber, L., Gajdon, G.K.: Technical intelligence in animals: the kea model. Anim. Cogn. **9**(4), 295–305 (2006)
425. Tebbich, S., Taborsky, M., Winkler, H.: Social manipulation causes cooperation in keas. Anim. Behav. **52**, 1–10 (1996)
426. Werdenich, D., Huber, L.: A case of quick problem solving in birds: string-pulling in keas (Nestor notabilis). Anim. Behav. **71**(4), 855–863 (2006)
427. Taylor, A.H., Medina, F.S., Holzhaider, J.C., Hearne, L.J., Hunt, G.R., Gray, R.D.: An investigation into the cognition behind spontaneous string pulling in New Caledonian crows. PLoS One. **5**(2), e9345 (2010)
428. Völter, C.J., Call, J.: Problem solving in great apes (Pan paniscus, Pan troglodytes, Gorilla gorilla, and Pongo abelii): the effect of visual feedback. Anim. Cogn. **15**(5), 923–936 (2012)
429. Taylor, A.H., Knaebe, B., Gray, R.D.: An end to insight? New Caledonian crows can spontaneously solve problems without planning their actions. Proc. R. Soc. B Biol. Sci. **279**(1749), 4977–4981 (2012)
430. Lorenz, K.Z., Tinbergen, N.: Taxis und Instinkthandlung in der Eirollbewegung der Graugans. Z. Tierpsychol. **2**, 1–29 (1938)
431. Thorndike, E.L.: Animal Intelligence. Macmillan, New York (1911)
432. Willatts, P.: Development of means–end behavior in young infants: pulling a support to retrieve a distant object. Dev. Psychol. **35**(3), 651–667 (1999)
433. Schmidt, G.F., Cook, R.G.: Mind the gap: means–end discrimination by pigeons. Anim. Behav. **71**(3), 599–608 (2006)
434. Herrmann, E., Wobber, V., Call, J.: Great apes' (Pan troglodytes, Pan paniscus, Gorilla gorilla, Pongo pygmaeus) understanding of tool functional properties after limited experience. J. Comp. Psychol. **122**(2), 220 (2008)
435. Auersperg, A.M.I., Gajdon, G.K., Huber, L.: Kea (Nestor notabilis) consider spatial relationships between objects in the support problem. Biol. Lett. **5**(4), 455–458 (2009)
436. Bird, C.D., Emery, N.J.: Rooks use stones to raise the water level to reach a floating worm. Curr. Biol. **19**, 1410 (2009)
437. Schwing, R., Weiss, F., Tichy, A., Gajdon, G.: Testing the causal understanding of water displacement by kea (Nestor notabilis). Behaviour. **156**(5–8), 447–478 (2019)
438. Cheke, L.G., Bird, C.D., Clayton, N.S.: Tool-use and instrumental learning in the Eurasian jay (Garrulus glandarius). Anim. Cogn. **14**(3), 441–455 (2011)
439. Taylor, A.H., Elliffe, D.M., Hunt, G.R., Emery, N.J., Clayton, N.S., Gray, R.D.: New Caledonian crows learn the functional properties of novel tool types. PLoS One. **6**(12), e26887 (2011)
440. Jelbert, S.A., Taylor, A.H., Cheke, L.G., Clayton, N.S., Gray, R.D.: Using the Aesop's fable paradigm to investigate causal understanding of water displacement by New Caledonian crows. PLoS One. **9**(3), e92895 (2014)
441. Logan, C.J., Jelbert, S.A., Breen, A.J., Gray, R.D., Taylor, A.H.: Modifications to the Aesop's fable paradigm change New Caledonian crow performances. PLoS One. **9**(7), e103049 (2014)
442. Miller, R., Jelbert, S.A., Taylor, A.H., Cheke, L.G., Gray, R.D., Loissel, E., Clayton, N.S.: Performance in object-choice Aesop's fable tasks are influenced by object biases in New Caledonian crows but not in human children. PLoS One. **11**(12), e0168056 (2016)
443. Stanton, L., Davis, E., Johnson, S., Gilbert, A., Benson-Amram, S.: Adaptation of the Aesop's fable paradigm for use with raccoons (Procyon lotor): considerations for future application in non-avian and non-primate species. Anim. Cogn. **20**(6), 1147–1152 (2017)

444. Logan, C.J., Harvey, B.D., Schlinger, B.A., Rensel, M.: Western scrub-jays do not appear to attend to functionality in Aesop's fable experiments. PeerJ. **4**, e1707 (2016)
445. Cheke, L.G., Loissel, E., Clayton, N.S.: How do children solve Aesop's fable? PLoS One. **7**(7), e40574 (2012)
446. Loissel, E., Cheke, L.G., Clayton, N.S.: Exploring the relative contributions of reward-history and functionality information to children's acquisition of the Aesop's fable task. PLoS One. **13**(2), e0193264 (2018)
447. Weir, A.A., Kacelnik, A.: A New Caledonian crow (Corvus moneduloides) creatively re-designs tools by bending or unbending aluminium strips. Anim. Cogn. **9**(4), 317 (2006)
448. Klump, B.C., Sugasawa, S., St Clair, J.J., Rutz, C.: Hook tool manufacture in New Caledonian crows: behavioural variation and the influence of raw materials. BMC Biol. **13**(1), 97 (2015)
449. Rutz, C., Sugasawa, S., Van der Wal, J.E., Klump, B.C., St Clair, J.J.: Tool bending in New Caledonian crows. R. Soc. Open Sci. **3**(8), 160439 (2016)
450. Sugasawa, S., Klump, B.C., St Clair, J.J., Rutz, C.: Causes and consequences of tool shape variation in New Caledonian crows. Curr. Biol. **27**(24), 3885–3890.e4 (2017)
451. Klump, B.C., Cantat, M., Rutz, C.: Raw-material selectivity in hook-tool-crafting New Caledonian crows. Biol. Lett. **15**(2), 20180836 (2019)
452. O'Connor, S., Ono, R., Clarkson, C.: Pelagic fishing at 42,000 years before the present and the maritime skills of modern humans. Science. **334**(6059), 1117–1121 (2011)
453. St Clair, J.J., Klump, B.C., Sugasawa, S., Higgott, C.G., Colegrave, N., Rutz, C.: Hook innovation boosts foraging efficiency in tool-using crows. Nat. Ecol. Evol. **2**(3), 441–444 (2018)
454. Emery, N.J., Clayton, N.S.: The mentality of crows: convergent evolution of intelligence in corvids and apes. Science. **306**(5703), 1903–1907 (2004)
455. Mioduszewska, B.M., O'Hara, M.C., Haryoko, T., Auersperg, A.M.I., Huber, L., Prawiradilaga, D.M.: Notes on ecology of wild Goffin's cockatoo in the late dry season with emphasis on feeding ecology. Treubia. **45**, 85–102 (2019)
456. Hansell, M., Ruxton, G.D.: Setting tool use within the context of animal construction behaviour. Trends Ecol. Evol. **23**(2), 73–78 (2008)
457. Parker, S.T., Gibson, K.R.: Object manipulation, tool use, and sensorimotor intelligence as feeding adaptations in cebus monkeys and great apes. J. Hum. Evol. **6**, 623–641 (1977)
458. King, B.J.: Extractive foraging and the evolution of primate intelligence. Hum. Evol. **1**(4), 361–372 (1986)
459. Parker, S.T.: Re-evaluating the extractive foraging hypothesis. New Ideas Psychol. **37**, 1–12 (2015)
460. Huber, L., O'Hara, M.: The extractive foraging hypothesis (Parker and Gibson 1997, 2015). In: Shackelford, T.K., Weekes-Shackelford, V.A. (eds.) Encyclopedia of Evolutionary Psychological Science. Springer, Cham (2016)
461. Auersperg, A.M.I., van Horik, J.O., Bugnyar, T., Kacelnik, A., Emery, N.J., von Bayern, A.M.P.: Combinatory actions during object play in psittaciformes (Diopsittaca nobilis, Pionites melanocephala, Cacatua goffini) and corvids (Corvus corax, C. monedula, C. moneduloides). J. Comp. Psychol. **129**(1), 62–71 (2015)
462. Hayashi, M., Matsuzawa, T.: Cognitive development in object manipulation by infant chimpanzees. Anim. Cogn. **6**(4), 225–233 (2003)
463. Bjorklund, D.F., Gardiner, A.K.: Object play and tool use: developmental and evolutionary perspectives. In: The Oxford Handbook of the Development of Play, pp. 153–171. Oxford University Press, New York (2011)
464. Auersperg, A.M.I., Szabo, B., von Bayern, A.M.P., Kacelnik, A.: Spontaneous innovation in tool manufacture and use in a Goffin's cockatoo. Curr. Biol. **22**(21), R903–R904 (2012)
465. Auersperg, A.M.I., Kacelnik, A., von Bayern, A.M.P.: Explorative learning and functional inferences on a five-step means-means-end problem in Goffin's cockatoos (Cacatua goffini). PLoS One. **8**(7), e68979 (2013)
466. Auersperg, A.M.I., von Bayern, A.M.I., Weber, S., Szabadvari, A., Bugnyar, T., Kacelnik, A.: Social transmission of tool use and tool manufacture in Goffin cockatoos (Cacatua goffini). Proc. R. Soc. B Biol. Sci. **281**(1793), 20140972 (2014)

467. Auersperg, A.M.I., Borasinski, S., Laumer, I., Kacelnik, A.: Goffin's cockatoos make the same tool type from different materials. Biol. Lett. **12**(11), 20160689 (2016)
468. Auersperg, A.M.I., Kock, C., O'Hara, M., Huber, L.: Tool making cockatoos adjust the lengths but not the widths of their tools to function. PLoS One. **13**(11), e0205429 (2018)
469. Cutting, N., Apperly, I.A., Beck, S.R.: Why do children lack the flexibility to innovate tools? J. Exp. Child Psychol. **109**(4), 497–511 (2011)
470. Laumer, I.B., Bugnyar, T., Reber, S.A., Auersperg, A.M.I.: Can hook-bending be let off the hook? Bending/unbending of pliant tools by cockatoos. Proc. R. Soc. B Biol. Sci. **284**(1862), 20171026 (2017)
471. Laumer, I.B., Call, J., Bugnyar, T., Auersperg, A.M.I.: Spontaneous innovation of hook-bending and unbending in orangutans (Pongo abelii). Sci. Rep. **8**, 16518 (2018)
472. Collias, N.E., Collias, E.C.: Nest Building and Bird Behaviour. Princeton University Press, Princeton (1984)
473. O'Hara, M., Mioduszewska, B., Haryoko, T., Prawiradilaga, D.M., Huber, L., Auersperg, A.: Extraction without tooling around – the first comprehensive description of the foraging- and socio-ecology of wild Goffin's cockatoos (Cacatua goffiniana). Behaviour. **156**(5–8), 661–690 (2019)
474. O'Hara, M., Mioduszewska, B., Mundry, R., Yohanna, Haryoko, T., Rachmatika, R., Prawiradilaga, D.M., Huber, L., Auersperg, A.M.I.: Wild Goffin's cockatoos flexibly manufacture and use tool sets. Curr. Biol. **31**(20), 4512–4520.e6 (2021)
475. Brewer, S.M., McGrew, W.: Chimpanzee use of a tool-set to get honey. Folia Primatol. **54**(1–2), 100–104 (1990)
476. Nishihara, T., Suzuki, S., Kuroda, S.: Tool-set for termite-fishing by chimpanzees in the Ndoki Forest, Congo. Behaviour. **132**(3–4), 219–235 (1995)
477. Gajdon, G.K., Fijn, N., Huber, L.: Limited spread of innovation in a wild parrot, the kea (Nestor notabilis). Anim. Cogn. **9**(3), 173–181 (2006)
478. Klump, B.C., Martin, J.M., Wild, S., Hörsch, J.K., Major, R.E., Aplin, L.M.: Innovation and geographic spread of a complex foraging culture in an urban parrot. Science. **373**(6553), 456–460 (2021)
479. Klump, B.C., Major, R.E., Farine, D.R., Martin, J.M., Aplin, L.M.: Is bin-opening in cockatoos leading to an innovation arms race with humans? Curr. Biol. **32**(17), R910–R911 (2022)
480. Whiten, A.: The burgeoning reach of animal culture. Science. **372**(6537), eabe6514 (2021)
481. Bischof-Köhler, D.: Kinder auf Zeitreise: Theory of Mind, Zeitverständnis und Handlungsorganisation. Huber, Bern, Göttingen, Toronto, Seattle (2000)
482. Bischof, N.: Comparative ethology of incest avoidance. In: Fox, R. (ed.) Biosocial Anthropology, pp. 37–68. Malaby Press, London (1975)
483. Suddendorf, T., Corballis, M.C.: Mental time travel and the evolution of the human mind. Genet. Soc. Gen. Psychol. Monogr. **123**(2), 133–167 (1997)
484. Shettleworth, S.J.: Clever animals and killjoy explanations in comparative psychology. Trends Cogn. Sci. **14**(11), 477–481 (2010)
485. Zentall, T.R.: Mental time travel in animals: a challenging question. Behav. Process. **72**(2), 173–183 (2006)
486. Rutz, C., Bluff, L.A., Weir, A.A., Kacelnik, A.: Video cameras on wild birds. Science. **318**(5851), 765 (2007)
487. Klump, B.C., van der Wal, J.E., St Clair, J.J., Rutz, C.: Context-dependent "safekeeping" of foraging tools in New Caledonian crows. Proc. R. Soc. B Biol. Sci. **2015**(282), 20150278 (1808)
488. Klump, B.C., St Clair, J.J., Rutz, C.: New Caledonian crows keep "valuable" hooked tools safer than basic non-hooked tools. ELife. **10**, e64829 (2021)
489. Auersperg, A.M.I., Kock, C., Pledermann, A., O'Hara, M., Huber, L.: Safekeeping of tools in Goffin's cockatoos, Cacatua goffiniana. Anim. Behav. **128**, 125–133 (2017)
490. Boesch, C., Boesch, H.: Mental map in wild chimpanzees: an analysis of hammer transports for nut cracking. Primates. **25**(2), 160–170 (1984)

491. Anderson, J., Henneman, M.-C.: Solutions to a tool-use problem in a pair of Cebus apella. Mammalia. **58**(3), 351–362 (1994)
492. Hihara, S., Obayashi, S., Tanaka, M., Iriki, A.: Rapid learning of sequential tool use by macaque monkeys. Physiol. Behav. **78**(3), 427–434 (2003)
493. Bunn, H.T.: Archaeological evidence for meat-eating by Plio-Pleistocene hominids from Koobi Fora and Olduvai Gorge. Nature. **291**(5816), 574 (1981)
494. deBeaune, S., Davidson, I., Hardy, B., McGrew, W., Marchant, L., Reader, S., Stout, D., Vauclair, J., deBeaune, S.: The invention of technology: prehistory and cognition. Curr. Anthropol. **45**(2), 139–162 (2004)
495. Matsuzawa, T. (ed.): Primate Origins of Human Cognition and Behavior. Springer, Tokyo (2001)
496. Mischel, W.: Der Marshmallow-Test: Willensstärke, Belohnungsaufschub und die Entwicklung der Persönlichkeit. Siedler Verlag, München (2015)
497. Flessert, M., Beran, M.J.: Delayed gratification. In: Encyclopedia of Animal Cognition and Behavior, pp. 1–7. Springer, Cham (2018)
498. Osvath, M., Osvath, H.: Chimpanzee (Pan troglodytes) and orangutan (Pongo abelii) forethought: self-control and pre-experience in the face of future tool use. Anim. Cogn. **11**(4), 661–674 (2008)
499. Tulving, E.: Episodic memory and autonoesis: uniquely human? In: Terrace, H.S., Metcalfe, J. (eds.) The Missing Link in Cognition: Origins of Self-Reflective Consciousness, pp. 3–56. Oxford University Press, New York, Oxford (2005)
500. Gilbert, D.T., Wilson, T.D.: Prospection: experiencing the future. Science. **317**(5843), 1351–1354 (2007)
501. Suddendorf, T., Corballis, M.C.: The evolution of foresight: what is mental time travel, and is it unique to humans? Behav. Brain Sci. **30**(3), 299–313 (2007).; discussion 313–351
502. Tulving, E.: Episodic and semantic memory. In: Tulving, E., Donaldson, W. (eds.) Organisation of Memory, pp. 381–403. Academic Press, New York (1972)
503. Buckner, R.L., Carroll, D.C.: Self-projection and the brain. Trends Cogn. Sci. **11**(2), 49–57 (2007)
504. Szpunar, K.K., Watson, J.M., McDermott, K.B.: Neural substrates of envisioning the future. Proc. Natl. Acad. Sci. USA. **104**(2), 642–647 (2007)
505. Rosenbaum, R.S., Kohler, S., Schacter, D.L., Moscovitch, M., Westmacott, R., Black, S.E., Gao, F., Tulving, E.: The case of K.C.: contributions of a memory-impaired person to memory theory. Neuropsychologia. **43**(7), 989–1021 (2005)
506. Atance, C.M., O'Neill, D.K.: Episodic future thinking. Trends Cogn. Sci. **5**(12), 533–539 (2001)
507. Raichle, M.E., MacLeod, A.M., Snyder, A.Z., Powers, W.J., Gusnard, D.A., Shulman, G.L.: A default mode of brain function. Proc. Natl. Acad. Sci. USA. **98**(2), 676–682 (2001)
508. Mantini, D., Gerits, A., Nelissen, K., Durand, J.-B., Joly, O., Simone, L., Sawamura, H., Wardak, C., Orban, G.A., Buckner, R.L.: Default mode of brain function in monkeys. J. Neurosci. **31**(36), 12954–12962 (2011)
509. Lu, H., Zou, Q., Gu, H., Raichle, M.E., Stein, E.A., Yang, Y.: Rat brains also have a default mode network. Proc. Natl. Acad. Sci. USA. **109**(10), 3979–3984 (2012)
510. Barks, S.K., Parr, L.A., Rilling, J.K.: The default mode network in chimpanzees (Pan troglodytes) is similar to that of humans. Cereb. Cortex. **25**(2), 538–544 (2013)
511. Kyathanahally, S.P., Jia, H., Pustovyy, O.M., Waggoner, P., Beyers, R., Schumacher, J., Barrett, J., Morrison, E.E., Salibi, N., Denney, T.S., Vodyanoy, V.J., Deshpande, G.: Anterior-posterior dissociation of the default mode network in dogs. Brain Struct. Funct. **220**(2), 1063–1076 (2015)
512. Boyer, P.: Evolutionary economics of mental time travel? Trends Cogn. Sci. **12**(6), 219–224 (2008)
513. Osvath, M.: Planning Primates. A Search for Episodic Foresight Lund University Cognitive Studies. Lund University, Lund (2010)

514. Johansson, P., Hall, L., Sikstrom, S., Olsson, A.: Failure to detect mismatches between intention and outcome in a simple decision task. Science. **310**(5745), 116–119 (2005)
515. Rensch, B.: Gedächtnis, Begriffsbildung und Planhandlungen bei Tieren. Parey, Berlin, Hamburg (1973)
516. Köhler, W.: Zur Psychologie des Schimpansen. Psychol. Forsch. **1**, 2–46 (1921)
517. Suddendorf, T.: Discovery of the fourth dimension: mental time travel and human evolution. Master's thesis, University of Waikato, New Zealand (1994)
518. Suddendorf, T., Busby, J.: Making decisions with the future in mind: developmental and comparative identification of mental time travel. Learn. Motiv. **36**(2), 110–125 (2005)
519. Correia, S.P., Dickinson, A., Clayton, N.S.: Western scrub-jays anticipate future needs independently of their current motivational state. Curr. Biol. **17**(10), 856–861 (2007)
520. Mulcahy, N.J., Call, J.: Apes save tools for future use. Science. **312**(5776), 1038–1040 (2006)
521. Dufour, V., Sterck, E.H.: Chimpanzees fail to plan in an exchange task but succeed in a tool-using procedure. Behav. Process. **79**(1), 19–27 (2008)
522. Osvath, M.: Spontaneous planning for future stone throwing by a male chimpanzee. Curr. Biol. **19**(5), R190–R191 (2009)
523. Osvath, M., Karvonen, E.: Spontaneous innovation for future deception in a male chimpanzee. PLoS One. **7**(5), e36782 (2012)
524. Visalberghi, E., Sirianni, G., Fragaszy, D., Boesch, C.: Percussive tool use by Tai Western chimpanzees and Fazenda Boa Vista bearded capuchin monkeys: a comparison. Philos. Trans. R. Soc. Lond. Ser. B Biol. Sci. **370**(1682), 20140351 (2015)
525. Redshaw, J., Suddendorf, T.: Children's and apes' preparatory responses to two mutually exclusive possibilities. Curr. Biol. **26**(13), 1758–1762 (2016)
526. Suddendorf, T., Crimston, J., Redshaw, J.: Preparatory responses to socially determined, mutually exclusive possibilities in chimpanzees and children. Biol. Lett. **13**(6), 20170170 (2017)
527. Redshaw, J., Suddendorf, T., Neldner, K., Wilks, M., Tomaselli, K., Mushin, I., Nielsen, M.: Young children from three diverse cultures spontaneously and consistently prepare for alternative future possibilities. Child Dev. **90**(1), 51–61 (2019)
528. Janmaat, K.R., Polansky, L., Ban, S.D., Boesch, C.: Wild chimpanzees plan their breakfast time, type, and location. Proc. Natl. Acad. Sci. USA. **111**(46), 16343–16348 (2014)
529. Clayton, N.S., Dickinson, A.: Episodic-like memory during cache recovery by scrub jays. Nature. **395**(6699), 272–274 (1998)
530. Griffiths, D., Dickinson, A., Clayton, N.: Episodic memory: what can animals remember about their past? Trends Cogn. Sci. **3**(2), 74–80 (1999)
531. Clayton, N.S., Griffiths, D.P., Dickinson, A.: Declarative and episodic-like memory in animals: personal musings of a scrub jay. In: Heyes, C., Huber, L. (eds.) The Evolution of Cognition, pp. 273–288. MIT Press, Cambridge, MA (2000)
532. Clayton, N.S., Griffiths, D.P., Emery, N.J., Dickinson, A.: Elements of episodic-like memory in animals. Philos. Trans. R. Soc. Lond. Ser. B Biol. Sci. **356**(1413), 1483–1491 (2001)
533. Raby, C.R., Alexis, D.M., Dickinson, A., Clayton, N.S.: Planning for the future by western scrub-jays. Nature. **445**(7130), 919–921 (2007)
534. Clayton, N.S., Dickinson, A.: Motivational control of caching behaviour in the scrub jay, Aphelocoma coerulescens. Anim. Behav. **57**(2), 435–444 (1999)
535. Suddendorf, T., Corballis, M.C.: Behavioural evidence for mental time travel in nonhuman animals. Behav. Brain Res. **215**(2), 292–298 (2010)
536. Kabadayi, C., Osvath, M.: Ravens parallel great apes in flexible planning for tool-use and bartering. Science. **357**(6347), 202–204 (2017)
537. Redshaw, J., Taylor, A.H., Suddendorf, T.: Flexible planning in ravens? Trends Cogn. Sci. **21**(11), 821–822 (2017)
538. Suddendorf, T., Nielsen, M., Von Gehlen, R.: Children's capacity to remember a novel problem and to secure its future solution. Dev. Sci. **14**(1), 26–33 (2011)
539. Martin-Ordas, G., Atance, C.M., Call, J.: Remembering in tool-use tasks in children and apes: the role of the information at encoding. Memory. **22**(1), 129–144 (2014)

540. Martin-Ordas, G., Berntsen, D., Call, J.: Memory for distant past events in chimpanzees and orangutans. Curr. Biol. **23**(15), 1438–1441 (2013)
541. Byrne, R.W., Sanz, C.M., Morgan, D.B.: Chimpanzees plan their tool use. In: Sanz, C.M., Call, J., Boesch, C. (eds.) Tool Use in Animals: Cognition and Ecology, pp. 48–63. Cambridge University Press, Cambridge (2013)
542. Kwan, D., Craver, C.F., Green, L., Myerson, J., Boyer, P., Rosenbaum, R.S.: Future decision-making without episodic mental time travel. Hippocampus. **22**(6), 1215–1219 (2012)
543. Roberts, W.A.: Are animals stuck in time? Psychol. Bull. **128**(3), 473–489 (2002)
544. Osvath, M., Martin-Ordas, G.: The future of future-oriented cognition in non-humans: theory and the empirical case of the great apes. Philos. Trans. R. Soc. Lond. Ser. B Biol. Sci. **369**(1655), 20130486 (2014)
545. Tulving, E.: Episodic memory: from mind to brain. Annu. Rev. Psychol. **53**, 1–25 (2002)
546. Schacter, D.L., Addis, D.R., Buckner, R.L.: Remembering the past to imagine the future: the prospective brain. Nat. Rev. Neurosci. **8**(9), 657–661 (2007)
547. Wheeler, M.A.: Episodic memory and autonoetic awareness. In: Tulving, E., Craik, F.I.M. (eds.) The Oxford Handbook of Memory, pp. 597–608. Oxford University Press, New York (2000)
548. Bednekoff, P.A., Balda, R.P., Kamil, A.C., Hile, A.G.: Long-term spatial memory in four seed-caching corvid species. Anim. Behav. **53**(2), 335–341 (1997)
549. Olson, D.J., Kamil, A.C., Balda, R.P., Nims, P.J.: Performance of four seed-caching corvid species in operant tests of nonspatial and spatial memory. J. Comp. Psychol. **109**(2), 173–181 (1995)
550. Lind, J., Enquist, M., Ghirlanda, S.: Animal memory: a review of delayed matching-to-sample data. Behav. Process. **117**, 52–58 (2015)
551. Garcia, J., Ervin, F.R., Koelling, R.A.: Learning with prolonged delay of reinforcement. Psychon. Sci. **5**(3), 121–122 (1966)
552. Garcia, J., Koelling, R.A.: Relation of cue to consequence in avoidance learning. Psychon. Sci. **4**(1), 123–124 (1966)
553. Garcia, J., Lasiter, P.S., Bermudez-Rattoni, F., Deems, D.A.: A general theory of aversion learning. Ann. N. Y. Acad. Sci. **443**(1), 8–21 (1985)
554. Denny, M.R.: Retention of SR in the midst of the cognitive invasion. In: Kendrick, D.F., Rilling, M.E., Denny, M.R. (eds.) Theories of Animal Memory, pp. 35–50. Psychology Press, Hove (1986)
555. Immelmann, K.: Sexual and other long-term aspects of imprinting in birds and other species. Adv. Study Behav. **4**, 147–174 (1972)
556. Oetting, S., Pröve, E., Bischof, H.-J.: Sexual imprinting as a two-stage process: mechanisms of information storage and stabilization. Anim. Behav. **50**(2), 393–403 (1995)
557. Kendrick, D.F., Rilling, M.E., Denny, M.R.: Theories of Animal Memory. Comparative Cognition and Neuroscience Series. Psychology Press, Hove (2014)
558. Dunnett, S.B., Martel, F.L.: Proactive interference effects on short-term memory in rats: I. Basic parameters and drug effects. Behav. Neurosci. **104**(5), 655–665 (1990)
559. Sloan, H.L., Good, M., Dunnett, S.B.: Double dissociation between hippocampal and prefrontal lesions on an operant delayed matching task and a water maze reference memory task. Behav. Brain Res. **171**(1), 116–126 (2006)
560. Suddendorf, T., Busby, J.: Mental time travel in animals? Trends Cogn. Sci. **7**(9), 391–396 (2003)
561. Overman, W.H.J., Doty, R.W.: Prolonged visual memory in macaques and man. Neuroscience. **5**, 1825–1831 (1980)
562. Clayton, N.S., Yu, K.S., Dickinson, A.: Scrub jays (Aphelocoma coerulescens) form integrated memories of the multiple features of caching episodes. J. Exp. Psychol. Anim. Behav. Process. **27**(1), 17–29 (2001)
563. Zentall, T.R.: Animals may not be stuck in time. Learn. Motiv. **36**(2), 208–225 (2005)
564. Tulving, E., Lepage, M.: Where in the brain is the awareness of one's past. In: Schacter, D., Scarry, E. (eds.) Memory, Brain, and Belief, pp. 208–230. Harvard University Press, Cambridge, MA (2000)

565. Gardiner, J.M., Java, R.I.: Forgetting in recognition memory with and without recollective experience. Mem. Cogn. **19**(6), 617–623 (1991)
566. Schacter, D.L., Verfaillie, M., Anes, M.D.: Illusory memories in amnesic patients: conceptual and perceptual false recognition. Neuropsychology. **11**, 331–342 (1997)
567. Tulving, E.: Elements of Episodic Memory. Oxford University Press, New York (1983)
568. Meltzoff, A.N., Moore, M.K.: Imitation in newborn infants: exploring the range of gestures imitated and the underlying mechanisms. Dev. Psychol. **25**(6), 954–962 (1989)
569. Zentall, T.R., Clement, T.S., Bhatt, R.S., Allen, J.: Episodic-like memory in pigeons. Psychon. Bull. Rev. **8**(4), 685–690 (2001)
570. Zentall, T.R.: Animals represent the past and the future. Evol. Psychol. **11**(3), 573–590 (2013)
571. Fugazza, C., Pogany, A., Miklosi, A.: Recall of others' actions after incidental encoding reveals episodic-like memory in dogs. Curr. Biol. **26**(23), 3209–3213 (2016)
572. Fugazza, C., Pogany, A., Miklosi, A.: Do as I … Did! Long-term memory of imitative actions in dogs (Canis familiaris). Anim. Cogn. **19**(2), 263–269 (2016)
573. Mercado, E., Murray, S.O., Uyeyama, R.K., Pack, A.A., Herman, L.M.: Memory for recent actions in the bottlenosed dolphin (Tursiops truncatus): repetition of arbitrary behaviors using an abstract rule. Anim. Learn. Behav. **26**(2), 210–218 (1998)
574. Wagenaar, W.A.: My memory: a study of autobiographical memory over six years. Cogn. Psychol. **18**(2), 225–252 (1986)
575. Menzel Jr., E.W.: Cognitive mapping in chimpanzees. In: Hulse, S.H., Fowler, H., Honig, W.K. (eds.) Cognitive Processes in Animal Behavior, pp. 375–422. Lawrence Erlbaum, Hillsdale (1978)
576. Savage-Rumbaugh, S., McDonald, K., Sevcik, R.A., Hopkins, W.D., Rubert, E.: Spontaneous symbol acquisition and communicative use by pygmy chimpanzees (Pan paniscus). J. Exp. Psychol. Gen. **115**(3), 211 (1986)
577. Beran, M.J., Rumbaugh, D.M., Savage-Rumbaugh, E.: Chimpanzee (Pan troglodytes) counting in a computerized testing paradigm. Psychol. Rec. **48**, 3–19 (1998)
578. Menzel, C.R.: Unprompted recall and reporting of hidden objects by a chimpanzee (Pan troglodytes) after extended delays. J. Comp. Psychol. **113**(4), 426 (1999)
579. Schwartz, B.L., Colon, M.R., Sanchez, I.C., Rodriguez, I.A., Evans, S.: Single-trial learning of "what" and "who" information in a gorilla (Gorilla gorilla gorilla): implications for episodic memory. Anim. Cogn. **5**(2), 85–90 (2002)
580. Jozet-Alves, C., Bertin, M., Clayton, N.S.: Evidence of episodic-like memory in cuttlefish. Curr. Biol. **23**(23), R1033–R1035 (2013)
581. Billard, P., Clayton, N.S., Jozet-Alves, C.: Cuttlefish retrieve whether they smelt or saw a previously encountered item. Sci. Rep. **10**(1), 1–7 (2020)
582. Billard, P., Schnell, A.K., Clayton, N.S., Jozet-Alves, C.: Cuttlefish show flexible and future-dependent foraging cognition. Biol. Lett. **16**(2), 20190743 (2020)
583. Eacott, M.J., Easton, A.: Episodic memory in animals: remembering which occasion. Neuropsychologia. **48**(8), 2273–2280 (2010)
584. Eacott, M.J., Easton, A.: Remembering the past and thinking about the future: is it really about time? Learn. Motiv. **43**(4), 200–208 (2012)
585. Friedman, W.J.: Memory for the time of past events. Psychol. Bull. **113**, 44–66 (1993)
586. Dally, J.M., Emery, N.J., Clayton, N.S.: Food-caching western scrub-jays keep track of who was watching when. Science. **312**(5780), 1662–1665 (2006)
587. Babb, S.J., Crystal, J.D.: Discrimination of what, when, and where is not based on time of day. Learn. Behav. **34**(2), 124–130 (2006)
588. Crystal, J.D., Smith, A.E.: Binding of episodic memories in the rat. Curr. Biol. **24**(24), 2957–2961 (2014)
589. Roberts, W.A., Feeney, M.C., Macpherson, K., Petter, M., McMillan, N., Musolino, E.: Episodic-like memory in rats: is it based on when or how long ago? Science. **320**(5872), 113–115 (2008)
590. Standing, L., Conezio, J., Haber, R.N.: Perception and memory for pictures: single-trial learning of 2500 visual stimuli. Psychon. Sci. **19**(2), 73–74 (1970)

591. Eacott, M.J., Easton, A.: On familiarity and recall of events by rats. Hippocampus. **17**(9), 890–897 (2007)
592. April, L.B., Bruce, K., Galizio, M.: The magic number 70 (plus or minus 20): variables determining performance in the rodent odor span task. Learn. Motiv. **44**(3), 143–158 (2013)
593. Slotnick, B.M., Schoonover, F.W.: Olfactory pathways and the sense of smell. Neurosci. Biobehav. Rev. **16**(4), 453–472 (1992)
594. Panoz-Brown, D., Corbin, H.E., Dalecki, S.J., Gentry, M., Brotheridge, S., Sluka, C.M., Wu, J.E., Crystal, J.D.: Rats remember items in context using episodic memory. Curr. Biol. **26**(20), 2821–2826 (2016)
595. Dudchenko, P.A., Wood, E.R., Eichenbaum, H.: Neurotoxic hippocampal lesions have no effect on odor span and little effect on odor recognition memory but produce significant impairments on spatial span, recognition, and alternation. J. Neurosci. **20**(8), 2964–2977 (2000)
596. Zola-Morgan, S., Squire, L.R., Amaral, D.G.: Human amnesia and the medial temporal region: enduring memory impairment following a bilateral lesion limited to field CA1 of the hippocampus. J. Neurosci. **6**(10), 2950–2967 (1986)
597. Rattenborg, N.D., Martinez-Gonzalez, D.: A bird-brain view of episodic memory. Behav. Brain Res. **222**(1), 236–245 (2011)
598. Dede, A.J., Frascino, J.C., Wixted, J.T., Squire, L.R.: Learning and remembering real-world events after medial temporal lobe damage. Proc. Natl. Acad. Sci. USA. **113**(47), 13480–13485 (2016)
599. Panoz-Brown, D., Iyer, V., Carey, L.M., Sluka, C.M., Rajic, G., Kestenman, J., Gentry, M., Brotheridge, S., Somekh, I., Corbin, H.E., Tucker, K.G., Almeida, B., Hex, S.B., Garcia, K.D., Hohmann, A.G., Crystal, J.D.: Replay of episodic memories in the rat. Curr. Biol. **28**(10), 1628–1634.e7 (2018)
600. Foster, D.J., Wilson, M.A.: Reverse replay of behavioural sequences in hippocampal place cells during the awake state. Nature. **440**(7084), 680–683 (2006)
601. Block, N.: Two neural correlates of consciousness. Trends Cogn. Sci. **9**(2), 46–52 (2005)
602. Kiani, R., Shadlen, M.N.: Representation of confidence associated with a decision by neurons in the parietal cortex. Science. **324**(5928), 759–764 (2009)
603. Kepecs, A., Uchida, N., Zariwala, H.A., Mainen, Z.F.: Neural correlates, computation and behavioural impact of decision confidence. Nature. **455**, 227 (2008)
604. Komura, Y., Nikkuni, A., Hirashima, N., Uetake, T., Miyamoto, A.: Responses of pulvinar neurons reflect a subject's confidence in visual categorization. Nat. Neurosci. **16**, 749 (2013)
605. Smith, J.D., Schull, J., Strote, J., McGee, K., Egnor, R., Erb, L.: The uncertain response in the bottlenosed dolphin (Tursiops truncatus). J. Exp. Psychol. Gen. **124**(4), 391–408 (1995)
606. Smith, J.D., Shields, W.E., Schull, J., Washburn, D.A.: The uncertain response in humans and animals. Cognition. **62**(1), 75–97 (1997)
607. Hampton, R.R.: Rhesus monkeys know when they remember. Proc. Natl. Acad. Sci. USA. **98**(9), 5359–5362 (2001)
608. Metcalfe, J.: Evolution of metacognition. In: Bjork, J.D.R.A. (ed.) Handbook of Memory Awareness and Memory, pp. 29–46. Psychology Press, New York (2008)
609. Nelson, T.O.: Consciousness and metacognition. Am. Psychol. **51**(2), 102–116 (1996)
610. Squire, L.R., Zola-Morgan, S.: The medial temporal lobe memory system. Science. **253**(5026), 1380–1386 (1991)
611. Clark, R.E., Manns, J.R., Squire, L.R.: Classical conditioning, awareness, and brain systems. Trends Cogn. Sci. **6**(12), 524–531 (2002)
612. Templer, V.L., Hampton, R.R.: Rhesus monkeys (Macaca mulatta) show robust evidence for memory awareness across multiple generalization tests. Anim. Cogn. **15**(3), 409–419 (2012)
613. Jozefowiez, J., Staddon, J.E.R., Cerutti, D.T.: Metacognition in animals: how do we know that they know? Comp. Cogn. Behav. Rev. **4**, 29–39 (2009)
614. Le Pelley, M.E.: Metacognitive monkeys or associative animals? Simple reinforcement learning explains uncertainty in nonhuman animals. J. Exp. Psychol. Learn. Mem. Cogn. **38**(3), 686–708 (2012)

615. Basile, B.M., Hampton, R.R., Suomi, S.J., Murray, E.A.: An assessment of memory awareness in tufted capuchin monkeys (Cebus apella). Anim. Cogn. **12**(1), 169–180 (2009)
616. Sutton, J.E., Shettleworth, S.J.: Memory without awareness: pigeons do not show metamemory in delayed matching to sample. J. Exp. Psychol. Anim. Behav. Process. **34**(2), 266–282 (2008)
617. Hampton, R.R.: Multiple demonstrations of metacognition in nonhumans: converging evidence or multiple mechanisms? Comp. Cogn. Behav. Rev. **4**, 17–28 (2009)
618. Kornell, N.: Where is the "meta" in animal metacognition? J. Comp. Psychol. **128**(2), 143–149 (2014)
619. Kornell, N., Son, L.K., Terrace, H.S.: Transfer of metacognitive skills and hint seeking in monkeys. Psychol. Sci. **18**(1), 64–71 (2007)
620. Morgan, G., Kornell, N., Kornblum, T., Terrace, H.S.: Retrospective and prospective metacognitive judgments in rhesus macaques (Macaca mulatta). Anim. Cogn. **17**(2), 249–257 (2014)
621. Call, J.: Do apes know that they could be wrong? Anim. Cogn. **13**(5), 689–700 (2010)
622. Hampton, R.R., Zivin, A., Murray, E.A.: Rhesus monkeys (Macaca mulatta) discriminate between knowing and not knowing and collect information as needed before acting. Anim. Cogn. **7**(4), 239–246 (2004)
623. Smith, J.D.: The study of animal metacognition. Trends Cogn. Sci. **13**(9), 389–396 (2009)
624. Carruthers, P.: How we know our own minds: the relationship between mindreading and metacognition. Behav. Brain Sci. **32**(2), 121–138 (2009).; discussion 138–182
625. Sakai, K.: Task set and prefrontal cortex. Annu. Rev. Neurosci. **31**, 219–245 (2008)
626. Shenhav, A., Botvinick, M.M., Cohen, J.D.: The expected value of control: an integrative theory of anterior cingulate cortex function. Neuron. **79**(2), 217–240 (2013)
627. Fernandez-Duque, D., Baird, J.A., Posner, M.I.: Executive attention and metacognitive regulation. Conscious. Cogn. **9**(2 Pt 1), 288–307 (2000)
628. Yeung, N.: Conflict monitoring and cognitive control. In: Ochsner, K., Kosslyn, S.M. (eds.) Oxford Handbook of Cognitive Neuroscience. Oxford University Press, Oxford (2013)
629. Shea, N., Boldt, A., Bang, D., Yeung, N., Heyes, C., Frith, C.D.: Supra-personal cognitive control and metacognition. Trends Cogn. Sci. **18**(4), 186–193 (2014)
630. Evans, J.S., Stanovich, K.E.: Dual-process theories of higher cognition: advancing the debate. Perspect. Psychol. Sci. **8**(3), 223–241 (2013)
631. Kahneman, D.: A perspective on judgment and choice: mapping bounded rationality. Am. Psychol. **58**(9), 697–720 (2003)
632. Bahrami, B., Olsen, K., Latham, P.E., Roepstorff, A., Rees, G., Frith, C.D.: Optimally interacting minds. Science. **329**(5995), 1081–1085 (2010)
633. Fusaroli, R., Bahrami, B., Olsen, K., Roepstorff, A., Rees, G., Frith, C., Tylen, K.: Coming to terms: quantifying the benefits of linguistic coordination. Psychol. Sci. **23**(8), 931–939 (2012)
634. Premack, D., Woodruff, G.: Does the chimpanzee have a theory of mind? Behav. Brain Sci. **1**, 515–526 (1978)
635. King, S.L., Connor, R.C., Krützen, M., Allen, S.J.: Cooperation-based concept formation in male bottlenose dolphins. Nat. Commun. **12**(1), 2373 (2021)
636. Gallup Jr., G.G.: Self-awareness and the emergence of mind in primates. Am. J. Primatol. **2**(3), 237–248 (1982)
637. Prinz, W.: Selbst im Spiegel. Die soziale Konstruktion von Subjektivität. Suhrkamp, Berlin (2013)
638. Frith, U., Happé, F.: Theory of mind and self-consciousness: what is it like to be autistic? Mind Lang. **14**(1), 82–89 (1999)
639. Frith, C., Frith, U.: Theory of mind. Curr. Biol. **15**(17), R644–R646 (2005)
640. Bartsch, K., Wellman, H.M.: Children Talk About the Mind, viii, 234 p. Oxford University Press, New York, Oxford (1995)
641. Heider, F., Simmel, M.: An experimental study of apparent behavior. Am. J. Psychol. **57**(2), 243–259 (1944)

642. Adolphs, R.: Social cognition and the human brain. Trends Cogn. Sci. **3**(12), 469–479 (1999)
643. Klin, A.: Attributing social meaning to ambiguous visual stimuli in higher-functioning autism and Asperger syndrome: The Social Attribution Task. J. Child Psychol. Psychiatry. **41**(7), 831–846 (2000)
644. Perner, J.: Understanding the Representational Mind Learning, Development, and Conceptual Change, XIV, 348 S. MIT Press, Cambridge, MA (1991)
645. Wellman, H.M., Cross, D., Watson, J.: Meta-analysis of theory-of-mind development: the truth about false belief. Child Dev. **72**(3), 655–684 (2001)
646. Flavell, J.H., Botkin, P.T., Fry, C.L., Wright, J.W., Jarvis, P.E.: The Development of Role-Taking and Communication Skills in Children. Wiley, New York (1968)
647. Flavell, J.H.: Cognitive development: children's knowledge about the mind. Annu. Rev. Psychol. **50**, 21–45 (1999)
648. Wimmer, H., Perner, J.: Beliefs about beliefs: representation and constraining function of wrong beliefs in young children's understanding of deception. Cognition. **13**(1), 103–128 (1983)
649. Perner, J., Frith, U., Leslie, A.M., Leekam, S.R.: Exploration of the autistic child's theory of mind: knowledge, belief, and communication. Child Dev. **60**(3), 688–700 (1989)
650. Baron-Cohen, S., Leslie, A.M., Frith, U.: Does the autistic child have a "theory of mind"? Cognition. **21**(1), 37–46 (1985)
651. Baron-Cohen, S., Tager-Flusberg, H., Lombardo, M.: Understanding Other Minds: Perspectives from Developmental Social Neuroscience, 3rd edn, xvii, 498 p. Oxford University Press, Oxford (2013)
652. Stone, V.E., Baron-Cohen, S., Knight, R.T.: Frontal lobe contributions to theory of mind. J. Cogn. Neurosci. **10**(5), 640–656 (1998)
653. Stuss, D.T., Gallup Jr., G.G., Alexander, M.P.: The frontal lobes are necessary for "theory of mind". Brain. **124**(Pt 2), 279–286 (2001)
654. Frith, C.D., Wolpert, D.M. (eds.): The Neuroscience of Social Interaction: Decoding, Imitating and Influencing the Actions of Others. Oxford University Press, Oxford (2004)
655. Jolly, A.: Lemur social behavior and primate intelligence. Science. **153**, 501–506 (1966)
656. Kummer, H.: Primate Societies. Harlan Davidson, Arlington Heights (1971)
657. Humphrey, N.K.: The social function of intellect. In: Bateson, P.P.G., Hinde, R.A. (eds.) Growing Points in Ethology, pp. 303–317. Cambridge University Press, Cambridge (1976)
658. Humphrey, N.K.: Nature's psychologists. In: Josephson, B., Ramachandran, V.S. (eds.) Consciousness and the Physical World, pp. 57–80. Pergamon, London (1980)
659. Humphrey, N.K.: The Inner Eye. Faber & Faber, London (1986)
660. Stone, T., Davies, M.: The mental simulation debate: a progress report. In: Carruthers, P., Smith, P.K. (eds.) Theories of Theories of Mind, pp. 119–137. Cambridge University Press, Cambridge, MA (1996)
661. Goldman, A.I.: Simulating Minds: The Philosophy, Psychology and Neuroscience of Mindreading. Oxford University Press, Oxford (2006)
662. Krebs, J.R., Dawkins, M.S.: Animal signals: mindreading and manipulation. In: Krebs, J.R., Davies, N.B. (eds.) Behavioural Ecology: An Evolutionary Approach, pp. 380–401. Blackwell Scientific Publications, Oxford (1984)
663. Carruthers, P., Smith, P.K.: Theories of Theories of Mind. Cambridge University Press, Cambridge (1996)
664. Byrne, R.W., Whiten, A.: Computation and mindreading in primate tactical deception. In: Whiten, A. (ed.) Natural Theories of Mind. Evolution, Development and Simulation of Everyday Mindreading, pp. 127–142. Basil Blackwell, Cambridge (1991)
665. Byrne, R.W., Whiten, A.: Cognitive evolution in primates: evidence from tactical deception. Man. **27**(3), 609–627 (1992)
666. Whiten, A., Byrne, R.W.: The Machiavellian intelligence hypotheses: editorial. In: Byrne, R.W., Whiten, A. (eds.) Machiavellian Intelligence. Social Expertise and the Evolution of the Intellect in Monkeys, Apes, and Humans, pp. 1–9. Clarendon, Oxford (1988)

667. Whiten, A.: The emergence of mindreading: steps towards an interdisciplinary enterprise. In: Whiten, A. (ed.) Natural Theories of Mind. Evolution, Development and Simulation of Everyday Mindreading, pp. 319–332. Basil Blackwell, Cambridge (1991)
668. Premack, D.: "Does the chimpanzee have a theory of mind" revisited. In: Byrne, R., Whiten, A. (eds.) Machiavellian Intelligence, pp. 160–179. Oxford University Press, Oxford (1988)
669. Povinelli, D.J., Nelson, K.E., Boysen, S.T.: Inferences about guessing and knowing by chimpanzees (Pan troglodytes). J. Comp. Psychol. **104**(3), 203–210 (1990)
670. Premack, D.: Gavagai! Or the Future History of the Animal Language Controversy The MIT Press Series in Learning, Development, and Conceptual Change, 164 p. MIT Press, Cambridge, MA, London (1986)
671. Woodruff, G., Premack, D.: Primative mathematical concepts in the chimpanzee: proportionality and numerosity. Nature. **293**(5833), 568–570 (1981)
672. Krupenye, C.: Guesser-knower paradigm. In: Vonk, J., Shackelford, T.K. (eds.) Encyclopedia of Animal Cognition and Behavior. Springer International Publishing AG, Cham (2017)
673. Povinelli, D.J., deBlois, S.: Young children's (Homo sapiens) understanding of knowledge formation in themselves and others. J. Comp. Psychol. **106**(3), 228–238 (1992)
674. Povinelli, D.J., Parks, K.A., Novak, M.A.: Do rhesus monkeys attribute knowledge and ignorance to others? J. Comp. Psychol. **105**, 318–325 (1991)
675. Flombaum, J.I., Santos, L.R.: Rhesus monkeys attribute perceptions to others. Curr. Biol. **15**(5), 447–452 (2005)
676. Perrett, D.I., Smith, P.A., Potter, D.D., Mistlin, A.J., Head, A.S., Milner, A.D., Jeeves, M.A.: Visual cells in the temporal cortex sensitive to face view and gaze direction. Proc. R. Soc. Lond. B Biol. Sci. **223**(1232), 293–317 (1985)
677. Schwab, C., Huber, L.: Obey or not obey? Dogs (Canis familiaris) behave differently in response to attentional states of their owners. J. Comp. Psychol. **120**(3), 169–175 (2006)
678. Wallis, L.J., Range, F., Muller, C.A., Serisier, S., Huber, L., Viranyi, Z.: Training for eye contact modulates gaze following in dogs. Anim. Behav. **106**, 27–35 (2015)
679. Schmidjell, T., Range, F., Huber, L., Viranyi, Z.: Do owners have a Clever Hans effect on dogs? Results of a pointing study. Front. Psychol. **3**(558), 1–15 (2012)
680. Huber, L., Racca, A., Scaf, B., Virányi, Z., Range, F.: Discrimination of familiar human faces in dogs (Canis familiaris). Learn. Motiv. **44**(4), 258–269 (2013)
681. Müller, C.A., Schmitt, K., Barber, A.L.A., Huber, L.: Dogs can discriminate emotional expressions of human faces. Curr. Biol. **25**, 1–5 (2015)
682. Huber, L.: How dogs perceive and understand us. Curr. Dir. Psychol. Sci. **25**(5), 339–344 (2016)
683. Maginnity, M.E., Grace, R.C.: Visual perspective taking by dogs (Canis familiaris) in a Guesser-Knower task: evidence for a canine theory of mind? Anim. Cogn. **17**(6), 1375–1392 (2014)
684. Povinelli, D.J., Eddy, T.J.: What young chimpanzees know about seeing. Monogr. Soc. Res. Child Dev. **61**(3), i–vi (1996)., 1–152; discussion 153–191
685. Kuroshima, H., Fujita, K., Adachi, I., Iwata, K., Fuyuki, A.: A Capuchin monkey (Cebus apella) recognizes when people do and do not know the location of food. Anim. Cogn. **6**(4), 283–291 (2003)
686. Held, S., Mendl, M., Devereux, C., Byrne, R.: Behaviour of domestic pigs in a visual perspective taking task.pdf. Behaviour. **138**(11–12), 1337–1354 (2001)
687. Pfungst, O.: Das Pferd des Herrn von Osten (Der Kluge Hans): Ein Beitrag zur experimentellen Tier- und Menschenpsychologie. Johann Ambrosius Barth, Leipzig (1907)
688. Catala, A., Mang, B., Wallis, L., Huber, L.: Dogs demonstrate perspective taking based on geometrical gaze following in a Guesser–Knower task. Anim. Cogn. **20**(4), 581–589 (2017)
689. Tomasello, M., Hare, B., Agnetta, B.: Chimpazees, Pan troglodytes, follow gaze direction geometrically. Anim. Behav. **58**, 769–777 (1999)
690. Povinelli, D.J., Eddy, T.J.: Chimpanzees: joint visual attention. Psychol. Sci. **7**(3), 129–135 (1996)

691. Bensky, M.K., Gosling, S.D., Sinn, D.L.: The world from a dog's point of view: a review and synthesis of dog cognition research. Adv. Study Behav. **45**, 209–406 (2013)
692. Kaminski, J., Marshall-Pescini, S. (eds.): The Social Dog: Behaviour and Cognition, xiv, 404 p. Academic Press/Elsevier, San Diego (2014)
693. Wynne, C.D.L.: What is special about dog cognition? Curr. Dir. Psychol. Sci. **25**(5), 345–350 (2016)
694. Met, A., Miklósi, A., Lakatos, G.: Gaze-following behind barriers in domestic dogs. Anim. Cogn. **17**(6), 1401–1405 (2014)
695. Brauer, J., Keckeisen, M., Pitsch, A., Kaminski, J., Call, J., Tomasello, M.: Domestic dogs conceal auditory but not visual information from others. Anim. Cogn. **16**(3), 351–359 (2013)
696. Kaminski, J., Pitsch, A., Tomasello, M.: Dogs steal in the dark. Anim. Cogn. **16**(3), 385–394 (2013)
697. Whiten, A.: Humans are not alone in computing how others see the world. Anim. Behav. **86**(2), 213–221 (2013)
698. Heyes, C.: Cues, convergence and a curmudgeon: a reply to Povinelli. Anim. Behav. **48**(1), 242–244 (1994)
699. Udell, M.A., Wynne, C.D.: Reevaluating canine perspective-taking behavior. Learn. Behav. **39**(4), 318–323 (2011)
700. Udell, M.A., Dorey, N.R., Wynne, C.D.: Can your dog read your mind?: understanding the causes of canine perspective taking. Learn. Behav. **39**(4), 289–302 (2011)
701. Buckner, C.: The semantic problem(s) with research on animal mind-reading. Mind Lang. **29**(5), 566–589 (2014)
702. Penn, D.C., Povinelli, D.J.: On the lack of evidence that non-human animals possess anything remotely resembling a "theory of mind". Philos. Trans. R. Soc. Lond. Ser. B Biol. Sci. **362**(1480), 731–744 (2007)
703. Tomasello, M., Hare, B., Fogleman, T.: The ontogeny of gaze following in chimpanzees, Pantroglodytes, and rhesus macaques, Macaca mulatta. Anim. Behav. **61**, 335–343 (2001)
704. Schmelz, M., Call, J.: Theory of mind in primates and humans, Comparative evolution of. In: The International Encyclopedia of Anthropology, pp. 1–10. American Cancer Society (2018)
705. Povinelli, D.J., Eddy, T.J.: Factors influencing young chimpanzees' (Pan troglodytes) recognition of attention. J. Comp. Psychol. **110**, 336–345 (1996)
706. Call, J., Tomasello, M.: Does the chimpanzee have a theory of mind? 30 years later. Trends Cogn. Sci. **12**(5), 187–192 (2008)
707. Schmelz, M., Call, J.: The psychology of primate cooperation and competition: a call for realigning research agendas. Philos. Trans. R. Soc. Lond. Ser. B Biol. Sci. **371**(1686), 20150067 (2016)
708. Hare, B., Call, J., Agnetta, B., Tomasello, M.: Chimpanzees know what conspecifics do and do not see. Anim. Behav. **59**(4), 771–785 (2000)
709. Hare, B., Call, J., Tomasello, M.: Do chimpanzees know what conspecifics know? Anim. Behav. **61**, 139–151 (2001)
710. Barton, R.A., Dunbar, R.I.M.: Evolution of the social brain. In: Whiten, A., Byrne, R.W. (eds.) Machiavellian Intelligence II. Extensions and Evaluations, pp. 240–263. Cambridge University Press, Cambridge (1997)
711. Emery, N.J., Clayton, N.S.: Evolution of the avian brain and intelligence. Curr. Biol. **15**(23), R946–R950 (2005)
712. Clayton, N.S., Emery, N.J.: The social life of corvids. Curr. Biol. **17**(16), R652–R656 (2007)
713. Güntürkün, O., Bugnyar, T.: Cognition without cortex. Trends Cogn. Sci. **20**(4), 291–303 (2016)
714. Jarvis, E.D., et al.: Avian brains and a new understanding of vertebrate brain evolution. Nat. Rev. Neurosci. **6**(2), 151–159 (2005)
715. Clayton, N.S., Dally, J.M., Emery, N.J.: Social cognition by food-caching corvids. The western scrub-jay as a natural psychologist. Philos. Trans. R. Soc. Lond. Ser. B Biol. Sci. **362**(1480), 507–522 (2007)

716. Bugnyar, T.: Social cognition in ravens. Comp. Cogn. Behav. Rev. **8**, 1–12 (2013)
717. Bond, A.B., Kamil, A.C., Balda, R.P.: Social complexity and transitive inference in corvids. Anim. Behav. **65**, 479–487 (2003)
718. de Waal, F., Tyack, P.: Animal Social Complexity: Intelligence, Culture, and Individualized Societies, xiv, 616 p. Harvard University Press, Cambridge, MA (2003)
719. Tomback, D.F.: Foraging strategies of Clark's nutcracker. Living Bird. **16**, 123–161 (1978)
720. Vander Wall, S.B., Balda, R.P.: Co-adaptation of the Clark's nutcracker and pinyon pine for efficient seed harvest and dispersal. Ecol. Monogr. **47**, 89–111 (1977)
721. Balda, R.P., Kamil, A.C.: Long-term spatial memory in Clark's nutcracker, Nucifraga columbiana. Anim. Behav. **44**(4), 761–769 (1992)
722. Kamil, A.C., Balda, R.P., Good, S.: Patterns of movement and orientation during caching and recovery by Clark's nutcrackers, Nucifraga columbiana. Anim. Behav. **57**(6), 1327–1335 (1999)
723. Bugnyar, T., Kotrschal, K.: Observational learning and the raiding of food caches in ravens, Corvus corax: is it "tactical deception"? Anim. Behav. **64**, 185–195 (2002)
724. Heinrich, B., Pepper, J.W.: Influence of competitors on caching behaviour in the common raven, Corvus corax. Anim. Behav. **56**, 1083–1090 (1998)
725. Bednekoff, P.A., Balda, R.P.: Observational spatial memory in Clark's nutcrackers and Mexican jays. Anim. Behav. **52**(4), 833–839 (1996)
726. Watanabe, S., Clayton, N.S.: Observational visuospatial encoding of the cache locations of others by western scrub-jays (Aphelocoma californica). J. Ethol. **25**(3), 271–279 (2007)
727. Bugnyar, T., Heinrich, B.: Ravens, Corvus corax, differentiate between knowledgeable and ignorant competitors. Proc. Biol. Sci. **272**(1573), 1641–1646 (2005)
728. Dally, J.M., Emery, N.J., Clayton, N.S.: Cache protection strategies by western scrub-jays, Aphelocoma californica: implications for social cognition. Anim. Behav. **70**(6), 1251–1263 (2005)
729. Dally, J.M., Emery, N.J., Clayton, N.S.: Cache protection strategies by western scrub-jays (Aphelocoma californica): hiding food in the shade. Proc. Biol. Sci. **271**(Suppl 6), S387–S390 (2004)
730. Dally, J.M., Emery, N.J., Clayton, N.S.: The social suppression of caching in western scrub-jays (Aphelocoma californica). Behaviour. **142**, 961–977 (2005)
731. Emery, N.J., Clayton, N.S.: Effects of experience and social context on prospective caching strategies by scrub jays. Nature. **414**(6862), 443–446 (2001)
732. Emery, N.J., Clayton, N.S.: Comparing the complex cognition of birds and primates. In: Rogers, L.J., Kaplan, G.S. (eds.) Comparative vertebrate cognition, pp. 3–55. Kluwer Academic Press, The Hague (2004)
733. Bugnyar, T., Kotrschal, K.: Scrounging tactics in free-ranging ravens. Ethology. **108**, 993–1009 (2002)
734. Bugnyar, T., Kotrschal, K.: Leading a conspecific away from food in ravens (Corvus corax)? Anim. Cogn. **7**(2), 69–76 (2004)
735. Bugnyar, T., Heinrich, B.: Pilfering ravens, Corvus corax, adjust their behaviour to social context and identity of competitors. Anim. Cogn. **9**(4), 369–376 (2006)
736. Bugnyar, T.: Knower-guesser differentiation in ravens: others' viewpoints matter. Proc. R. Soc. B Biol. Sci. **278**(1705), 634–640 (2011)
737. Heyes, C.M.: Theory of mind in nonhuman primates. Behav. Brain Sci. **21**(1), 101–148 (1998)
738. Bugnyar, T., Stowe, M., Heinrich, B.: Ravens, Corvus corax, follow gaze direction of humans around obstacles. Proc. Biol. Sci. **271**(1546), 1331–1336 (2004)
739. Bugnyar, T., Stöwe, M., Heinrich, B.: The ontogeny of caching in ravens, Corvus corax. Anim. Behav. **74**(4), 757–767 (2007)
740. Schloegl, C., Kotrschal, K., Bugnyar, T.: Gaze following in common ravens, Corvus corax: ontogeny and habituation. Anim. Behav. **74**(4), 769–778 (2007)
741. Tomasello, M., Call, J., Hare, B.: Chimpanzees understand psychological states – the question is which ones and to what extent. Trends Cogn. Sci. **7**(4), 153–156 (2003)

742. Gerrans, P.: The theory of mind module in evolutionary psychology. Biol. Philos. **17**(3), 305–321 (2002)
743. Heyes, C.: Animal mindreading: what's the problem? Psychon. Bull. Rev. **22**(2), 313–327 (2015)
744. Heyes, C.: Submentalizing: I am not really reading your mind. Perspect. Psychol. Sci. **9**(2), 131–143 (2014)
745. Bugnyar, T., Reber, S.A., Buckner, C.: Ravens attribute visual access to unseen competitors. Nat. Commun. **7**, 10506 (2016)
746. Kaminski, J., Call, J., Tomasello, M.: Chimpanzees know what others know, but not what they believe. Cognition. **109**(2), 224–234 (2008)
747. Karg, K., Schmelz, M., Call, J., Tomasello, M.: Differing views: can chimpanzees do level 2 perspective-taking? Anim. Cogn. **19**(3), 555–564 (2016)
748. Lohmann, H., Tomasello, M.: The role of language in the development of false belief understanding: a training study. Child Dev. **74**(4), 1130–1144 (2003)
749. Call, J., Tomasello, M.: A nonverbal false belief task: the performance of children and great apes. Child Dev. **70**(2), 381–395 (1999)
750. Flavell, J.H., Shipstead, S.G., Croft, K.: Young children's knowledge about visual perception: hiding objects from others. Child Dev. **49**(4), 1208–1211 (1978)
751. O'Connell, S., Dunbar, R.: A test for comprehension of false belief in chimpanzees. Evol. Cogn. **9**(2), 131–140 (2003)
752. Krachun, C., Carpenter, M., Call, J., Tomasello, M.: A competitive nonverbal false belief task for children and apes. Dev. Sci. **12**(4), 521–535 (2009)
753. Karg, K., Schmelz, M., Call, J., Tomasello, M.: Chimpanzees strategically manipulate what others can see. Anim. Cogn. **18**(5), 1069–1076 (2015)
754. Crockford, C., Wittig, R.M., Mundry, R., Zuberbuhler, K.: Wild chimpanzees inform ignorant group members of danger. Curr. Biol. **22**(2), 142–146 (2012)
755. Krachun, C., Call, J.: Chimpanzees (Pan troglodytes) know what can be seen from where. Anim. Cogn. **12**(2), 317–331 (2009)
756. Karg, K., Schmelz, M., Call, J., Tomasello, M.: All great ape species (Gorilla gorilla, Pan paniscus, Pan troglodytes, Pongo abelii) and two-and-a-half-year-old children (Homo sapiens) discriminate appearance from reality. J. Comp. Psychol. **128**(4), 431–439 (2014)
757. Vonk, J., Povinelli, D.J.: Social and physical reasoning in human-reared chimpanzees: preliminary studies. In: Eilan, N., Lerman, H., Roessler, J. (eds.) Perception, Causation, and Objectivity. Issues in Philosophy and Psychology, pp. 342–352. Oxford University Press, Oxford (2011)
758. Meltzoff, A.N., Brooks, R.: Self-experience as a mechanism for learning about others: a training study in social cognition. Dev. Psychol. **44**(5), 1257–1265 (2008)
759. Povinelli, D.J., Vonk, J.: Chimpanzee minds: suspiciously human? Trends Cogn. Sci. **7**(4), 157–160 (2003)
760. Karg, K., Schmelz, M., Call, J., Tomasello, M.: The goggles experiment: can chimpanzees use self-experience to infer what a competitor can see? Anim. Behav. **105**, 211–221 (2015)
761. Hare, B., Tomasello, M.: Chimpanzees are more skillful in competitive than in cooperative tasks. Anim. Behav. **68**, 571–581 (2004)
762. Krupenye, C., Kano, F., Hirata, S., Call, J., Tomasello, M.: Great apes anticipate that other individuals will act according to false beliefs. Science. **354**(6308), 110–114 (2016)
763. Southgate, V., Senju, A., Csibra, G.: Action anticipation through attribution of false belief by 2-year-olds. Psychol. Sci. **18**(7), 587–592 (2007)
764. Perner, J., Ruffman, T.: Infants' insight into the mind: how deep? Science. **308**(5719), 214–216 (2005)
765. Baillargeon, R., Scott, R.M., He, Z.: False-belief understanding in infants. Trends Cogn. Sci. **14**(3), 110–118 (2010)
766. Senju, A., Southgate, V., Snape, C., Leonard, M., Csibra, G.: Do 18-month-olds really attribute mental states to others? A critical test. Psychol. Sci. **22**(7), 878–880 (2011)

767. Kano, F., Krupenye, C., Hirata, S., Tomonaga, M., Call, J.: Great apes use self-experience to anticipate an agent's action in a false-belief test. Proc. Natl. Acad. Sci. USA. **116**(42), 20904–20909 (2019)
768. Buttelmann, D., Carpenter, M., Tomasello, M.: Eighteen-month-old infants show false belief understanding in an active helping paradigm. Cognition. **112**(2), 337–342 (2009)
769. Call, J., Tomasello, M.: Distinguishing intentional from accidental actions in orangutans (Pongo pygmaeus), chimpanzees (Pan troglodytes), and human children (Homo sapiens). J. Comp. Psychol. **112**(2), 192–206 (1998)
770. Warneken, F., Tomasello, M.: Altruistic helping in human infants and young chimpanzees. Science. **311**, 1301–1303 (2006)
771. Buttelmann, D., Buttelmann, F., Carpenter, M., Call, J., Tomasello, M.: Great apes distinguish true from false beliefs in an interactive helping task. PLoS One. **12**(4), e0173793 (2017)
772. Huber, L., Lamm, C.: Understanding dog cognition by functional magnetic resonance imaging. Learn. Behav. **45**, 1–2 (2017)
773. Karl, S., Boch, M., Viranyi, Z., Lamm, C., Huber, L.: Training pet dogs for eye-tracking and awake fMRI. Behav. Res. Methods. **52**(2), 838–856 (2019)
774. Boch, M., Karl, S., Sladky, R., Huber, L., Lamm, C., Wagner, I.C.: Tailored haemodynamic response function increases detection power of fMRI in awake dogs (Canis familiaris). NeuroImage. **224**, 117414 (2021)
775. Karl, S., Boch, M., Zamansky, A., van der Linden, D., Wagner, I.C., Völter, C., Lamm, C., Huber, L.: Exploring the dog-human relationship using a multi-method approach: combining fMRI, eye-tracking and behavioural measures. Sci. Rep. **10**, 22273 (2020)
776. Karl, S., Sladky, R., Lamm, C., Huber, L.: Neural responses of pet dogs witnessing their caregiver's positive interactions with a conspecific: an fMRI study. Cereb. Cortex Commun. **2**, 1–12 (2021)
777. Lonardo, L., Völter, C., Lamm, C., Huber, L.: Dogs follow human misleading suggestions more often when the informant has a false belief. Proc. R. Soc. Lond. B. **2021**(288), 20210906 (1955)
778. Andrews, K.: Apes track false beliefs but might not understand them. Learn. Behav. **46**(1), 3–4 (2018)
779. Andrews, K.: Do apes attribute beliefs to predict behavior? Harv. Rev. Philos. **25**, 89–110 (2018)
780. Andrews, K.: Do Apes Read Minds? Toward a New Folk Psychology. MIT Press, Cambridge, MA (2012)
781. Kano, F., Krupenye, C., Hirata, S., Call, J., Tomasello, M.: Submentalizing cannot explain belief-based action anticipation in apes. Trends Cogn. Sci. **21**(9), 633–634 (2017)
782. Krupenye, C., Kano, F., Hirata, S., Call, J., Tomasello, M.: A test of the submentalizing hypothesis: apes' performance in a false belief task inanimate control. Commun. Integr. Biol. **10**(4), e1343771 (2017)
783. Heyes, C.: Apes submentalise. Trends Cogn. Sci. **21**(1), 1–2 (2017)
784. Kano, F., Call, J.: Great apes generate goal-based action predictions: an eye-tracking study. Psychol. Sci. **25**(9), 1691–1698 (2014)
785. Von Frisch, K.: The Dancing Bees: An Account of the Life and Senses of the Honey Bee. Harvest Books, New York (1953)
786. Von Frisch, K.: The Dance Language and Orientation of Bees. Harvard University Press, Cambridge, MA (1967)
787. Dyer, F.C.: The biology of the dance language. Annu. Rev. Entomol. **47**(1), 917–949 (2002)
788. Su, S., Cai, F., Si, A., Zhang, S., Tautz, J., Chen, S.: East learns from West: Asiatic honeybees can understand dance language of European honeybees. PLoS One. **3**(6), e2365 (2008)
789. Plato: Cratylus, Parmenides, Greater Hippias, Lesser Hippias. With translation by Harold N. Fowler. Loeb Classical Library 167. Harvard Univ. Press (originally published 1926)
790. Bühler, K.: Sprachtheorie: Die Darstellungsfunktion der Sprache. Gustav Fischer, Jena (1934)

791. Herrnstein, R.J., Loveland, D.H., Cable, C.: Natural concepts in pigeons. J. Exp. Psychol. Anim. Behav. Process. **2**(4), 285–302 (1976)
792. Smith, E.E., Medin, D.L.: Categories and Concepts. Harvard University Press, Cambridge, MA (1981)
793. Margolis, E., Laurence, S. (eds.): Concepts. MIT Press, Cambridge, MA (1999)
794. Fitch, W.T.: The Evolution of Language, xii, 610 p. Cambridge University Press, Cambridge, New York (2010)
795. Wich, S.A., Swartz, K.B., Hardus, M.E., Lameira, A.R., Stromberg, E., Shumaker, R.W.: A case of spontaneous acquisition of a human sound by an orangutan. Primates. **50**(1), 56–64 (2009)
796. von Humboldt, W.: Über die Verschiedenheit des menschlichen Sprachbaues und ihren Einfluss auf die geistige Entwicklung des Menschengeschlechts. F. Dümmler, Berlin (1836)
797. Hauser, M.D., Chomsky, N., Fitch, W.T.: The faculty of language: what is it, who has it, and how did it evolve? Science. **298**(5598), 1569–1579 (2002)
798. Fitch, W.T.: The evolution of language: a comparative review. Biol. Philos. **20**(2–3), 193–230 (2005)
799. Riedl, R.: A systems-analytical approach to macroevolutionary phenomena. Q. Rev. Biol. **52**(4), 351–370 (1977)
800. Gould, S.J., Lewontin, R.C.: The spandrels of San Marco and the Panglossian paradigm: a critique of the adaptationist programme. Proc. R. Soc. Lond. B Biol. Sci. **205**(1161), 581–598 (1979)
801. Maynard Smith, J.: The Theory of Evolution. Cambridge University Press, Cambridge (1993)
802. Riedl, R.: Order in Living Organisms. A Systems Analysis of Evolution. Wiley, Chichester, New York, Brisbane, Toronto (1978)
803. Chomsky, N.: Knowledge of Language. Its Nature, Origin, and Use Convergence, XXIX, 307 S. Praeger, Westport (1986)
804. Kellogg, W.N.: Communication and language in the home-raised chimpanzee. Science. **162**, 423–438 (1968)
805. Gardner, R.A., Gardner, B.T.: Teaching sign language to a chimpanzee. Science. **187**, 644–672 (1969)
806. Premack, D.: On the assessment of language-competence in the chimpanzee. In: Schrier, A.M., Stollnitz, F. (eds.) Behavior of Nonhuman Primates, pp. 186–228. Academic Press, New York (1971)
807. Rumbaugh, D.M.: Language Learning by a Chimpanzee. Academic Press, New York (1977)
808. Gardner, R.A., Gardner, B.T., Van Cantfort, T.E.: Teaching Sign Language to Chimpanzees. SUNY Press, Albany (1989)
809. Miles, H.L.: Language acquisition in apes and children. In: Peng, F.C.C. (ed.) Sign Language and Language Acquisition in Man and Ape, pp. 108–120. Westview Press, Boulder (1978)
810. Patterson, F.: Linguistic capabilities of a lowland gorilla. In: Peng, F.C.C. (ed.) Sign Language and Language Acquisition in Man and Ape, pp. 161–201. Westview Press, Boulder (1978)
811. Pepperberg, I.M.: Animal language studies: what happened? Psychon. Bull. Rev. **24**, 181–185 (2017)
812. Hauser, M.D.: The Evolution of Communication, xii, 760 p. MIT Press, Cambridge, MA, London (1996)
813. Corballis, M.C.: From Hand to Mouth: The Origins of Language. Princeton University Press, Princeton (2003)
814. Seyfarth, R.M., Cheney, D.L.: Signalers and receivers in animal communication. Annu. Rev. Psychol. **54**, 145–173 (2003)
815. Rendall, D., Owren, M.J., Ryan, M.J.: What do animal signals mean? Anim. Behav. **78**(2), 233–240 (2009)
816. Rumbaugh, D.M.: Language Learning by a Chimpanzee: The Lana Project. Academic Press, New York (2014)

817. Savage-Rumbaugh, E.S.: Ape Language: From Conditioned Response to Symbol. Columbia University Press, New York (1986)
818. Savage-Rumbaugh, S., Shanker, S.G., Taylor, T.J.: Apes, Language and the Human Mind. Oxford University Press, New York (1998)
819. Herman, L.M., Richards, D.G., Wolz, J.P.: Comprehension of sentences by bottlenosed dolphins. Cognition. **16**, 129–219 (1984)
820. Terrace, H.S.: Nim, X, 303 p. Knopf, New York (1979)
821. Terrace, H.S., Petitto, L.A., Sanders, R.J., Bever, T.G.: Ape language. Science. **211**(4477), 87–88 (1981)
822. Terrace, H.S.: Animal cognition: thinking without language. In: Weiskrantz, L. (ed.) Animal Intelligence, pp. 113–128. Clarendon Press, Oxford (1985)
823. Sebeok, T.A., Rosenthal, R.: The Clever Hans Phenomenon: Communication with Horses, Whales, Apes, and People Annals of the New York Academy of Sciences, VIII, 311 p. New York Academy of Sciences (Annals), New York (1980)
824. Pepperberg, I.M.: Grey parrot numerical competence: a review. Anim. Cogn. **9**(4), 377–391 (2006)
825. Matsuzawa, T.: Symbolic representation of number in chimpanzees. Curr. Opin. Neurobiol. **19**, 92–98 (2009)
826. Boysen, S.T., Capaldi, E.J.: The Development of Numerical Competence: Animal and Human Models. Psychology Press, New York (2014)
827. Schusterman, R.J., Krieger, K.: Artificial language comprehension and size transposition by a California sea lion (Zalophus californianus). J. Comp. Psychol. **100**(4), 348–355 (1986)
828. Herman, L.M.: Cognition and language competencies of bottlenosed dolphins. In: Schusterman, R.J., Thomas, J.A., Wood, F.G. (eds.) Dolphin Cognition and Behavior: A Comparative Approach, pp. 221–252. Erlbaum, Hillsdale (1986)
829. Schusterman, R.J., Gisiner, R.C.: Artificial language comprehension in dolphins and sea lions: the essential cognitive skills. Psychol. Rec. **38**, 311–348 (1988)
830. Kastak, C.R., Schusterman, R.J.: Sea lions and equivalence: expanding classes by exclusion. J. Exp. Anal. Behav. **78**, 449–465 (2002)
831. Gentner, T.Q., Fenn, K.M., Margoliash, D., Nusbaum, H.C.: Recursive syntactic pattern learning by songbirds. Nature. **440**, 1204–1207 (2006)
832. Fitch, W.T., Hauser, M.D.: Computational constraints on syntactic processing in a nonhuman primate. Science. **303**(5656), 377–380 (2004)
833. Zuberbühler, K.: Referential signaling in non-human primates: cognitive precursors and limitations for the evolution of language. Adv. Study Behav. **33**, 265–307 (2003)
834. Searcy, W.A., Nowicki, S.: The Evolution of Animal Communication: Reliability and Deception in Signaling System. Princeton University Press, Princeton (2005)
835. Marler, P.R.: Avian and primate communication: the problem of natural categories. Neurosci. Biobehav. Rev. **6**, 87–94 (1982)
836. Endler, J.A.: Some general comments on the evolution and design of animal communication systems. Philos. Trans. R. Soc. Lond. Ser. B Biol. Sci. **340**, 215–225 (1993)
837. Tinbergen, N.: On aims and methods of ethology. Z. Tierpsychol. **20**(4), 410–433 (1963)
838. Bateson, P., Laland, K.N.: Tinbergen's four questions: an appreciation and an update. Trends Ecol. Evol. **28**(12), 712–718 (2013)
839. Hauser, M.D., Konishi, M.: The Design of Animal Communication, xi, 701 p. MIT Press, Cambridge, MA (1999)
840. Rosenqvist, G., Admundsen, T., Esprnark, Y. (eds.): Animal Signals: Signalling and Signal Design in Animal Communication. Tapir Akademika Publishing Press, Trondheim (2000)
841. Bradbury, J.W., Vehrencamp, S.L.: Principles of Animal Communication, 2nd edn. Sinauer Association, Sunderland (2011)
842. Kaplan, G.: Animal communication. Wiley Interdiscip. Rev. Cogn. Sci. **5**(6), 661–677 (2014)
843. Todt, D., Naguib, M.: Vocal interactions in birds: the use of song as a model in communication. In: Advances in the Study of Behavior, pp. 247–296. Elsevier (2000)

844. Naguib, M., Janik, V., Clayton, N., Zuberbuhler, K.: Vocal Communication in Birds and Mammals. Academic Press, Oxford (2009)
845. Beecher, M.D., Akçay, Ç.: Friends and enemies: how social dynamics shape communication and song learning in song sparrows. Anim. Behav. **3**, 33–61 (2014)
846. Pepperberg, I.M.: Vocal communication in nonhuman animals: view from the wings. Anim. Behav. Cogn. **7**(2), 95–100 (2020)
847. Tyack, P.L.: Dolphins whistle a signature tune. Science. **289**(5483), 1310–1311 (2000)
848. Herman, L.M.: The multiple functions of male song within the humpback whale (Megaptera novaeangliae) mating system: review, evaluation, and synthesis. Biol. Rev. **92**(3), 1795–1818 (2017)
849. Wyatt, T.D.: Pheromones and Animal Behavior: Chemical Signals and Signatures, 2nd edn. Cambridge University Press, Cambridge (2014)
850. Higham, J.P., Hebets, E.A.: An introduction to multimodal communication. Behav. Ecol. Sociobiol. **67**(9), 1381–1388 (2013)
851. Preininger, D., Boeckle, M., Freudmann, A., Starnberger, I., Sztatecsny, M., Hödl, W.: Multimodal signaling in the small torrent frog (Micrixalus saxicola) in a complex acoustic environment. Behav. Ecol. Sociobiol. **67**(9), 1449–1456 (2013)
852. Zuberbühler, K.: Survivor signals: the biology and psychology of animal alarm calling. Adv. Study Behav. **40**, 277–322 (2009)
853. Seyfarth, R.M., Cheney, D.L., Marler, P.: Monkey responses to three different alarm calls: evidence of predator classification and semantic communication. Science. **210**(4471), 801–803 (1980)
854. Seyfarth, R.M., Cheney, D.L., Marler, P.: Vervet monkey alarm calls: semantic communication in a free-ranging primate. Anim. Behav. **28**, 1070–1094 (1980)
855. Cäsar, C., Byrne, R.W., Hoppitt, W., Young, R.J., Zuberbühler, K.: Evidence for semantic communication in titi monkey alarm calls. Anim. Behav. **84**(2), 405–411 (2012)
856. Cäsar, C., Byrne, R., Young, R.J., Zuberbühler, K.: The alarm call system of wild black-fronted titi monkeys, Callicebus nigrifrons. Behav. Ecol. Sociobiol. **66**(5), 653–667 (2012)
857. Cäsar, C., Zuberbühler, K., Young, R.J., Byrne, R.W.: Titi monkey call sequences vary with predator location and type. Biol. Lett. **9**(5), 20130535 (2013)
858. Schlenker, P., Chemla, E., Cäsar, C., Ryder, R., Zuberbühler, K.: Titi semantics: context and meaning in titi monkey call sequences. Nat. Lang. Linguist. Theory. **35**(1), 271–298 (2017)
859. Schlenker, P., Chemla, E., Zuberbühler, K.: Semantics and pragmatics of monkey communication. In: Oxford Research Encyclopedia of Linguistics. Oxford University Press, New York (2017)
860. Zuberbuhler, K.: Language evolution: the origin of meaning in primates. Curr. Biol. **16**(4), R123–R125 (2006)
861. Zuberbühler, K.: The origins of meaning: language in the light of evolution. J. Hum. Evol. **55**(1), 186–186 (2008)
862. Schlenker, P., Chemla, E., Schel, A.M., Fuller, J., Gautier, J.-P., Kuhn, J., Veselinović, D., Arnold, K., Cäsar, C., Keenan, S.: Formal monkey linguistics. Theoret. Linguist. **42**(1–2), 1–90 (2016)
863. Schlenker, P., Chemla, E., Zuberbühler, K.: What do monkey calls mean? Trends Cogn. Sci. **20**(12), 894–904 (2016)
864. Zuberbühler, K.: Evolutionary roads to syntax. Anim. Behav. **151**, 259–265 (2019)
865. Ouattara, K., Lemasson, A., Zuberbühler, K.: Campbell's monkeys use affixation to alter call meaning. PLoS One. **4**(11), e7808 (2009)
866. Engesser, S., Crane, J.M., Savage, J.L., Russell, A.F., Townsend, S.W.: Experimental evidence for phonemic contrasts in a nonhuman vocal system. PLoS Biol. **13**(6), e1002171 (2015)
867. Bowling, D.L., Fitch, W.T.: Do animal communication systems have phonemes? Trends Cogn. Sci. **19**(10), 555–557 (2015)
868. Marler, P.: Origins of music and speech: insights from animals. In: Wallin, N.L., Merker, B., Brown, S. (eds.) The Origins of Music, pp. 31–48. MIT Press, Cambridge, MA (2001)

869. Fitch, W.T.: The biology and evolution of music: a comparative perspective. Cognition. **100**(1), 173–215 (2006)
870. Janik, V.M., Sayigh, L.S., Wells, R.S.: Signature whistle shape conveys identity information to bottlenose dolphins. Proc. Natl. Acad. Sci. USA. **103**, 8293–8297 (2006)
871. Sayigh, L.S., Tyack, P.L., Wells, R.S., Solow, A.R., Scott, M.D., Irvine, A.B.: Individual recognition in wild bottlenose dolphins: a field test using playback experiments. Anim. Behav. **57**(1), 41–50 (1999)
872. Herman, L.M.: Intelligence and rational behavior in the bottlenosed dolphin. In: Hurley, S., Nudds, M. (eds.) Rational Animals? pp. 439–467. Oxford University Press, Oxford (2006)
873. Cheney, D.L., Seyfarth, R.M.: Baboon Metaphysics. The Evolution of a Social Mind. University of Chicago Press, Chicago (2007)
874. Seyfarth, R.M., Cheney, D.L., Bergman, T.J.: Primate social cognition and the origins of language. Trends Cogn. Sci. **9**(6), 264–266 (2005)
875. Seyfarth, R.M., Cheney, D.L.: The evolution of language from social cognition. Curr. Opin. Neurobiol. **28**, 5–9 (2014)
876. Seyfarth, R.M., Cheney, D.L.: The Social Origins of Language. Princeton University Press, Princeton (2017)
877. Hurford, J.: The role of expression and representation in language evolution. In: Wray, A. (ed.) The Transition to Language, pp. 311–334. Oxford University Press, New York (2002)
878. Hespos, S.J., Spelke, E.: Conceptual precursors to language. Nature. **430**, 453–456 (2004)
879. Engh, A.L., Hoffmeier, R.R., Cheney, D.L., Seyfarth, R.M.: Who, me? Can baboons infer the target of vocalizations? Anim. Behav. **71**, 381–387 (2006)
880. Slocombe, K.E., Zuberbuhler, K.: Chimpanzees modify recruitment screams as a function of audience composition. Proc. Natl. Acad. Sci. USA. **104**(43), 17228–17233 (2007)
881. Wittig, R.M., Crockford, C., Seyfarth, R.M., Cheney, D.L.: Vocal alliances in Chacma baboons (Papio hamadryas ursinus). Behav. Ecol. Sociobiol. **61**(6), 899–909 (2007)
882. Wittig, R.M., Crockford, C., Wikberg, E., Seyfarth, R.M., Cheney, D.L.: Kin-mediated reconciliation substitutes for direct reconciliation in female baboons. Proc. R. Soc. B Biol. Sci. **274**(1613), 1109–1115 (2007)
883. Seyfarth, R.M., Cheney, D.L.: Animal cognition: chimpanzee alarm calls depend on what others know. Curr. Biol. **22**(2), R51–R52 (2012)
884. Crockford, C., Wittig, R.M., Zuberbühler, K.: Vocalizing in chimpanzees is influenced by social-cognitive processes. Sci. Adv. **3**(11), e1701742 (2017)
885. Enard, W., Przeworski, M., Fisher, S.E., Lai, C.S., Wiebe, V., Kitano, T., Monaco, A.P., Paabo, S.: Molecular evolution of FOXP2, a gene involved in speech and language. Nature. **418**(6900), 869–872 (2002)
886. Haesler, S., Wada, K., Nshdejan, A., Morrisey, E.E., Lints, T., Jarvis, E.D., Scharff, C.: FoxP2 expression in avian vocal learners and non-learners. J. Neurosci. **24**(13), 3164–3175 (2004)
887. Tomasello, M.: A Natural History of Human Thinking, 178 S. Harvard University Press, Cambridge, MA (2014)
888. Davidson, D. (ed.): Inquiries into Truth and Interpretation. Clarendon Press, Oxford (1984)
889. Riedl, R.: Kultur – Spätzündung der Evolution? R. Piper & Co, München (1987)
890. Kattmann, U.: Der Mensch … ein besonderes Lebewesen. Die Neue Brehmbücherei, Magdeburg (2020)
891. Freud, S.: The Interpretation of Dreams (orig. Die Traumdeutung, 1900, Deuticke). MacMillan Co (1913)
892. James, W.: Principles of Psychology. Holt, New York (1890)
893. Schlicht, T.: Experiencing organisms: from mineness to subject of experience. Philos. Stud. **175**(10), 2447–2474 (2018)
894. Block, N.: On a confusion about a function of consciousness. Behav. Brain Sci. **18**(2), 227–247 (1995)
895. Tononi, G.: Consciousness as integrated information: a provisional manifesto. Biol. Bull. **215**(3), 216–242 (2008)

896. Seth, A.K., Dienes, Z., Cleeremans, A., Overgaard, M., Pessoa, L.: Measuring consciousness: relating behavioural and neurophysiological approaches. Trends Cogn. Sci. **12**(8), 314–321 (2008)
897. Bayne, T., Hohwy, J., Owen, A.M.: Are there levels of consciousness? Trends Cogn. Sci. **20**(6), 405–413 (2016)
898. Chalmers, D.J.: The Conscious Mind: in Search of a Fundamental Theory. Philosophy of Mind Series, XVII, 414 S. Oxford University Press, New York (1996)
899. Brown, R., Lau, H., LeDoux, J.E.: Understanding the higher-order approach to consciousness. Trends Cogn. Sci. **23**(9), 754–768 (2019)
900. Nagel, T.: What is it like to be a bat? Philos. Rev. **83**(4), 435–450 (1974)
901. Dawkins, M.: Animal welfare and the paradox of animal consciousness. In: Naguib, M., et al. (eds.) Advances in the Study of Behavior, pp. 5–38. Academic Press, San Diego (2015)
902. Gennaro, R.J. (ed.): Higher-Order Theories of Consciousness: An Anthology. John Benjamins, Philadelphia, Amsterdam (2004)
903. Rosenthal, D.: Consciousness and Mind. Oxford University Press, New York (2005)
904. Seager, W.: A cold look at HOT theory. In: Gennaro, R.J. (ed.) Higher-Order Theories of Consciousness: An Anthology, pp. 255–275. John Benjamins, Philadelphia, Amsterdam (2004)
905. Panksepp, J.: The periconscious substrates of consciousness: affective states and the evolutionary origins of the SELF. J. Conscious. Stud. **5**, 566–582 (1998)
906. Panksepp, J.: Affective Neuroscience: the Foundations of Human and Animal Emotions Series in Affective Science, xii, 466 p. Oxford University Press, New York (1998)
907. Merker, B.: Consciousness without a cerebral cortex: a challenge for neuroscience and medicine. Behav. Brain Sci. **30**(1), 63–81 (2007).; discussion 81–134
908. Morin, A.: Levels of consciousness and self-awareness: a comparison and integration of various neurocognitive views. Conscious. Cogn. **15**(2), 358–371 (2006)
909. Zelazo, P.D.: The development of conscious control in childhood. Trends Cogn. Sci. **8**(1), 12–17 (2004)
910. Stuss, D.T., Anderson, V.: The frontal lobes and theory of mind: developmental concepts from adult focal lesion research. Brain Cogn. **55**(1), 69–83 (2004)
911. Newen, A., Vogeley, K.: Self-representation: searching for a neural signature of self-consciousness. Conscious. Cogn. **12**(4), 529–543 (2003)
912. Damasio, A.R.: Descartes' Error: Emotion, Reason, and the Human Brain, xix, 312 p. Putnam, New York (1994)
913. Von Bertalanffy, L.: The theory of open systems in physics and biology. Science. **111**(2872), 23–29 (1950)
914. Edelman, G.M.: The Remembered Present: A Biological Theory of Consciousness, xx, 346 p. Basic Books, New York (1989)
915. Edelman, G.M.: Bright Air, Brilliant Fire: On the Matter of the Mind, xvi, 280 p. Basic Books, New York (1992)
916. Baars, B.J.: A Cognitive Theory of Consciousness, xxiii, 424 p. Cambridge University Press, New York (1988)
917. Neisser, U.: The roots of self-knowledge: perceiving self, it, and thou. In: Snodgrass, J.G., Thompson, R.L. (eds.) The Self Across Psychology: Self-Recognition, Self-Awareness, and the Self-Concept, vol. 818, pp. 18–33. New York Academy of Sciences, New York (1997)
918. Wild, M.: Bewusstsein. In: Ach, J.S., Borchers, D. (eds.) Handbuch Tierethik. Grundlagen – Kontexte – Perspektiven, pp. 35–40. J. B. Metzler, Berlin (2018)
919. Dennett, D.C.: Intentional systems in cognitive ethology: the "Panglossian paradigm" defended. Behav. Brain Sci. **6**, 343–390 (1983)
920. Grice, P.: Meaning. Philos. Rev. **64**, 377–388 (1957)
921. Sperber, D., Wilson, D.: Relevance: Communication and Cognition. Basil Blackwell, Oxford (1986)
922. Tomasello, M.: Origins of Human Communication The Jean Nicod Lectures. MIT Press, Cambridge, MA (2008)

923. Moore, R.: A common intentional framework for ape and human communication. Curr. Anthropol. **56**(1), 71–72 (2015)
924. Townsend, S.W., et al.: Exorcising Grice's ghost: an empirical approach to studying intentional communication in animals. Biol. Rev. **92**(3), 1427–1433 (2017)
925. Gennaro, R.J.: Consciousness. In: Vonk, J., Shackelford, T. (eds.) Encyclopedia of Animal Cognition and Behavior, pp. 1–14. Springer International Publishing, Cham (2018)
926. Dennett, D.C.: Consciousness Explained. Little, Brown and Co, Boston (1991)
927. Frith, D., Perry, R., Lumer, E.: The neural correlates of conscious experience: an experimental framework. Trends Cogn. Sci. **3**, 105–114 (1999)
928. Cowey, A., Stoerig, P.: Blindsight in monkeys. Nature. **373**, 247–249 (1995)
929. Griffin, D.R., Speck, G.B.: New evidence of animal consciousness. Anim. Cogn. **7**, 5–18 (2004)
930. Tononi, G., Koch, C.: The neural correlates of consciousness – an update. Ann. N. Y. Acad. Sci. **1124**, 239–261 (2008)
931. Butler, A.B.: Evolution of brains, cognition, and consciousness. Brain Res. Bull. **75**(2–4), 442–449 (2008)
932. Seth, A.K., Baars, B.J., Edelman, D.B.: Criteria for consciousness in humans and other mammals. Conscious. Cogn. **14**(1), 119–139 (2005)
933. Baars, B.J., Banks, W.P., Newman, J. (eds.): Essential Sources in the Scientific Study of Consciousness. MIT Press, Cambridge, MA (2003)
934. Roth, G.: The Long Evolution of Brains and Minds. Springer, Dordrecht, Heidelberg (2013)
935. Deacon, T.W.: Rethinking mammalian brain evolution. Am. Zool. **30**(3), 629–705 (1990)
936. Teffer, K., Semendeferi, K.: Human prefrontal cortex: evolution, development, and pathology. In: Progress in Brain Research, pp. 191–218. Elsevier (2012)
937. Roth, G., Dicke, U.: Evolution of the brain and intelligence. Trends Cogn. Sci. **9**(5), 250–257 (2005)
938. Azevedo, F.A., Carvalho, L.R., Grinberg, L.T., Farfel, J.M., Ferretti, R.E., Leite, R.E., Filho, W.J., Lent, R., Herculano-Houzel, S.: Equal numbers of neuronal and nonneuronal cells make the human brain an isometrically scaled-up primate brain. J. Comp. Neurol. **513**(5), 532–541 (2009)
939. Herculano-Houzel, S.: Numbers of neurons as biological correlates of cognitive capability. Curr. Opin. Behav. Sci. **16**, 1–7 (2017)
940. Herculano-Houzel, S.: The remarkable, yet not extraordinary, human brain as a scaled-up primate brain and its associated cost. Proc. Natl. Acad. Sci. USA. **109**(Supplement 1), 10661–10668 (2012)
941. Herculano-Houzel, S., Catania, K., Manger, P.R., Kaas, J.H.: Mammalian brains are made of these: a dataset of the numbers and densities of neuronal and nonneuronal cells in the brain of Glires, Primates, Scandentia, Eulipotyphlans, Afrotherians and Artiodactyls, and their relationship with body mass. Brain Behav. Evol. **86**(3–4), 145–163 (2015)
942. Fodor, J.: The Mind Doesn't Work that Way. MIT Press, Cambridge, MA (2000)
943. Kaas, J.H.: Evolution of Nervous Systems. Academic Press, San Diego (2016)
944. Changizi, M.A., Shimojo, S.: Parcellation and area-area connectivity as a function of neocortex size. Brain Behav. Evol. **66**(2), 88–98 (2005)
945. Hart, B.L., Hart, L.A., Pinter-Wollman, N.: Large brains and cognition: where do elephants fit in? Neurosci. Biobehav. Rev. **32**(1), 86–98 (2008)
946. Tye, M.: Tense Bees and Shell-Shocked Crabs. Are Animals Conscious? Oxford University Press, Oxford (2016)
947. Dugas-Ford, J., Rowell, J.J., Ragsdale, C.W.: Cell-type homologies and the origins of the neocortex. Proc. Natl. Acad. Sci. USA. **109**(42), 16974–16979 (2012)
948. Güntürkün, O.: Evolution of cognitive neural structures. Psychol. Res. **76**, 212–219 (2012)
949. Montiel, J.F., Vasistha, N.A., Garcia-Moreno, F., Molnar, Z.: From sauropsids to mammals and back: new approaches to comparative cortical development. J. Comp. Neurol. **524**(3), 630–645 (2016)

950. Karten, H.J.: Evolutionary developmental biology meets the brain: the origins of mammalian cortex. Proc. Natl. Acad. Sci. USA. **94**(7), 2800–2804 (1997)
951. Ito, H., Yamamoto, N.: Non-laminar cerebral cortex in teleost fishes? Biol. Lett. **5**(1), 117–121 (2009)
952. Wilkinson, A., Chan, H.M., Hall, G.: Spatial learning and memory in the tortoise (Geochelone carbonaria). J. Comp. Psychol. **121**(4), 412–418 (2007)
953. Wilkinson, A., Coward, S., Hall, G.: Visual and response-based navigation in the tortoise (Geochelone carbonaria). Anim. Cogn. **12**(6), 779–787 (2009)
954. Lohman, A.H., Smeets, W.J.: The dorsal ventricular ridge and cortex of reptiles in historical and phylogenetic perspective. In: The Neocortex, pp. 59–74. Springer, New York (1991)
955. Shimizu, T., Karten, H.J.: Multiple origins of neocortex: contributions of the dorsal ventricular ridge. In: The Neocortex, pp. 75–86. Springer, New York (1991)
956. Prinz, J.: Attention, working memory, and animal consciousness. In: Andrews, K., Beck, J. (eds.) The Routledge Handbook of Philosophy of Animal Minds, pp. 185–195. Routledge, Abingdon (2018)
957. Godfrey-Smith, P.: Other Minds: The Octopus, the Sea, and the Deep Origins of Consciousness. Farrar, Straus and Giroux, New York (2016)
958. Trestman, M.: The Cambrian explosion and the origins of embodied cognition. Biol. Theory. **8**, 80–92 (2013)
959. Merker, B.: The liabilities of mobility: a selection pressure for the transition to consciousness in animal evolution. Conscious. Cogn. **14**, 89–114 (2005)
960. Klein, C., Barron, A.: Insects have the capacity for subjective experience. Anim. Sentience. **1**, 1–19 (2016)
961. Ito, K., Shinomiya, K., Ito, M., Armstrong, J.D., Boyan, G., Hartenstein, V., Harzsch, S., Heisenberg, M., Homberg, U., Jenett, A., Keshishian, H., Restifo, L.L., Rössler, W., Simpson, J.H., Strausfeld, N.J., Strauss, R., Vosshall, L.B.: A systematic nomenclature for the insect brain. Neuron. **81**(4), 755–765 (2014)
962. Putnam, H.: Psychological predicates. In: Capitan, W.H., Merrill, D.D. (eds.) Art, Mind, and Religion, pp. 37–48. University of Pittsburgh Press, Pittsburgh (1967)
963. Noesselt, T., Hillyard, S.A., Woldorff, M.G., Schoenfeld, A., Hagner, T., Jancke, L., Tempelmann, C., Hinrichs, H., Heinze, H.J.: Delayed striate cortical activation during spatial attention. Neuron. **35**(3), 575–587 (2002)
964. Soon, C.S., Brass, M., Heinze, H.J., Haynes, J.D.: Unconscious determinants of free decisions in the human brain. Nat. Neurosci. **11**(5), 543–545 (2008)
965. Libet, B., Wright Jr., E.W., Feinstein, B., Pearl, D.K.: Subjective referral of the timing for a conscious sensory experience: a functional role for the somatosensory specific projection system in man. Brain. **102**(1), 193–224 (1979)
966. Libet, B.: Conscious vs neural time. Nature. **352**(6330), 27–28 (1991)
967. Libet, B.: The neural time factor in conscious and unconscious events. Ciba Found. Symp. **174**, 123–137 (1993).; discussion 137–146
968. Cleeremans, A.: Computational correlates of consciousness. Prog. Brain Res. **150**, 81–98 (2005)
969. Bode, S., Bogler, C., Soon, C.S., Haynes, J.D.: The neural encoding of guesses in the human brain. NeuroImage. **59**(2), 1924–1931 (2012)
970. Logothetis, N.K., Pauls, J., Augath, M., Trinath, T., Oeltermann, A.: Neurophysiological investigation of the basis of the fMRI signal. Nature. **412**(6843), 150–157 (2001)
971. Engel, A.K., König, P., Kreiter, A.K., Schillen, T.B., Singer, W.: Temporal coding in the visual cortex: new vistas on integration in the nervous system. Trends Neurosci. **15**(6), 218–226 (1992)
972. Crick, F., Koch, C.: A framework for consciousness. Nat. Neurosci. **6**(2), 119–126 (2003)
973. Taylor, K., Mandon, S., Freiwald, W.A., Kreiter, A.K.: Coherent oscillatory activity in monkey area v4 predicts successful allocation of attention. Cereb. Cortex. **15**(9), 1424–1437 (2005)

974. Tononi, G., Sporns, O., Edelman, G.M.: Reentry and the problem of integrating multiple cortical areas: simulation of dynamic integration in the visual system. Cereb. Cortex. **2**(4), 310–335 (1992)
975. Tononi, G., Edelman, G.M.: Consciousness and complexity. Science. **282**(5395), 1846–1851 (1998)
976. Lamme, V.A., Roelfsema, P.R.: The distinct modes of vision offered by feedforward and recurrent processing. Trends Neurosci. **23**(11), 571–579 (2000)
977. Edelman, G.M., Tononi, G.: A Universe of Consciousness How Matter Becomes Imagination, 1st edn, xiii, 274 p. Basic Books, New York (2000)
978. Roelfsema, P.R., Lamme, V.A., Spekreijse, H.: Synchrony and covariation of firing rates in the primary visual cortex during contour grouping. Nat. Neurosci. **7**(9), 982–991 (2004)
979. Edelman, G.M., Tononi, G.: Consciousness: How Matter Becomes Imagination. Penguin Books Limited, London (2013)
980. Edelman, D.B., Seth, A.K.: Animal consciousness: a synthetic approach. Trends Neurosci. **32**(9), 476–484 (2009)
981. Logothetis, N.K.: Single units and conscious vision. Philos. Trans. R. Soc. Lond. Ser. B Biol. Sci. **353**(1377), 1801–1818 (1998)
982. Rattenborg, N.C., Martinez-Gonzalez, D., Lesku, J.A.: Avian sleep homeostasis: convergent evolution of complex brains, cognition and sleep functions in mammals and birds. Neurosci. Biobehav. Rev. **33**(3), 253–270 (2009)
983. Butler, A.B., Manger, P.R., Lindahl, B.I., Arhem, P.: Evolution of the neural basis of consciousness: a bird-mammal comparison. BioEssays. **27**(9), 923–936 (2005)
984. Luo, M., Perkel, D.J.: A GABAergic, strongly inhibitory projection to a thalamic nucleus in the zebra finch song system. J. Neurosci. **19**(15), 6700–6711 (1999)
985. Llinas, R.R., Steriade, M.: Bursting of thalamic neurons and states of vigilance. J. Neurophysiol. **95**(6), 3297–3308 (2006)
986. Bullock, T.H., Basar, E.: Comparison of ongoing compound field potentials in the brains of invertebrates and vertebrates. Brain Res. **472**(1), 57–75 (1988)
987. Leadbeater, E., Chittka, L.: Social learning in insects – from miniature brains to consensus building. Curr. Biol. **17**(16), R703–R713 (2007)
988. Philippi, C.L., Feinstein, J.S., Khalsa, S.S., Damasio, A., Tranel, D., Landini, G., Williford, K., Rudrauf, D.: Preserved self-awareness following extensive bilateral brain damage to the insula, anterior cingulate, and medial prefrontal cortices. PLoS One. **7**(8), e38413 (2012)
989. Damasio, A., Damasio, H., Tranel, D.: Persistence of feelings and sentience after bilateral damage of the insula. Cereb. Cortex. **23**(4), 833–846 (2013)
990. Alkire, M.T., Hudetz, A.G., Tononi, G.: Consciousness and anesthesia. Science. **322**(5903), 876–880 (2008)
991. Merker, B.: The efference cascade, consciousness, and its self: naturalizing the first person pivot of action control. Front. Psychol. **4**, 501 (2013)
992. Damasio, A., Carvalho, G.B.: The nature of feelings: evolutionary and neurobiological origins. Nat. Rev. Neurosci. **14**(2), 143–152 (2013)
993. Mashour, G.A., Alkire, M.T.: Evolution of consciousness: phylogeny, ontogeny, and emergence from general anesthesia. Proc. Natl. Acad. Sci. USA. **110**(Suppl 2), 10357–10364 (2013)
994. McHaffie, J.G., Stanford, T.R., Stein, B.E., Coizet, V., Redgrave, P.: Subcortical loops through the basal ganglia. Trends Neurosci. **28**(8), 401–407 (2005)
995. Sparks, D.L.: Neural cartography: sensory and motor maps in the superior colliculus. Brain Behav. Evol. **31**(1), 49–56 (1988)
996. von Holst, E.: The principle of reafference: interactions between the central nervous system and the peripheral organs. Die Naturwissenschften. **37**, 464 (1950)
997. Revonsuo, A.: Consciousness, dreams and virtual realities. Philos. Psychol. **8**(1), 35–58 (1995)
998. Edelman, G.M.: Naturalizing consciousness: a theoretical framework. Proc. Natl. Acad. Sci. USA. **100**(9), 5520–5524 (2003)

999. Parvizi, J., Damasio, A.: Consciousness and the brainstem. Cognition. **79**(1–2), 135–160 (2001)
1000. Panksepp, J.: The affective brain and core consciousness: how does neural activity generate emotional feelings? In: Lewis, M., Haviland-Jones, J.M., Barrett, L.F. (eds.) Handbook of Emotions, pp. 47–67. Guilford Publications, New York (2008)
1001. Barron, A.B., Klein, C.: What insects can tell us about the origins of consciousness. Proc. Natl. Acad. Sci. USA. **113**(18), 4900–4908 (2016)
1002. Strausfeld, N.J.: Arthropod Brains: Evolution, Functional Elegance, and Historical Significance, xvi, 830 p. Harvard University Press, Cambridge, MA, London (2012)
1003. Seelig, J.D., Jayaraman, V.: Neural dynamics for landmark orientation and angular path integration. Nature. **521**(7551), 186–191 (2015)
1004. Paulk, A.C., Stacey, J.A., Pearson, T.W., Taylor, G.J., Moore, R.J., Srinivasan, M.V., van Swinderen, B.: Selective attention in the honeybee optic lobes precedes behavioral choices. Proc. Natl. Acad. Sci. USA. **111**(13), 5006–5011 (2014)
1005. Christoff, K., Cosmelli, D., Legrand, D., Thompson, E.: Specifying the self for cognitive neuroscience. Trends Cogn. Sci. **15**(3), 104–112 (2011)
1006. Northoff, G., Heinzel, A., de Greck, M., Bermpohl, F., Dobrowolny, H., Panksepp, J.: Self-referential processing in our brain – a meta-analysis of imaging studies on the self. NeuroImage. **31**(1), 440–457 (2006)
1007. Thompson, E.: Mind in Life: Biology, Phenomenology, and the Sciences of Mind. Harvard University Press, Cambridge, MA (2007)
1008. Legrand, D., Ruby, P.: What is self-specific? Theoretical investigation and critical review of neuroimaging results. Psychol. Rev. **116**(1), 252–282 (2009)
1009. Tononi, G., Boly, M., Massimini, M., Koch, C.: Integrated information theory: from consciousness to its physical substrate. Nat. Rev. Neurosci. **17**(7), 450–461 (2016)
1010. Dehaene, S., Lau, H., Kouider, S.: What is consciousness, and could machines have it? Science. **358**(6362), 486–492 (2017)
1011. Dehaene, S., Kerszberg, M., Changeux, J.P.: A neuronal model of a global workspace in effortful cognitive tasks. Proc. Natl. Acad. Sci. USA. **95**(24), 14529–14534 (1998)
1012. Dehaene, S., Naccache, L.: Towards a cognitive neuroscience of consciousness: basic evidence and a workspace framework. Cognition. **79**(1–2), 1–37 (2001)
1013. Panagiotaropoulos, T.I., Deco, G., Kapoor, V., Logothetis, N.K.: Neuronal discharges and gamma oscillations explicitly reflect visual consciousness in the lateral prefrontal cortex. Neuron. **74**(5), 924–935 (2012)
1014. Pincham, H.L., Bowman, H., Szucs, D.: The experiential blink: mapping the cost of working memory encoding onto conscious perception in the attentional blink. Cortex. **81**, 35–49 (2016)
1015. Marti, S., Sigman, M., Dehaene, S.: A shared cortical bottleneck underlying attentional blink and psychological refractory period. NeuroImage. **59**(3), 2883–2898 (2012)
1016. Dunlosky, J., Metcalfe, J.: Metacognition. SAGE Publications, Thousand Oaks (2008)
1017. Cleeremans, A.: Connecting conscious and unconscious processing. Cogn. Sci. **38**(6), 1286–1315 (2014)
1018. Peña-Ayala, A.: Metacognition: Fundaments, Applications, and Trends: A Profile of the Current State-of-the-Art. Springer International Publishing, Cham (2014)
1019. Birch, J., Schnell, A.K., Clayton, N.S.: Dimensions of animal consciousness. Trends Cogn. Sci. **24**(10), 789–801 (2002)
1020. Nieder, A., Wagener, L., Rinnert, P.: A neural correlate of sensory consciousness in a corvid bird. Science. **369**(6511), 1626–1629 (2020)
1021. Travers, E., Frith, C.D., Shea, N.: Learning rapidly about the relevance of visual cues requires conscious awareness. Q. J. Exp. Psychol. **71**(8), 1698–1713 (2018)
1022. Clark, R.E., Squire, L.R.: Classical conditioning and brain systems: the role of awareness. Science. **280**, 77–81 (1998)
1023. Balasko, M., Cabanac, M.: Behavior of juvenile lizards (Iguana iguana) in a conflict between temperature regulation and palatable food. Brain Behav. Evol. **52**(6), 257–262 (1998)

1024. Appel, M., Elwood, R.W.: Motivational trade-offs and potential pain experience in hermit crabs. Appl. Anim. Behav. Sci. **119**(1–2), 120–124 (2009)
1025. Fraser, D., Nicol, C.J.: Preference and motivation research. In: Appleby, M.C., Hughes, B.O., Mench, J.A. (eds.) Animal Welfare, pp. 183–199. CABI, Wallingford, Cambridge, MA (2011)
1026. Balleine, B.W., Dickinson, A.: Goal-directed instrumental action: contingency and incentive learning and their cortical substrates. Neuropharmacology. **37**(4–5), 407–419 (1998)
1027. Harding, E.J., Paul, E.S., Mendl, M.: Animal behaviour: cognitive bias and affective state. Nature. **427**(6972), 312 (2004)
1028. Crump, A., Arnott, G., Bethell, E.J.: Affect-driven attention biases as animal welfare indicators: review and methods. Animals. **8**(8), 136 (2018)
1029. Huber, A., Barber, A.L.A., Faragó, T., Müller, C.A., Huber, L.: Investigating emotional contagion in dogs (Canis familiaris) to emotional sounds of humans and conspecifics. Anim. Cogn. **20**, 703–715 (2017)
1030. Adriaense, J.E.C., Koski, S.E., Huber, L., Lamm, C.: Challenges in the comparative study of empathy and related phenomena in animals. Neurosci. Biobehav. Rev. **112**, 62–82 (2020)
1031. de Haan, E.H., Corballis, P.M., Hillyard, S.A., Marzi, C.A., Seth, A., Lamme, V.A., Volz, L., Fabri, M., Schechter, E., Bayne, T.: Split-brain: what we know now and why this is important for understanding consciousness. Neuropsychol. Rev. **30**, 224–233 (2020)
1032. Ortega, L.J., Stoppa, K., Güntürkün, O., Troje, N.F.: Limits of intraocular and interocular transfer in pigeons. Behav. Brain Res. **193**(1), 69–78 (2008)
1033. Dainton, B.: Temporal consciousness. In: Zalta, E.N. (ed.) The Stanford Encyclopedia of Philosophy. Stanford University, Stanford (2018)
1034. Kolers, P.A., von Grünau, M.: Shape and color in apparent motion. Vis. Res. **16**(4), 329–335 (1976)
1035. Michaelian, K.: Mental Time Travel: Episodic Memory and Our Knowledge of the Personal Past. MIT Press, Cambridge, MA (2016)
1036. Vandekerckhove, M.M.P.: Memory, autonoetic consciousness and the self: consciousness as a continuum of stages. Self Identity. **8**, 4–23 (2009)
1037. Povinelli, D.J.: The self: elevated in consciousness and extended in time. In: Moore, C., Lemmon, K. (eds.) The Self in Time: Developmental Perspectives, pp. 75–95. Erlbaum, Mahwah (2001)
1038. Suddendorf, T., Redshaw, J.: The development of mental scenario building and episodic foresight. Ann. N. Y. Acad. Sci. **1296**, 135–153 (2013)
1039. Tulving, E.: Episodic memory and common sense: how far apart? Philos. Trans. R. Soc. Lond. Ser. B Biol. Sci. **356**, 1505–1515 (2001)
1040. Fonagy, P., Gergely, G., Jurist, E.L., Target, M.: Affect Regulation, Mentalization, and the Development of the Self. Other Press, New York (2002)
1041. DeGrazia, D.: Self-awareness in animals. In: Lurz, R. (ed.) The Philosophy of Animal Minds, pp. 201–217. Cambridge University Press, New York (2009)
1042. Hurley, S.L.: Consciousness in Action. Harvard University Press, Cambridge, MA (1998)
1043. Boyle, A.: Mirror self-recognition and self-identification. Philos. Phenomenol. Res. **97**(2), 284–303 (2018)
1044. Gallup Jr., G.G.: Chimpanzees: self-recognition. Science. **167**(3914), 86–87 (1970)
1045. Anderson, J.R., Gallup Jr., G.G.: Mirror self-recognition: a review and critique of attempts to promote and engineer self-recognition in primates. Primates. **56**, 317–326 (2015)
1046. Mitchell, R.W.: Evidence of dolphin self-recognition and the difficulties of interpretation. Conscious. Cogn. **4**, 229–234 (1995)
1047. Rajala, A.Z., Reininger, K.R., Lancaster, K.M., Populin, L.C.: Rhesus monkeys (Macaca mulatta) do recognize themselves in the mirror: implications for the evolution of self-recognition. PLoS One. **5**, e12865 (2010)
1048. Lewis, M.: "Myself and me". In: Parker, S.T., Mitchell, R.W., Boccia, M.L. (eds.) Self-awareness in Animals and Humans, pp. 20–34. Cambridge University Press, Cambridge (1994)

1049. Schilhab, T.S.S.: What mirror self-recognition in nonhumans can tell us about aspects of self. Biol. Philos. **19**, 111–126 (2004)
1050. Gallup Jr., G.G., Anderson, J.R., Shillito, D.J.: The mirror test. In: Bekoff, M., Allen, C., Burghardt, G.M. (eds.) The Cognitive Animal. Empirical and Theoretical Perspectives on Animal Cognition, pp. 325–333. MIT Press, Cambridge, MA (2002)
1051. Heyes, C.M.: Reflections on self-recognition in primates. Anim. Behav. **47**(4), 909–919 (1994)
1052. Heyes, C.M.: Self-recognition in primates: further reflections create a hall of mirrors. Anim. Behav. **50**, 1533–1542 (1995)
1053. Amsterdam, B.: Mirror self-image reactions before age two. Dev. Psychobiol. **5**(4), 297–305 (1972)
1054. Nielsen, M., Dissanayake, C.: Pretend play, mirror self-recognition and imitation: a longitudinal investigation through the second year. Infant Behav. Dev. **27**(3), 342–365 (2004)
1055. Suddendorf, T.: The Gap: The Science of What Separates Us from Other Animals. Basic Books, New York (2013)
1056. Mitchell, R.W.: A comparison of the self-awareness and kinesthetic-visual matching theories of self-recognition: autistic children and others. Ann. N. Y. Acad. Sci. **818**, 39–62 (1997)
1057. Bard, K.A., Todd, B.K., Bernier, C., Love, J., Leavens, D.A.: Self-awareness in human and chimpanzee infants: what is measured and what is meant by the mark and mirror test? Infancy. **9**(2), 191–219 (2006)
1058. Posada, S., Colell, M.: Another gorilla (Gorilla gorilla gorilla) recognizes himself in a mirror. Am. J. Primatol. **69**(5), 576–583 (2007)
1059. Hauser, M.D., Kralik, J., Botto-Mahan, C., Garrett, M., Oser, J.: Self-recognition in primates: phylogeny and the salience of species-typical features. Proc. Natl. Acad. Sci. USA. **92**(23), 10811–10814 (1995)
1060. Hauser, M.D., Miller, C.T., Liu, K., Gupta, R.: Cotton-top tamarins (Saguinus oedipus) fail to show mirror-guided self-exploration. Am. J. Primatol. **53**, 131–137 (2001)
1061. Plotnik, J.M., de Waal, F.B.M., Reiss, D.: Self-recognition in an Asian elephant. Proc. Natl. Acad. Sci. USA. **103**(45), 17053–17057 (2006)
1062. Povinelli, D.J.: Failure to find self-recognition in Asian elephants (Elephas maximus) in contrast to their use of mirror cues to discover hidden food. J. Comp. Psychol. **103**(2), 122 (1989)
1063. Nissani, M., Hoefler-Nissani, D.: Absence of mirror self-referential behavior in two Asian elephants. J. Vet. Sci. **1**, 1 (2007)
1064. Dale, R., Plotnik, J.M.: Elephants know when their bodies are obstacles to success in a novel transfer task. Sci. Rep. **7**, 46309 (2017)
1065. Reiss, D., Marino, L.: Mirror self-recognition in the bottlenose dolphin: a case of cognitive convergence. Proc. Natl. Acad. Sci. USA. **98**(10), 5937–5942 (2001)
1066. Morrison, R., Reiss, D.: Precocious development of self-awareness in dolphins. PLoS One. **13**(1), e0189813 (2018)
1067. Kohda, M., Hotta, T., Takeyama, T., Awata, S., Tanaka, H., Asai, J.Y., Jordan, A.L.: If a fish can pass the mark test, what are the implications for consciousness and self-awareness testing in animals? PLoS Biol. **17**(2), e3000021 (2019)
1068. de Waal, F.B.M.: Fish, mirrors, and a gradualist perspective on self-awareness. PLoS Biol. **17**(2), e3000112 (2019)
1069. Bleiweiss, R.: Ultraviolet plumage reflectance distinguishes sibling bird species. Proc. Natl. Acad. Sci. USA. **101**, 16561–16564 (2014)
1070. Pepperberg, I.M., Garcia, S.E., Jackson, E.C., Marconi, S.: Mirror use by African grey parrots (Psittacus erithacus). J. Comp. Psychol. **109**(2), 182 (1995)
1071. Anderson, J.R.: Monkeys with mirrors: some questions for primate psychology. Int. J. Primatol. **5**(1), 81–98 (1984)
1072. Gallup Jr., G.G.: Toward a comparative psychology of self-awareness: species limitations and cognitive consequences. In: Strauss, J., Goethals, G.R. (eds.) The Self: Interdisciplinary Approaches, pp. 121–135. Springer-Verlag, New York (1991)

1073. van Buuren, M., Auersperg, A., Gajdon, G., Tebbich, S., von Bayern, A.: No evidence of mirror self-recognition in keas and Goffin's cockatoos. Behaviour. **156**(5–8), 763–786 (2019)
1074. Kusayama, T., Bischof, H.-J., Watanabe, S.: Responses to mirror-image stimulation in jungle crows (Corvus macrorhynchos). Anim. Cogn. **3**(1), 61–64 (2000)
1075. Medina, F.S., Taylor, A.H., Hunt, G.R., Gray, R.D.: New Caledonian crows' responses to mirrors. Anim. Behav. **82**(5), 981–993 (2011)
1076. Vanhooland, L.-C., Bugnyar, T., Massen, J.J.M.: Crows (Corvus corone ssp.) check contingency in a mirror yet fail the mirror-mark test. J. Comp. Psychol. **134**(2), 158–169 (2020)
1077. Soler, M., Pérez-Contreras, T., Peralta-Sánchez, J.M.: Mirror-mark tests performed on jackdaws reveal potential methodological problems in the use of stickers in avian mark-test studies. PLoS One. **9**(1), e86193 (2014)
1078. Clary, D., Kelly, D.M.: Graded mirror self-recognition by Clark's nutcrackers. Sci. Rep. **6**(1), 1–11 (2016)
1079. Clary, D., Stow, M.K., Vernouillet, A., Kelly, D.M.: Mirror-mediated responses of California scrub jays (Aphelocoma californica) during a caching task and the mark test. Ethology. **126**(2), 140–152 (2020)
1080. de Waal, F.B., Dindo, M., Freeman, C.A., Hall, M.J.: The monkey in the mirror: hardly a stranger. Proc. Natl. Acad. Sci. USA. **102**(32), 11140–11147 (2005)
1081. Povinelli, D.J., Rulf, A.B., Landau, K.R., Bierschwale, D.T.: Self-recognition in chimpanzees: distribution, ontogeny, and patterns of emergence. J. Comp. Psychol. **107**, 347–372 (1993)
1082. Mahovetz, L.M., Young, L.J., Hopkins, W.D.: The influence of AVPR1A genotype on individual differences in behaviors during a mirror self-recognition task in chimpanzees (Pan troglodytes). Genes Brain Behav. **15**(5), 445–452 (2016)
1083. Archer, J.: Ethology and Human Development. Harvester Wheatsheaf, Herfordshire (1992)
1084. Prior, H., Schwarz, A., Güntürkün, O.: Mirror-induced behavior in the magpie (Pica Pica): evidence of self-recognition. PLoS Biol. **6**(8), e202 (2008)
1085. Krachun, C., Lurz, R., Mahovetz, L.M., Hopkins, W.D.: Mirror self-recognition and its relationship to social cognition in chimpanzees. Anim. Cogn. **22**(6), 1171–1183 (2019)
1086. Marino, L., Reiss, D., Gallup, G.G.J.: Mirror self-recognition in bottlenose dolphins: implications for comparative investigations of highly dissimilar species. In: Parker, S.T., Mitchell, R.W., Boccia, M.L. (eds.) Self-Awareness in Animals and Humans, pp. 380–391. Cambridge University Press, Cambridge (1994)
1087. Swartz, K.B., Evans, S.: Not all chimpanzees (Pan troglodytes) show self-recognition. Primates. **32**, 483–496 (1991)
1088. Shumaker, R.W., Swartz, K.B.: When traditional methodologies fail: cognitive studies of great apes. In: Bekoff, M., Allen, C., Burghardt, G.M. (eds.) The Cognitive Animal. Empirical and Theoretical Perspectives on Animal Cognition, pp. 335–343. MIT Press, Cambridge, MA (2002)
1089. Rochat, P.: Five levels of self-awareness as they unfold early in life. Conscious. Cogn. **12**, 717–731 (2003)
1090. Hodos, W., Campbell, C.B.G.: Scala naturae: why there is no theory in comparative psychology. Psychol. Rev. **76**(4), 337–350 (1969)
1091. Lenkei, R., Faragó, T., Zsilák, B., Pongrácz, P.: Dogs (Canis familiaris) recognize their own body as a physical obstacle. Sci. Rep. **11**(1), 1–8 (2021)
1092. Hediger, H.: Zur Frage des Selbstbewußtseins beim Tier. In: Stamm, R.A. (ed.) Tierpsychologie. Die biologische Erforschung tierischen und menschlichen Verhaltens, pp. 280–291. Beltz, Weinheim, Basel (1978)
1093. Leary, M.R., Buttermore, N.R.: The evolution of the human self: tracing the natural history of self-awareness. J. Theory Soc. Behav. **33**(4), 365–404 (2003)
1094. Proske, U., Gandevia, S.C.: The kinaesthetic senses. J. Physiol. **587**(17), 4139–4146 (2009)
1095. Vogeley, K., Fink, G.R.: Neural correlates of the first-person-perspective. Trends Cogn. Sci. **7**(1), 38–42 (2003)

1096. Mitchell, R.W.: Multiplicities of self. In: Parker, S.T., Mitchell, R.W., Boccia, M.L. (eds.) Self-Awareness in Animals and Humans. Developmental Perspectives, pp. 81–107. Cambridge University Press, New York (1994)
1097. Mitchell, R.W.: Kinesthetic-visual matching and the self-concept as explanations of mirror-self-recognition. J. Theory Soc. Behav. **27**(1), 17–39 (1997)
1098. Prinz, W.: Self in the mirror. Conscious. Cogn. **22**(3), 1105–1113 (2013)
1099. Mitchell, R.W.: Mental models of mirror-self-recognition: two theories. New Ideas Psychol. **11**(3), 295–325 (1993)
1100. Mitchell, R.W.: Recognizing one's self in a mirror? A reply to Gallup and Povinelli, de Lannoy, Anderson, and Byrne. New Ideas Psychol. **11**, 351–377 (1993)
1101. Hart, D., Karmel, M.P.: Self-awareness and self-knowledge in humans, apes, and monkeys. In: Russon, A.E., Bard, K.A., Parker, S.T. (eds.) Reaching into Thought: The Minds of the Great Apes, pp. 325–347. Cambridge University Press, Cambridge, MA (1996)
1102. Fleming, S.M., Weil, R.S., Nagy, Z., Dolan, R.J., Rees, G.: Relating introspective accuracy to individual differences in brain structure. Science. **329**(5998), 1541–1543 (2010)
1103. Miyamoto, K., Osada, T., Setsuie, R., Takeda, M., Tamura, K., Adachi, Y., Miyashita, Y.: Causal neural network of metamemory for retrospection in primates. Science. **355**(6321), 188–193 (2017)
1104. Lau, H.C., Passingham, R.E.: Unconscious activation of the cognitive control system in the human prefrontal cortex. J. Neurosci. **27**(21), 5805–5811 (2007)
1105. Ruby, E., Maniscalco, B., Peters, M.A.: On a "failed" attempt to manipulate visual metacognition with transcranial magnetic stimulation to prefrontal cortex. Conscious. Cogn. **62**, 34–41 (2018)
1106. Gennaro, R.J.: The Consciousness Paradox: Consciousness, Concepts, and Higher-Order Thoughts Representation and Mind, x, 378 p. MIT Press, Cambridge, MA, London (2012)
1107. Gennaro, R.J.: Animal consciousness and higher-order thoughts. In: Andrews, K., Beck, J. (eds.) The Routledge Handbook of Philosophy of Animal Minds, pp. 196–205. Routledge, Abingdon (2018)
1108. Terrace, H.S., Metcalfe, J.: The Missing Link in Cognition: Origins of Self-Reflective Consciousness, ix, 364 p., 16 p. of plates. Oxford University Press, New York, Oxford (2005)
1109. Custers, R., Aarts, H.: The unconscious will: how the pursuit of goals operates outside of conscious awareness. Science. **329**(5987), 47–50 (2010)
1110. Logan, G.D., Crump, M.J.: Cognitive illusions of authorship reveal hierarchical error detection in skilled typists. Science. **330**(6004), 683–686 (2010)
1111. Terrace, H.S., Son, L.K.: Comparative metacognition. Curr. Opin. Neurobiol. **19**(1), 67–74 (2009)
1112. Huber, L.: The smart set. Inference Int. Rev. Sci. **3**, 1–10 (2018)
1113. Kattmann, U.: Sonderstellung oder Eigenart? Zur Stellung des Menschen innerhalb der Lebewesen. Prax. Naturwiss. Biol. **23**(10), 253–264 (1974)
1114. Balda, R.P., Pepperberg, I.M., Kamil, A.C.: Animal Cognition in Nature: The Convergence of Psychology and Biology in Laboratory and Field, ix, 465 p. Academic Press, San Diego (1998)
1115. Waal, F.B.M.: Are We Smart Enough to Know How Smart Animals Are?, 1st edn, 340 S. W. W. Norton & Company, New York, London (2016)
1116. Kamil, A.C.: On the proper definition of cognitive ethology. In: Balda, R.P., Pepperberg, I.M., Kamil, A.C. (eds.) Animal Cognition in Nature: The Convergence of Psychology and Biology in Laboratory and Field, pp. 1–28. Academic Press, San Diego (1998)
1117. Kamil, A.C.: A synthetic approach to the study of animal intelligence. In: Leger, D. (ed.) Comparative Perspectives in Modern Psychology: Nebraska Symposium on Motivation, pp. 230–257. University of Nebraska Press, Lincoln (1988)

1118. Alessandroni, N., Altschul, D., Bazhydai, M., Byers-Heinlein, K., Elsherif, M., Gjoneska, B., Huber, L., Mazza, V., Miller, R., Nawroth, C., Pronizius, E., Qadri, M.A.J., Šlipogor, V., Soderstrom, M., Stevens, J.R., Visser, I., Williams, M., Zettersten, M., Prétôt, L.: Comparative cognition needs big team science: how large-scale collaborations will unlock the future of the field. Comp. Cogn. Behav. Rev. (2024) https://doi.org/10.31234/osf.io/rynvu
1119. Farrar, B.G., Voudouris, K., Clayton, N.S.: Replications, comparisons, sampling and the problem of representativeness in animal cognition research. Anim. Behav. Cogn. **8**(2), 273 (2021)
1120. Tecwyn, E.C.: Doing reliable research in comparative psychology: challenges and proposals for improvement. J. Comp. Psychol. **135**(3), 291 (2021)
1121. Roberts, M.J.: Integrating the Mind: Domain General Versus Domain Specific Processes in Higher Cognition. Psychology Press, Hove, New York (2008)
1122. Hauser, M.D.: Wild Minds: What Animals Really Think, 1st edn. Henry Holt, New York (2000)
1123. MacLean, E.L., et al.: How does cognition evolve? Phylogenetic comparative psychology. Anim. Cogn. **15**(2), 223–238 (2012)
1124. Fitch, W.T., Huber, L., Bugnyar, T.: Social cognition and the evolution of language: constructing cognitive phylogenies. Neuron. **65**(6), 795–814 (2010)
1125. Barkow, J., Cosmides, L., Tooby, J. (eds.): The Adapted Mind: Evolutionary Psychology and the Generation of Culture. Oxford University Press, New York (1992)
1126. Huber, L.: Früchte vom Baum der Erkenntnis. Zur Evolution von kognitiven und moralanalogen Fähigkeiten. In: Herzog, E.M., Bauer, H.-C., Überriegler, K.P. (eds.) Blickpunkt: Intelligenz. Ein Erfolgsmodell der Evolution? Books on Demand, Norderstedt (2014)
1127. Huber, L., Wilkinson, A.: Cognitive evolution: a comparative approach. In: Barth, F.G., Giampieri-Deutsch, P., Klein, H.-D. (eds.) Sensory Perception: Mind and Matter, pp. 137–154. Springer, Wien, New York (2012)
1128. Wilkinson, A., Huber, L.: Cold-blooded cognition: reptilian cognitive abilities. In: Vonk, J., Shackelford, T.K. (eds.) The Oxford Handbook of Comparative Evolutionary Psychology, pp. 129–143. Oxford University Press, New York (2012)
1129. Wilkinson, A., Kuenstner, K., Mueller, J., Huber, L.: Social learning in a non-social reptile (Geochelone carbonaria). Biol. Lett. **6**(5), 614–616 (2010)
1130. Kis, A., Huber, L., Wilkinson, A.: Social learning by imitation in a reptile (Pogona vitticeps). Anim. Cogn. **18**(1), 325–331 (2015)
1131. Bshary, R., Grutter, A.S.: Image scoring and cooperation in a cleaner fish mutualism. Nature. **441**(7096), 975–978 (2006)
1132. Bshary, R., Grutter, A.S., Willener, A.S., Leimar, O.: Pairs of cooperating cleaner fish provide better service quality than singletons. Nature. **455**(7215), 964–966 (2008)
1133. Bshary, R., Wickler, W., Fricke, H.: Fish cognition: a primate's eye view. Anim. Cogn. **5**(1), 1–13 (2002)
1134. Salwiczek, L.H., Pretot, L., Demarta, L., Proctor, D., Essler, J., Pinto, A.I., Wismer, S., Stoinski, T., Brosnan, S.F., Bshary, R.: Adult cleaner wrasse outperform capuchin monkeys, chimpanzees and orang-utans in a complex foraging task derived from cleaner–client reef fish cooperation. PLoS One. **7**(11), e49068 (2012)
1135. Mather, J.A., Anderson, R.C., Wood, J.B.: Octopus: the ocean's Intelligent Invertebrate. Timber Press, Portland (2013)
1136. Darmaillacq, A.-S., Dickel, L., Mather, J.: Cephalopod Cognition. Cambridge University Press, Cambridge (2014)
1137. Chittka, L.: Bee cognition. Curr. Biol. **27**, R1037–R1059 (2017)
1138. Avarguès-Weber, A., Deisig, N., Giurfa, M.: Visual cognition in social insects. Annu. Rev. Entomol. **56**, 423–443 (2011)
1139. Giurfa, M.: Cognition with few neurons: higher-order learning in insects. Trends Neurosci. **36**(5), 285–294 (2013)

1140. Giurfa, M.: Learning and cognition in insects. Wiley Interdiscip. Rev. Cogn. Sci. **6**(4), 383–395 (2015)
1141. Menzel, R., Giurfa, M.: Dimensions of cognition in an insect, the honeybee. Behav. Cogn. Neurosci. Rev. **5**(1), 24–40 (2006)
1142. Loukola, O.J., Perry, C.J., Coscos, L., Chittka, L.: Bumblebees show cognitive flexibility by improving on an observed complex behavior. Science. **355**(6327), 833–836 (2017)
1143. Solvi, C., Al-Khudhairy, S.G., Chittka, L.: Bumble bees display cross-modal object recognition between visual and tactile senses. Science. **367**(6480), 910–912 (2020)
1144. Farrar, B., Ostojic, L.: The illusion of science in comparative cognition. PsyArXiv. (2019) https://psyarxiv.com/hduyx/
1145. Stevens, J.R.: Replicability and reproducibility in comparative psychology. Front. Psychol. **8**, 862 (2017)
1146. Ioannidis, J.P.A.: Why most published research findings are false. PLoS Med. **2**, e124 (2005)
1147. Farrar, B., Ostojic, L., Clayton, N.: The hidden side of animal cognition research: scientists' attitudes toward bias, replicability and scientific practice. PsyArXiv. (2021) https://doi.org/10.31234/osf.io/fwvpe
1148. Beran, M.J.: Replication and pre-registration in comparative psychology. Int. J. Comp. Psychol. **31**, 1–8 (2018) https://escholarship.org/uc/item/59f4z2nd
1149. Open Science Collaboration: Estimating the reproducibility of psychological science. Science. **349**, aac4716 (2015)
1150. Forscher, P.S., Wagenmakers, E.-J., Coles, N.A., Silan, M.A., Dutra, N., Basnight-Brown, D., IJzerman, H.: The benefits, barriers, and risks of big-team science. Perspect. Psychol. Sci. **18**(3), 607–623 (2023)
1151. Baumgartner, H.A., Alessandroni, N., Byers-Heinlein, K., Frank, M.C., Hamlin, J.K., Soderstrom, M., Voelkel, J.G., Willer, R., Yuen, F., Coles, N.A.: How to build up big team science: a practical guide for large-scale collaborations. R. Soc. Open Sci. **10**(6), 230235 (2023)
1152. ManyPrimates, et al.: Establishing an infrastructure for collaboration in primate cognition research. PLoS One. **14**(10), e0223675 (2019)
1153. Lambert, M.L., Farrar, B.G., Garcia-Pelegrin, E., Reber, S., Miller, R.: ManyBirds: a multi-site collaborative Open Science approach to avian cognition and behaviour research. PsyArXiv. (2021) https://doi.org/10.31234/osf.io/83xkt
1154. ManyBabies Consortium: Quantifying sources of variability in infancy research using the infant-directed-speech preference. Adv. Methods Pract. Psychol. Sci. **3**, 24–52 (2020)
1155. ManyDogs Project, Espinosa, J., Bray, E.E., Buchsbaum, D., Byosiere, S.-E., Byrne, M., Freeman, M.S., Gnanadesikan, G.E., Guran, C.-N.A., Horschler, D.J., Huber, L., Johnston, A.M., MacLean, E.L., Pelgrim, M.H., Santos, L., Silver, Z.A., Stevens, J.R., Völter, C.J., Zipperling, L.: ManyDogs 1: a multi-lab replication study of dogs' pointing comprehension. Anim. Behav. Cogn. **10**, 232–286 (2023)
1156. Kohse-Höinghaus, K., Fuchs, H., Zhang, T., Wu, Y.: Das Unbekannte erforschen – der Wert der Grundlagenforschung. Angew. Chem. **131**(50), 18048–18050 (2019)
1157. Russell, W.M.S., Burch, R.L.: The Principles of Humane Experimental Technique. Methuen, London (1959)
1158. Poucet, B., Benhamou, S.: The neuropsychology of spatial cognition in the rat. Crit. Rev. Neurobiol. **11**(2–3), 101–120 (1997)
1159. Bartal, I.B.-A., Decety, J., Mason, P.: Empathy and pro-social behavior in rats. Science. **334**(6061), 1427–1430 (2011)
1160. Bartal, I.B.-A., Rodgers, D.A., Sarria, M.S.B., Decety, J., Mason, P.: Pro-social behavior in rats is modulated by social experience. Elife. **3**, e01385 (2014)
1161. Precht, R.D.: Tiere denken. Vom Recht der Tiere und den Grenzen des Menschen, Originalausgabe, 1 Auflage, 509 S. Goldmann, München (2016)
1162. Höffe, O.: Lexikon der Ethik. Beck, München (1997)

1163. Duncan, I., Petherick, J.C.: The implications of cognitive processes for animal welfare. J. Anim. Sci. **69**(12), 5017–5022 (1991)
1164. Špinka, M.: How important is natural behaviour in animal farming systems? Appl. Anim. Behav. Sci. **100**(1–2), 117–128 (2006)
1165. Mendl, M., Paul, E.S.: Do animals live in the present?: current evidence and implications for welfare. Appl. Anim. Behav. Sci. **113**(4), 357–382 (2008)
1166. Bentham, J.: Introduction to the Principles of Morals and Legislation. Oxford University Press, Oxford (1789)
1167. Braithwaite, V.: Do Fish Feel Pain? Oxford University Press, Oxford (2010)
1168. Sneddon, L.: Pain perception in fish. J. Conscious. Stud. **18**(9–10), 209–229 (2011)
1169. Woodruff, M.L.: Sentience in fishes: more on the evidence. Anim. Sentience. **2**(13), 16 (2018)
1170. Elwood, R.W.: Evidence for pain in decapod crustaceans. Anim. Welf. **21**(S2), 23–27 (2012)
1171. Sneddon, L.U., Elwood, R.W., Adamo, S.A., Leach, M.C.: Defining and assessing animal pain. Anim. Behav. **97**, 201–212 (2014)
1172. Dawkins, M.S.: From an animal's point of view: motivation, fitness, and animal welfare. Behav. Brain Sci. **13**(1), 1–61 (1990)
1173. Dawkins, M.S.: Animal welfare with and without consciousness. J. Zool. **301**(1), 1–10 (2017)
1174. Singer, P.: Applied Ethics Oxford Readings in Philosophy, vi, 264 p. Oxford University Press, Oxford, New York (1986)
1175. Korff, W., Beck, L., Mikat, P. (eds.): Lexikon der Bioethik. Gütersloher Verlagshaus, Gütersloh (2000)
1176. Gray, H.M., Gray, K., Wegner, D.M.: Dimensions of mind perception. Science. **315**(5812), 619–619 (2007)
1177. Foer, J.S.: Eating Animals, 1st edn, vii, 341 p. Little, Brown and Company, New York (2009)
1178. Bastian, B., Loughnan, S., Haslam, N., Radke, H.R.: Don't mind meat? The denial of mind to animals used for human consumption. Personal. Soc. Psychol. Bull. **38**(2), 247–256 (2012)
1179. Clarke, C., Dzieciolowski, R.: Feral pigs in the northern South Island, New Zealand: I. Origin, distribution, and density. J. R. Soc. N. Z. **21**(3), 237–247 (1991)
1180. D'Eath, R.B., Turner, S.P.: The natural behaviour of the pig. In: Marchant-Fonde, J.N. (ed.) The Welfare of Pigs, pp. 13–45. Springer, Berlin (2009)
1181. Wondrak, M., Conzelmann, E., Veit, A., Huber, L.: Pigs (Sus scrofa domesticus) categorize pictures of human heads. Appl. Anim. Behav. Sci. **205**, 19–27 (2018)
1182. Veit, A., Wondrak, M., Huber, L.: Object movement re-enactment in free-ranging Kune Kune piglets. Anim. Behav. **132**, 49–59 (2017)
1183. Held, S., Mendl, M., Devereux, C., Byrne, R.W.: Social tactics of pigs in a competitive foraging task: the "informed forager" paradigm. Anim. Behav. **59**(3), 569–576 (2000)
1184. Held, S.D.E., Byrne, R.W., Jones, S., Murphy, E., Friel, M., Mendl, M.T.: Domestic pigs, Sus scrofa, adjust their foraging behaviour to whom they are foraging with. Anim. Behav. **79**(4), 857–862 (2010)
1185. Nawroth, C., Langbein, J., Coulon, M., Gabor, V., Oesterwind, S., Benz-Schwarzburg, J., von Borell, E.: Farm animal cognition—linking behavior, welfare and ethics. Front. Vet. Sci. **6**, 24 (2019)
1186. Mendl, M., Held, S., Byrne, R.W.: Pig cognition. Curr. Biol. **20**(18), R796–R798 (2010)
1187. Rault, J.-L.: Be kind to others: prosocial behaviours and their implications for animal welfare. Appl. Anim. Behav. Sci. **210**, 113–123 (2019)
1188. de Waal, F.: The tower of morality. In: de Waal, F. (ed.) Primates and Philosophers: How Morality Evolved, pp. 161–181. Princeton University Press, Princeton (2006)
1189. Yamamoto, S., Humle, T., Tanaka, M.: Chimpanzees help each other upon request. PLoS One. **4**(10), e7416 (2009)
1190. de Waal, F.B., Suchak, M.: Prosocial primates: selfish and unselfish motivations. Philos. Trans. R. Soc. Lond. Ser. B Biol. Sci. **365**(1553), 2711–2722 (2010)

1191. de Waal, F.B.M., Preston, S.D.: Mammalian empathy: behavioural manifestations and neural basis. Nat. Rev. Neurosci. **18**, 498 (2017)
1192. Rogers, L.J., Kaplan, G.: All animals are not equal. The interface between scientific knowledge and legislation for animal rights. In: Sunstein, C.R., Nussbaum, M.C. (eds.) Animal Rights. Current Debates and New Directions, pp. 175–202. Oxford University Press, New York (2004)
1193. Goodall, J., Bekoff, M.: The Ten Trusts. What We Must Do to Care for the Animals We Love, 200 S. Harper Collins, San Francisco (2002)
1194. Goodall, J., Berman, P.L.: Reason for Hope: A Spiritual Journey, xv, 280 p. Warner Books, New York (1999)
1195. Lindsey, J., Goodall, J.: Jane Goodall: 40 Years at Gombe. Abrams Books, New York (1999)

Person Index

L. Huber, *The Rational Animal*, https://doi.org/10.1007/978-3-031-60803-2

Subject Index

L. Huber, *The Rational Animal*, https://doi.org/10.1007/978-3-031-60803-2

Taxonomic Index

L. Huber, *The Rational Animal*, https://doi.org/10.1007/978-3-031-60803-2

The manufacturer's authorised representative in the EU is Springer Nature Customer Service Centre GmbH, Europaplatz 3, 69115 Heidelberg, Germany. If you have any concerns regarding our products, please contact ProductSafety@springernature.com

Printed and bound by CPI Group (UK) Ltd, Croydon, CR0 4YY
01/10/2026
02225897-0014